LEAD TIME

LEAD TIME

A Journalist's Education

GARRY WILLS

A Mariner Book
Houghton Mifflin Company
Boston / New York

First Mariner Books edition 2004
Copyright © 1983, 2004 by Literary Research, Inc.
Preface to the Mariner Edition © 2004 by Garry Wills

Visit our Web site: www.houghtonmifflinbooks.com.

LIBRARY OF CONGRESS CATALOGING-IN-PUBLICATION DATA
Wills, Garry, 1934–
Lead time.
Includes index.
1. United States — Politics and government — 1945–
— Addresses, essays, lectures. 2. United States —
Civilization — 1970– Addresses, essays, lectures.
I. Title.
[E839.5.W55 1984] 973.9 83-25031
ISBN 0-618-44690-7 (pbk.)

Printed in the United States of America

MP 10 9 8 7 6 5 4 3 2 1

Excerpt from *The Armies of the Night* by Norman Mailer: copyright © Norman Mailer, 1968; used by permission of the author and his agent, Scott Meredith Literary Agency, Inc. Excerpt from Introduction to *Scoundrel Time* by Lillian Hellman: Introduction copyright © Garry Wills, 1976; reprinted by permission.

Grateful acknowledgment is made to the following magazines and newspapers for permission to reprint material that originally appeared in their pages. *New York,* copyright © *New York* magazine, 1972, 1978, 1979. *Harper's Magazine:* Pieces from the October 1974 and April 1975 issues; copyright © *Harper's Magazine,* 1974, 1975. *New York Times,* copyright © The New York Times Company, 1972, 1974, 1976, 1977. *Atlantic Monthly,* copyright © The Atlantic Monthly Company, 1976. *Columbia Journalism Review,* copyright © *Columbia Journalism Review,* 1980. *Sojourners,* P.O. Box 29272, Washington, D.C. 20017; copyright © *Sojourners,* 1982. *Family Weekly,* copyright © *Family Weekly,* 1982. *Playboy:* Pieces from the May 1971, November 1972, and December 1973 issues; copyright © *Playboy* magazine, 1971, 1972, 1973. *Esquire:* "Dr. King on the Case," "The Devil," "Beverly Sills," "Stained-Glass Watergate," and "Truman"; copyright © Esquire Associates, 1968, 1973, 1974, 1975, 1976. *New York Review of Books:* "Alger Hiss," "Summer of '74," "The Senate," "The House," "Bobby Baker," "Bert Lance," "Jerry Brown," "1974 Miniconvention," "Democrats '76," "Republicans '76," "Miniconvention '78," "Johnson," and "Ford," *Quest:* "The Best Reporter." *Esquire* and *Soho News:* "Reagan." *New York Review of Books:* "Blood Sport."

To John
who put it together

Contents

Preface to the Mariner Edition

As a sedentary old man, sitting at a desk and writing, I find it amusing and odd to read about a younger self with the energy to chase around reporting on demonstrations, riots, communes, conventions, and political brawls. Part of that old energy just rubbed off from the times I was living in, the chronologically stretchable period called the sixties, which have seemed in retrospect so frenetic. I was surprised by the era as I covered it, but not really disoriented. My classical training and religious faith prepared me for the effects of human frailty and original sin, as well as for the redeeming qualities of sacrifice and idealism.

It is sometimes forgotten now how central was religion to the sixties. The civil rights movement arose in that part of America that is most religious, the black community, and was led by the most respected figures in that community, its ministers. I recall going to interview Andrew Young, after this book appeared, when he was the mayor of Atlanta. I told the black cab driver who picked me up at the airport that I wanted to go to city hall. "Whew," he said, "I don't like that place!" I asked him why. "Too many preachers there." When I went in to the mayor's office, I recounted this conversation, and Young said, "Well, he has that right," and named off members of his staff who were clergymen, men and women who had forged their political identities in the civil rights movement. Congressman John Lewis happened to be visiting at the same time, and Young said, "John here is a preacher, too." The civil rights movement could not have begun or lasted or prevailed without these leaders. My own Catholic fellows, and the Jewish and Protestant communities, also furnished believing allies to the black ministers' cause—as well as to the antiwar and feminist movements. My personal guides at the time included the Catholics Dorothy Day and Elizabeth McAlister, Philip Berrigan and Daniel Berrigan.

Some look back on the sixties as the point at which America broke down, lost its cohesiveness, its sense of purpose. I hope it will be clear from this collection why I believe that the country went deeper into itself, under the stress of events, and reinvented itself. While others call World War II's men (*men,* of course) "the greatest generation," I side rather with my friend Studs Terkel, who calls the generation of the sixties "the greatest." That is when the equality of half the human race was taken seriously for the first time in history, and when human rights were enforced for all kinds of minorities, religious and ethnic and sexual. That time introduced a great axial shift, as Peter Steinfels argues, in attitudes toward women, gender, the family, reproduction, professionalism, urbanism, and globalism—a shift comparable with the scientific revolution and the democratic revolution.[1] Human rights acquired a whole new context, with the result that some people now complain about our society stressing rights *too much*—an odd charge to bring against the background of a world history plagued by constant violation of human rights.

For me it is a pleasure to renew acquaintance with the people in this book. I have a slightly higher regard for some of them now than I had then—the late senator Pat Moynihan, for instance, about whom while he lived I wrote what I think a better-balanced assessment later on (though my view of his prose style did not change).[2] For others I have a slightly lower opinion now—especially Pope John Paul II. But I change nothing in the book, since reflecting the original impact of events and people is what matters in this kind of writing. One thing, however, I must not so much change as deplore—my 1975 treatment of Muhammad Ali, which has the effect (though not quite the aim) of glorifying boxing.

We are all the victims, as well as the beneficiaries, of our upbringing. My father was a Golden Gloves competitor, a part-time boxing coach at Albion College, and a judge at Golden Glove matches. He took me, as a young boy, to sit with him in the front row of the matches he judged, so I grew up knowing boxers and boxing fairly well, and wanting to know more. I read Nat Fleischer's *Ring* magazine religiously, and clipped from it pictures of my boyhood heroes (Tony Zale, say, or Willie Pep) to stick on the bedroom wall, my only athletic "pinups." Joe Louis was my greatest hero, and I won money almost every time he fought, since my father, the victim of his upbringing, had the prejudices of his Virginia forebears and was always looking for a "white hope." Billy Conn was his favorite for that role.

With this background, I was bound to be awed when Muhammad Ali came along, and so were my sons. Even my pacifist wife, who rejects every other kind of violence, admired Ali. In fact, during our first year of marriage, when I was still a graduate student, I came home late from class one day, climbed to our two-room apartment under the eaves of a house in Wallingford, Connecticut, and was greeted with the words "I have just seen the most beautiful body in the world." She had watched, on the black-and-white TV of the time, as young Cassius Clay won his gold medal in the 1960 Olympics.

That background explains the encomium on Ali in this book and the implicit glorification of boxing—a glorification that I cannot let stand. Even as I wrote it, I was beginning to have misgivings about boxing, partly from my reading of Saint Augustine's denunciation of gladiatorial sports, partly from knowing what human wrecks many boxers ended up as. But it was only as I read things more scholarly than Nat Fleischer ever published that I realized how degrading is the sport of boxing. At this point, I found that even the defenses of boxing helped turn me against it—the macho strut of an Elliott Gorn, the rather sick adulation of a Joyce Carol Oates. The work of Jeffrey Sammons did much to open my eyes. So I felt impelled, in 1988, to write a palinode (the classical term for a recantation) to my Ali article, in the same place where that piece had been published. I include it as a corrective appendix to this collection. We all have to outgrow some aspects of our upbringing, as well as benefit from the more nourishing aspects of it. I am glad to see that my sons have also stopped watching boxing.

NOTES

1. Peter Steinfels, *A People Adrift* (Simon & Schuster, 2003), pp. 266–80.
2. Garry Wills, "Honorable Man," *New York Review of Books,* November 16, 2000.

After the Fact

Journalism was a way to get other people to pay for my education. I had led a very sheltered life until the year 1967, when I became (for a while) a full-time journalist. After going from a Jesuit seminary to graduate school, and from there straight into a classroom, I suddenly found myself in strip joints, police helicopters, black nationalist headquarters. And the years that followed took me to places—from backstage at the Met to locked up in a cell—I never expected to see.

At first I was sent where a good editor—Harold Hayes of *Esquire*—thought I might learn something. Later, when I had more say in the matter, curiosity helped me choose some of my learning spots. They were all worth a visit, it turned out, because I was not merely a visitor but, perforce, an observer. We look with different eyes at things we have to write about later—one of the side benefits of having to write for a living. To attend events at some remove from them does not decrease participation; it swivels the head in a kind of double take that does, indeed, take in more. I was commissioned not only to observe but to check my observations, on the spot, with the reactions of editors and future readers in mind. How could I convince them, make them present with me, seeing what I saw? Human consciousness is most intense when it is both "photographing" an event and simultaneously (what the camera cannot do) studying the photograph with others' knowledge in mind. Standing off from things is the best way to plunge into them.

The duty of warily circling one's own perceptions of a thing *and* the thing itself was doubly enforced on magazine writers in the sixties. All magazine editors have to cope with lead time, the gap between assignment of an article and its actual appearance on the newsstand or in the mail box—a gap often ranging from several weeks to several months. The timelier an editor aspires to be, the more troubling this barrier between conception and completion. He must send a writer out after the facts, acutely aware that the reader will be judging "after the fact" in a variety of senses. Not only has the event receded in time, but intervening accounts of it—on TV and radio, in the papers and newsweeklies —have come and gone. Can there be anything more to say?

This burden of the lead time stimulated some of the best and worst magazine writing of the fast-moving sixties. Editors tried to assign writers who would have things to say when the event had been "exhausted"—whence Clay Felker's dispatch of Norman Mailer to the 1960 Democratic convention. This could, just as well, lead to fanciful or forced assignments—whence Harold Hayes's use of Jean Genet and William Burroughs to cover the 1968 Democratic convention. Much of what was called "the new journalism" in the sixties had less to do with literary vision than with the mechanics of the lead time. That is what brought novelists like Mailer and Genet into journalism.

The best editors made a virtue of necessity—they learned to stand off from the flow of discrete items filling daily newspapers, to look for longer trends, subtler evidence. They developed an instinct for the things a daily reporter runs too fast to notice. They were saying, in effect, "Lead, time." Despite a couple of famous excesses—the risks of this fast game—Harold Hayes was a model of editorial virtues developed under stress. He lived in several times at once. He always had a magazine just hitting the stands—past history for him, something worried at for months, now barely noticeable. He had writers out on future stories, whose progress he tracked on the phone and through other editors at the magazine. And he had an issue taking shape in frames on his wall—the issue dated six months from "now." By the time that issue's cover date became a reality, Harold's restless attention would still be half a year ahead, with the boxes filling up on his wall, out of synch with actual time, trying to "keep up" with things that had not happened yet.

The rapidity of events, which stimulated him, came at last to defeat him. He strained his attention farther and farther ahead, which just made things whizz past him at a faster pace. When, finally, he left the magazine, this expert in time travel forward seemed stuck in the past, still awed by the way things changed in the sixties, trying to tell and retell that story for a book, or a movie, or a TV series. Other journalists came out of the tumultuous sixties with this "war wound" of the editorial battles—a neck broken by turning one's head fast and faster to keep track of the explosion's fragments. But few had Harold's gift, amid chaos, for imagining the chaos.

It was sad to see the excitement yield to discouragement. An early sign was the August issue of 1968, which contained my piece on Martin

Luther King's funeral. Robert Kennedy had been killed during the lead time. Harold shook his head wearily: "What can you do when the coverage of one assassination comes out *after* the next one?" It weighs down the spirit, having to anticipate the next assassination—which had something to do, I suspect, with Harold's readiness to resign when *Esquire*'s business managers would not meet his terms for editing after Arnold Gingrich's retirement. One can, finally, get too far outside history's more obvious rhythms.

But, for a long time, Harold was good at nudging others out toward the edges of that decade's rapid stream, where they could *steer* instead of being carried along. His instinct was centrifugal—get to the sidelines and watch. When debate was hot in 1967 over new riot-control agents, he encouraged me to find the young inventor of Mace, who brought all the arguments down to human scale. I was more willing to try Mace on myself, and recommend it to others, when I found that its creator was humane. One of the reasons the domestic "arms race" of antiriot agents cooled off was the confidence Mace gave to security people who had been yearning toward machine guns.

Harold was quick to run up the expense account if it meant bringing a new character into the story. He wanted to hear from his writers every few days, as much from curiosity about the people they were meeting as to get a progress report on the number of words typed. And his curiosity piqued one's own. Harold seemed even more interested at long distance than I was in the very presence of the characters. What was I missing?

Like so many "Eastern" journalists, he was a descendant of Southern preachers; and, better than most, he retained an edge of country-boy wonder in that cynical city of miracles. The word I remember most, from our telephone conversations, was his pleasantly incredulous *"Really?"* Like others, I began to look for things that would elicit that particular inflection from him. His gift was to feel, ahead of time, what he wanted the magazine's readers to feel. The implicit attitude was, always, "Tell me *more*"—which made writers scurry about to find more things to tell him.

When he assigned me to Richard Nixon early in 1968, it was to write an article during the New Hampshire primary that would come out after the Wisconsin primary—in that time of fewer and more sharply defined primary races. He said at the outset I must work on the "worst case" assumption—that Nixon might be eliminated by the time the article appeared. (George Romney was still the front-runner for the Republicans' nomination.) The gist of his advice was: Go find something in the past that will have future interest, no matter what is occurring in the present of our May readers. In short: Shop around in time. Just arranging the time-layers of that article forced me to see Nixon from different angles; and I got one of Harold's most satisfying *really*s when I phoned with "news" from the past—that Father John Cronin had leaked FBI information to Nixon during the Hiss case, and had told Nixon about Hiss before the Chambers accusations were made public. The last point expressly contradicted Nixon's own account in *Six Crises*.

Father Cronin was a good example of the peripheral character Harold always felt was "out there"—the one who had valuable information no one else had asked for. Father Cronin would not have talked so freely at the time he was writing speeches for Vice-President Nixon. And he did not talk so openly after Nixon was elected President. But early in 1968 he was in that gray interval where gems of information look dull or valueless because no light of publicity picks them out. When he was newsworthy, Father Cronin would not have talked. When he was not newsworthy, no one talked to him. He talked to me only "after the fact," after Nixon's apparent demise.

Magazine writers have to sort through leftovers, when the television crews have packed up and left. But that is often the best time to talk with people. The witnesses to an event are, at first, rushed and overwhelmed. They feel they must guard the precious hoard of their knowledge. They might write a book themselves. The networks compete to put them on morning talk shows, and reporters vie for interviews. When the restless light of publicity travels on, people exposed overnight to its intensity often feel they have been dropped into darkness. They wonder if they *exist,* now that others do not notice them insistently; and they are glad to talk. This is one of the cruel side effects of modern journalism. Some people, blinded by the light, spend the rest of their lives blundering around trying to get back into its center—just think of Lee Harvey Oswald's mother.

Since magazine writers do not have to file daily, they are not tied to the campaign plane or the candidate's presence, like newspaper or television reporters. Several times, I watched a local person undergo a siege of questions and cameras and microphones, then saw the dozens of reporters leave, and found the person standing there alone, looking somehow violated. Never was a person readier to talk.

People talk more openly on their own turf, and a magazine reporter is often freer to go there. They especially talk more openly if you spend time with them. What might not be said to a stranger off the street will be forced on you after a day or two in their presence. I often tried to stay overnight with people I was interviewing for Harold. As Jimmy Carter realized when campaigning, this sets up a special tie of mutual candor.

Harold's advice to edge off and look around while other reporters were storming some center of "hot" news made me stay behind in Memphis when all the other reporters got on planes for the Atlanta funeral of Dr. King. The Southern Christian Leadership Conference (SCLC) leaders all took off, too, and left the garbage strikers, for whom Dr. King was working when he died, stranded without transportation to the funeral. I was the only journalist—the only white person —who watched their struggle to get there. It was impossible, traveling through Mississippi at night with these frightened people, not to share their fears, and admire their courage, even if duty seemed to be tugging me a different way—toward objectivity, toward "not becoming part of the story." If the journalist's task enforced a certain distance, the observer's double take, one had to resist becoming a *participant*.

Yet one could be *forced* to participate, or thought of as participating, even when one's attitude was entirely detached. I was searched by black militants three times before being allowed to talk with them—"put under discipline" to that extent. And in Memphis, as I followed an SCLC organizer from one small gathering to another, I delivered my first and only "witness" in a black Baptist church.

We were in the church basement, and spirits were high. Everyone was declaring, in turn, what Dr. King's death meant *personally*. I sat scribbling notes, not aware that the circle was reaching its completion where I sat next to my guide. When the person to my immediate left broke off in a sibilant wash of *Yes-Lord*s, silence fell; my pen stopped. I was afraid to look up, but the young man who had brought me said, "Well, Brother Wills?" There was no way to avoid my obvious duty without insulting everyone present as well as the memory of Dr. King. As the only white person there, I had a more attentive audience than I wanted. Yet I felt obliged to express white people's real sorrow, since I had been at the funeral parlor that morning when Dr. King's body was put on display and not a single white person filed by the casket except for out-of-town journalists.

Baptist preachers are the last great orators remaining to our nation, so I was in the big league of speechmaking. But my feeble effort was greeted with hearty *Amen*s and *Tell-it*s just because I said *something*.

Nonetheless, if not coerced by a kind of social pressure, I would not have risen. And I did not feel "co-opted" or taken up into the event: speaking out there was as natural a condition of attendance as are coat and tie in certain restaurants. To cover what goes on in such a place, one must meet the conditions of attendance.

Years later I had a similar experience while following a candidate for the Japanese Diet. Campaigning in that country is a very personal, almost intimate affair, little covered at the local level even by local journalists. My Western presence around the campaign headquarters caused so many lowered conversations and lifted glances that I was asked to explain myself in a brief speech to the campaign workers. Through an interpreter I said that sister democracies should learn more about the way their politics work, and I was grateful to them for letting me watch this edifying spectacle. God knows what the interpreter did with this brief sentiment; but I was cheered as if I had become a candidate myself. Once again, my "participation" did not affect the way I reported the campaign.

But in the sixties those who covered the civil rights movement were attacked as tools of that movement by Spiro Agnew and others. Reporters accused of "aiding and abetting" a black revolution were gloomily amused, since they encountered little but suspicion and resentment from the blacks they were supposed to be collaborating with. When I went to write an *Esquire* article about the Institute for Policy Studies, the only hostility I encountered was Ivanhoe Donaldson's. He accused me of drawing up a hit list for knocking off black leaders, since *Esquire* had published a chart of the most important black activists and their ties to each other. Such paranoia was common in "the movement," black or white; so journalists were suspect to both sides, rarely able to convince either their subjects or their audience that they really did want to observe events rather than shape them. There is no dispelling a suspicious person's suspicion, no matter what the evidence.

As soon as one slipped even slightly past the barriers of black distrust, white distrust took over. One of my early articles for *Esquire* involved an interview with the Pentagon general in charge of riot assistance to the cities. J. Edgar Hoover, though not an *Esquire* reader, saw a reprint of that piece in a Washington newspaper, and opened his jejune file on me in 1968. My offense seems to have been insufficient respect for the military.

In this interplay of paranoias, one could be saved from odd enemies by even odder allies. Later I would be arrested voluntarily; but in 1968

I had to step lively on several occasions to *escape* arrest. One night in Washington, I was rounded up with a group of demonstrators who had run from tear gas into a parking garage. I showed my press card to an officer who thought it less a recommendation than incrimination, and who pushed me in with the others to wait for the paddy wagon.

As I continued arguing, my friend John Waters came up—the underground filmmaker, who had his camera crew out looking for usable movie footage. Spiffily dressed as always, and with the unfreaky blond star of his early films beside him, he looked like the head of a TV crew. "He's with us," he assured the cop—who turned me loose, never guessing he had done the will of a man who was surely a smut king in J. Edgar's eyes. (One advantage of covering a demonstration with a TV crew is that policemen fear the camera, its record of any brutality.)

"Involvement" of those sorts was mechanical, and no challenge to the ideal of detachment. But other entanglements were more intimate. For one thing, even though the lead time made me fashion an article after the fact, I was still drawn back *after* the article. Much of what I wrote for Harold's *Esquire* is not in this book because it went into my *Nixon Agonistes, Bare Ruined Choirs, Jack Ruby,* and *The Second Civil War.* Harold encouraged or demanded more in the way of interviews and background than could fit into one article; and publishers in the sixties were always reading *Esquire* for book ideas (one of the indirect ways Harold shaped the perception of his times).

The book ideas were attractive because Harold exacted a heavy investment of time and effort in single articles. Though *Esquire* paid well by the standards of that day, the breakdown into man-hours expended on a single article made for discouraging figures. That is why so many of us were ready to develop our unused material for a second paycheck.

But working even *farther* after the fact made me realize there was more to learn. People I had interviewed once for an article taught me a great deal more when I went back to them—and some are teaching me still. Bill Willis, the drummer in Jack Ruby's strip joint, is a shrewd observer of Dallas and America and the world, and a friend whose opinion I have sought long after I finished the Ruby book. Leonard Hunter, who showed me what riots would mean in Chicago, was my guide to black leaders elsewhere. The demonstrators I met in Knoxville kept me informed on antiwar activists in Canada. I met one subject's mother, and acquired a second mother in the process.

These opportunities for "late learning" were not all tied to work on books. After covering Washington's Institute for Policy Studies for *Es-*

quire, I attended Institute seminars and consulted its fellows. At one seminar, I heard Daniel Ellsberg, then unknown, describe his conversion from Vietnam hawk to dove. Fellows like Marc Raskin and Dick Barnet profoundly affected my own thinking, and when they asked me to join the Institute's board of directors, I felt I had no right to turn them down. Education should be paid for in some way.

My deepest involvements thus came after I had written my articles; but even the first interviews have to establish some mutuality of trust. The trick is to keep the journalistic role foremost at that stage of the game. But, even from the outset, imaginative participation and detached judgment have a dialectical relationship. Journalists should not be so distant that all they can hear are shouts, nor so close that they become more conspirators than critics. I often delayed expressing any agreement with my subjects, or seeking them out as friends, till after my article was completed and appeared. (For one thing, they might not be so friendly after they read what I had written of them.) Yet, in a sense, no article was ever completed—as my Afterwords are meant to indicate in this volume. The publication of a piece was a beginning as well as an end. Most articles brought me new teachers, as people called or wrote to argue or agree with what I had said. During the sixties and seventies I deliberately kept a listed phone number, so I could not miss these opportunities. The cranks and "nut calls" were easily balanced by new sources of information and instruction.

Late learning, I found, is a never-ending task; it is only as good as the most recent thing you have learned. Journalism lives too much by deadlines—the newest report, the latest edition. But all learning worth the name is based on curiosity, and that is never slaked. It is fashionable with some to attack "revisionism"; but Samuel Johnson, in his criticism of anti-Stuart writings of his day, pointed out that all good history is in a constant state of revision—as fresh evidence comes in, old prejudice gives way; as the joint human endeavor after truth becomes more intense and orderly:

> When an opinion has once become popular, very few are willing to oppose it. Idleness is more willing to credit than enquire; cowardice is afraid of controversy, and vanity of answer.

Those who speak of "the wisdom of the ages" are often just canonizing yesterday's scholarship, or what they learned in school; not a *continuity* of *attention* over the years.

It is easier in magazine journalism than in daily newspaper work to see the lead problem as a late-learning opportunity. Gilbert Chesterton objected to the daily papers as giving us the end of many stories and the beginnings of none. We learn that Jones has died before we knew that he had lived. Something in the very format of such papers suggests that knowledge merely agglutinates—that you stick discrete new items onto an unchanged body of past knowledge. Actually, of course, new evidence entails the *repossession* of past knowledge, a repenetration and new sifting out. When I went from Harold's *Esquire* to less direct forms of reportage, I was lucky to find editors who thought of "mere" book reviewing as such a repenetration of our history. Bob Silvers, at the *New York Review,* treats the history of ideas as a series of news flashes from the ongoing battles of the past. He set me to read all around a new book, much as Harold wanted interviews with people at the edges of events. Bob encouraged me to cover a conference on the history of ideas as I would a political convention. He paired the reviewing of political books with coverage of the Watergate trials, or of elections; or combined a review of books on boxing with reporting on a championship bout.

Bob has been criticized for pitting books against each other in single reviews, and for indulging long descriptions of the background for such contests. But this seems to me another way of looking at the lead-time problem, or of turning it on its head. *Esquire* reported events after the fact—one had to imagine a future reader. Bob's journal starts with a current book, and goes backward to the body of knowledge it must complement or correct if it is to deserve the trees sacrificed to its printing.

So far as that was possible, I tried to "interview" books or past subjects in the way Harold taught me. To write about Jefferson, I spent as much time as I could in and near his own two homes (Monticello and Poplar Forest), used his own copies of books at the Library of Congress, read what he read, saw what he saw, touched what he touched. Though I was not at Jack Ruby's trial, I read the transcripts of other Dallas trials, and all Ruby's testimony in other forums, and his lawyer's books; and I met as many participants as I could, to ask them about particular passages in the transcript. When Tom Wolfe was putting together his anthology of "new journalists," he included my account of that trial. Then, to my surprise, when the anthology came out, the Ruby trial had been removed and my article on Dr. King substituted. Only when I read Wolfe's introduction did I understand why—he insisted on actual presence at the events being covered, and at first reading thought I had attended Ruby's trial. But participation can be imaginative as well as physical—though, once again, the two should interact so far as possible.

Moving forward in time toward anticipated readers, and moving backward to imagined "presents," are complementary actions, not contradictory. Chesterton, who deplored the discrete items delivered in the daily headlines, insisted as well that all real knowledge is "news," that ancient reality only reaches us in the freshest perceptions—as in the vision of his King Alfred:

> All things sprang at him, sun and weed,
> Till the grass grew to be grass indeed,
> And the tree was a tree at last.

That "at last" is the late learner's reward, the latest glance that unlocks the oldest information. St. Augustine lamented, *Sero te amavi . . . sero te amavi,* "Too late I came to love you, loved you too late." But in the area of knowledge that regretful *sero* is changed from "late" to "recent," or fresh; to Chesterton's "at last." All knowledge is the latest knowledge, as different from our knowledge yesterday as our body must be different when we wake up in the morning and find ourselves alive.

The lead-time problem of journalism is just one form of the revisionist problem of the scholars. We must move backward and forward in time simultaneously to get a fix on anything knowable. Caught in the flux of things, we can still partially chart our course, stand outside and above the flux to that extent, both participants and critics. We move in time while measuring time from a fixed standpoint. We edge out of the stream in order to steer down it, escape the flow in order to direct it. The mind moves most naturally not along a straight course but, like Stephen Leacock's hero, by "riding off in all directions."

Harold's office, with its articles being pushed ahead on multiple schedules, became for me a symbol of the knowing process. Our brains are such command centers for processing information. New things are constantly coming in, old things being revised, the two tested against each other. Fresh reports revive the past as well as challenge it, make one say "Really?" and reconsider. Everything is news. One can almost hear the newsboys from old movies shouting "Wuxtry! Wuxtry!" *Sero. Sero.*

I
Years of Turmoil

For a time assassins were setting the pace. President Kennedy was killed, and so was his killer. George Lincoln Rockwell, Fred Hampton, Malcolm X. And Robert Kennedy. And Dr. King.

Civil rights workers were killed—Medgar Evers, Herbert Lee, Louis Allen, James Chaney, Andrew Goodman, Michael Schwerner, Jimmie Lee Jackson, James Reeb, Viola Liuzzo. Black and white together, men and women. Even children: Addie Mae Collins, Denise McNair, Carol Robertson, Cynthia Wesley.

Rioting completed the harsh picture—in Watts, Detroit, Newark, Chicago, Miami, Memphis, Washington. Antirioters rode to the rescue, harsh as the rioters themselves—Bull Connor, Frank Rizzo, Spiro Agnew, George Wallace. Only the nonviolent victims gave us any hope of redemption. Primarily, Martin King.

1

War Protest: Commune

In a Canadian schoolyard, children are washing cars to make money for their class project. Two Americans—call them Peter and Mickie—gun in on ancient motorcycles. How much to wash a bike? "Fifty cents." They confer, come up with sixty cents, try to bargain with the kids for a two-bike package deal. They are turned down. "What next?" Mickie asks. "Nothing," says Peter. "Unless you want to rip off a kid to wash cars and dishes at the house."

A VW bus already dead and half-risen again, and a feline little sports car now on the last of its nine lives, join the scene and disgorge more Americans into the waning brilliance of Canada's autumn. A mock trial takes place to decide which bike, as dirtier, gets the fifty-cent treatment. Peter surreptitiously throws dirt on the one he rode. The result is a draw and the bikes remain unwashed, like their owners.

The Americans, with headbands keeping shoulder-length hair out of their eyes, could be rebel Indians breaking out of a reservation. Instead, this afternoon, they are a ragged touch-football team. Mickie, thin and loose-jointed, leads the way onto the dry field, caricaturing a drum majorette, knees almost hitting his chin with each prance and each pump of the great baton.

"Don't fuck off," Dusty shouts indignantly. "This is for the honor of the United States of fuckin' *America.*" Dusty left the Army in haste—he had been shipped back to America on suspicion of selling arms to the Vietcong ("A hundred and fifty dollars in scrip for an M-sixteen," he reminisces dreamily).

They come here every Sunday to play a collection of Canadian high school footballers, phys-ed teachers, and semipro castoffs—two-handed touch, Canadian rules (three downs a drive, no fair catch). The Americans are not high this time—they ran out of grass and money two days ago and are waiting for a hashish shipment to peddle. In fact, they are badly hung over; without money, all they can get is beer, charged to one of several accounts (all delinquent) at the grocery store. In the store they pretend not to recognize one another. "That guy? Just another fucker from America dodging the draft," Dusty tells the owner with contempt. He went to drama school before the Army got him.

"Siss," the Americans whistle, "boom," as the Canadians boot it, "baaahhh," as it settles into Big Al's hands. Al is the quiet one who holds the house together, puffing moodily, never drinking, writing poems and manifestos, reading Ché. He scampers well, fakes a lateral, then screams in pain—Canadian cleats have gouged away most of a big toenail; it dangles bloodily until Jimmy, Al's brother, twists the mangled thing off and wraps his own headband around the toe.

"An international incident!" Dusty trumpets. "Off the cleats! That's a non-fuckin'-negotiable demand." The Americans have taken the field in boots and sandals—all but Al, who is barefoot. Lladislaw, "our international diplomat," is chosen to lead a legation to the other side. Llad is a Hungarian defector to the Israeli Army who jumped ship with a large store of hashish in Montreal and worked his way inland selling the stuff. His prime qualification as diplomat is that Canadians cannot understand his accent. Eventually, everyone is shoeless, and it is first down Americans. Dwayne takes charge—"I'll run the option." It doesn't work, and no wonder. He had told me the night before how he "flunked Arson I."

"It was my first try and I was alone, so I thought I'd knock off the only wooden building on campus—just for practice, you know? It meant working right under a streetlight where campus police patrolled, but every other building looked so damn *strong*. This was hardly more than a shack. I soaked rags in gas, and spread them along the walls, leading out of a big gas drum and back into it. I had a roll of explosive fuse. So I got across the street and lit it. The silly fire just sat there and looked at me; it didn't go out, but it was smoldering away at a rate of about one inch every ten minutes—no light to it, just a little smoke, people walked right by it in the street, it was so damn sneaky and slow. Hell, I had bought *slow fuse!* I didn't want to spend the night watching it, so I split. It finally got there, I was told, and a little fire started. But it was put out. I figured it was time to retire. If I couldn't knock over a half-assed building like that, I couldn't bring down a goddamn *tent!*"

Big Al had done better. He had got an ROTC building before he crossed the border. He has designs on other U.S. buildings, and has lined up the dynamite; but he would rather wait for some *plastique:* "It's easier to get across the border. I'll make goddamn decorative *candles* of the stuff." Much of the dope dealt by the house—marijuana up across the border, hashish down—is transmitted inside the large candles they pour and sculpt. Now, crippled on the sidelines, Al unwraps his bloodied toe to appreciative oohs and ahs of the children. He is good with kids. One of his poems tells of *"the mirror-faces of the very young,"* and his notebooks say they are the reason he must risk further bombings.

The Canadians are scoring, it is 12–0. Nothing Dwayne can think of moves the ball from scrimmage. The big play so far was Jimmy's interception and twenty-yard runback of a pass. Dusty pounded his back. "MVP here, M-fuckin'-VP!"

"Great," Jimmy shouts. "What does the MVP get?"

Llad, Hungarian potatoes in his English diction, smiles, "He ball Dani first when she cured."

Dusty scowls at him—Dani is his chick, off to the city for her Monday morning gonorrhea treatment. "I *would* get one with the clap—but it's the last time she'll have it, you can bet. There's nothing she hates more than those two shots in the ass on Monday."

Soon the superior Canadians, having run their score up to 42, tire of the game—they are friendly but rather quiet; they like the Americans more for their theatrics than their football. (Each hard block brings weird cries and magical treatments. Even one yard gained from scrimmage calls for Mickie's stirring rendition of the American national anthem.) Sides are now rearranged, three Americans and two Canadians on each, and a stream of little kids pours onto the field—this is the moment they have been waiting for. They know all the "freaks" by name, and know they will be welcomed into the huddle. Even a passing group of high school girls is invited out to play.

"They're minnows," Mickie says. "Throw them back."

Jimmy: "But so *many* minnows—nothing like a whole *stream* of minnows to squirm in." The game disintegrates as the freaks manage to give each kid a turn at passing or receiving. This is the only quarterbacking Dwayne is good at—he has a two-year-old son back in the States.

Dusty breaks things off with, "I got to go to fuckin' *work.*"

"Cure him."

"Pop him a mescaline."

"Chant him an O-O-H-M."

"Bring on the medicine man."

But the Canadian who owns the football is leaving anyway, and the freaks are hungry.

Back at the house, strays and teenyboppers who passed out or bedded down late Saturday night are awake now, trying to find something in the refrigerator. "There's nothing but salad," one girl complains; she's a French-Canadian high schooler who comes every weekend, and is called Frou-Frou at the house.

"Make way for Big Al. It's pancake time." They have chaired him up the porch in their arms and told the women of his toenail sacrifice. But now they need a cook. They rose too late to eat and make it to the game. Each had grabbed a remedial bottle of beer to drink on the way to the field.

"Al, make a big supply of pancakes. What we don't eat we can use as Frisbees."

"Frisbees, hell! I saved one last week and put it on a stick for a flyswatter."

"Make me two big ones—I'll use them for snowshoes next month."

Llad has gone up to watch TV—he spends hours before the screen, giggling and picking up English. His favorite shows in Israel were American and English. "My friends were brokenhearted when 'The Saint' was canceled." Llad and his brother live in a different house, occupied by non-American defectors, but he comes over here every day for the TV.

Dusty calls Mickie into the front room to cut his hair and trim his beard—it dwindles to a matted goatee under the shears. "Who has a pair of pants?"

"I do," from Jimmy.

"Not your dungarees with fuckin' bell-bottoms. I mean *real* pants. I gotta look straight for this job." Dusty begins a temporary job as bouncer in a nearby tavern tonight—just till the shipment of hash arrives. By the time he gets into Al's pants—a foot too short and certain to split if he actually bounces anyone—the girls are giving him a Mr. America treatment, all of them judges with fake little notebooks:

"Nice ass on him."

"A grhayt beeg blownd *Greek God!*"

Frou-Frou applauds, "Yeah, but his swimsuit is too long—right down over his goddamn knees." Tina laughs but does not join in—she was a high school teacher last spring.

Al, looking round the corner from his stove, says, "You look like a French faggot."

"Yeah," Dusty agrees and goes up to shave the rest of his beard off. Llad, at the head of the stairs, shouts, "FLQ ripped off another!" Several people head for the TV.

"Mother-sweet-fuck*er!*" Dusty croons approvingly. The news is that the *Front de Libération de Québec* has kidnaped a second government official—the house admires the FLQ and has contacted it in the search for *plastique*.

"They're so much more together than American radicals," Al explains over his batter. "Wow! If they pull this off, the Panthers will bust every black man out of America's prisons."

A car door slams—Tony, back from taking Dani to the city. His hair is short, the army crew cut still growing out; his tanned thin arms are scribbled over with "good ole boy" sentimental tattoos. His eyes light up at the sight of the two motorcycles and he kicks one off into the field, wheels slipping as he bangs against thin deciduous trees, then races halfway up an incline, where the loose grass and leaves throw him, laughing crazily. The motor coughs itself to rest on the ground.

"Bombed out of his head," Al mutters. "He was supposed to deal some dope in the city, but he got high on the first batch. Well, it always happens. When people first come over the border, they have to stay high for a couple of weeks before they can get themselves together." Tony deserted last week, when his company was preparing to ship out for Vietnam. "That means we'll have nothing but rice and salad for dinner tonight."

Dusty is back downstairs, clean-shaven. Frou-Frou sees him first: "Look at the surf keed."

"Yeah," he moans, "Troy-fuckin'-Donahue." The pancakes are moving fast now, and taste good—stuffing welcome rags into their hunger. The only pause is when "Ohio" drops onto the phonograph (a machine fed continuously night and day), and the benches are scuffed back for everyone to stand, hand over heart. "That's our house anthem," Al whispers as Crosby, Stills, Nash and Young weave the lament, *"Four dead in Ohio."*

Llad knows these English words well: *"Soldiers are cutting us down."* As everyone sits down again, he says, "I left army because I could not . . . I can kill no one."

Jimmy disagrees: "There are some people I would kill with pleasure."

Dusty: "Hell, they're killing *us*. Sending us out to kill others. The marines are worst. I once saw them string plastic explosive on wires

from hut to hut in a Vietnam village, letting the people think it was dec-
oration, so they could get off on the way people touched it and played
with it and giggled—before they detonated it."

Al, who has been sampling his wares as he poured batter and flipped
pancakes, calls people away from the table: "We have to get the trial
started if Dusty is going to work."

"On with the trial!" They retire to the front room, rough and paneled
but kept very clean, with a well-polished rifle over the fireplace. Every
Sunday, violations of house decorum are assessed and punished. Affida-
vits have to be made up before Saturday midnight to keep the session
from becoming a cockpit of sudden hostilities.

"We don't want this to be too frivolous," Al explains carefully. "People can't live together if they are not all into the community, if some are
taking a free ride on it." The tone is facetious, but a tense trial last
week ended in the vote to purge one couple from the house.

Tina is on trial first, for waste. She took a bottle of beer, sipped from
it, did not finish it. Jimmy prosecutes—he found the nearly full bottle
next morning. Dusty defends—he argues it is the duty of others to
fuckin' *find* any bottles with beer left in them and drink the stuff. She is
voted guilty and made to wash dishes one extra time next week. "That's
all we do is wash dishes," she complains. "This house is rotten with
male chauvinism."

"Goddamn *right,*" Jimmy applauds—"The only thing we dig about
women's lib is no bras!" (Tina, alone of the girls in the house, wears a
bra.)

Al tells Jimmy to sit down: "The teenyboppers you let in don't have
enough yet to put bras around." It is clear that, though everyone votes,
Al is the real judge at trials.

Yet Al is also the next one convicted—of losing the house football on
a mountainside picnic (they climb a nearby mountain whenever they
have some particularly good stuff to smoke or drop—"We get off on the
trees"). Jimmy prosecutes again. Al had earlier sworn out an affidavit
against Jimmy for leaving the football in the rain—there are old conflicts
nagging at the brothers. Currents of serious criticism run beneath all the
banter of trial. Jimmy finds Al too officious.

Dwayne tries to bring Dusty to trial. "He shows a bourgeois posses-
siveness about Dani."

"Oh, she can fuckin' fuck anyone she wants," Dusty answers.

"But if she does, you might kill her—or one of us," Dwayne says.

"It's not an issue, she can't ball anyone now anyway, not till she fuckin' gets over the clap. But this is the time to get it all out front."

Mickie objects: "I thought who is balling who was not a matter for trial."

"Right," Dwayne answers. "But this is not about balling—only about Dusty trying to *prevent* people from balling freely with whoever they want; and that is a violation of the house code."

Dusty: "Why not wait till Dani is back to see if she *wants* to ball others?"

Dwayne: "No chick has ever been here without balling more than one."

"Sure they have. Remember Silvy?"

"But she didn't like it here; that's why she left."

"Balling should be nobody's business but those involved."

"Remember, this is a trial—one should speak, henceforth, of the ball*or* and the ball*ee*."

Mickie comes back, "If balling is nobody's business, why does everybody try to make the most noise possible? I sometimes think we're going to shake the goddamn house down. Everyone wants *cheerleaders* at the bedside, to see he's getting his."

"That's right," Tina spouts. "That *he* is getting *his*. The whole balling ethic here is male piggism."

"You seem to enjoy it," Jimmy leers.

"But I want to be more than a piece of *meat* for you to get off on—that's why I left you."

"But you came back."

"Just when I got too cold alone in bed."

Several kinds of hostility are out now, and naked—and sex dragged it all out, here as in any uptight suburb.

Al intervenes: "This is a trial, not a gross-off."

Dusty is acquitted. "Sometimes," he says, "this place reminds me of a fuckin' fraternity house."

Mickie, who graduated from an Ivy League school, lifts a maudlin tenor, *"We're poor little lambs who have gone astray. . . ."*

To lighten things, Al gets out some of his favorite bad poems and reads them melodramatically, Dwayne doing silent-movie chords and shakes on his guitar. The first poem is *The Highwayman*. "She blew off her tits," Jimmy says afterward. "No wonder he split." Then *The Face on the Barroom Floor*.

Dwayne, who has recorded several songs on an obscure label, airs his new compositions, all lovelorn with passionate boot-stompings and *gittar*-lashings. Mickie comes in at the end with *"Gentlemen songsters off on a spree. . . ."*

Al, turned serious again, reads from Evtushenko (how not people die, but worlds die in them) and his own notebooks (how his shadow glides at his side, the revolutionary in him, stalking him with accusation for worlds not brought to birth). Tina is nuzzling the house cat, a furry collection of crossed wires (there is LSD in its saucer on good nights). Jimmy flickers a dim flashlight on the ceiling. "Hell, who can get off that," he finally sighs, and goes for wood to make a fire.

Mickie softly rubs Tina awhile as she rubs the cat, and then they head for the stairs together. "What's up for tomorrow?" Al asks.

Jimmy: "Rifle practice."

Dwayne: "Up the mountain."

Tina: "There's no grass."

Dwayne: "There's only one thing I care about. Tomorrow the hash!"

Mickie brays his way upstairs, hugging Tina: *"Doomed from here to eternity. . . ."*

I ask Al what *will* happen tomorrow. "Tomorrow?" he says with theatrical pretentiousness—his only way, now, of preserving all the soured hopes: "Tomorrow the revolution!" And goes upstairs to write in his notebooks.

—*Playboy*, May 1971

AFTERWORD

I had met some members of this commune at an antiwar rally in Knoxville. They called occasionally with news of protest activity (and used my telephone charge card to make other calls to other people). When our phone malfunctioned in odd ways, my wife was convinced that the FBI was interested in these phone calls. The first thing she looked for, when I got my FBI files through the Freedom of Information Act, was a sign of interest in these communards. We should have known better. All of my dossier deals with opinions, not actions. The Bureau's energies were spent on the effort at policing *thought*—which left it little time to catch people who *did* anything illegal.

2

War Protest: Jail

Prison is the "in" thing this year. More than one person said nervously, "Think what Tom Wolfe could do with this!" We were, after all, going to jail with the Beautiful People, being taken into fashionable custody. Richard Avedon never had a hair out of place through the whole ordeal; I bet even his mug shot was flattering. We had the very den mother of radical chic, Felicia Bernstein, along with us. Large as her apartment is, she could not invite a whole jail in for the evening.

Everyone involved could recite Wolfe's objections by heart; and the night before, at the Dupont Plaza Hotel, almost everyone did. Singer Judy Collins said she had not called her friends together to see them jugged but to do something useful. Joe Papp, one of the friends she had called, said, "I like theater, *but* . . ." He may be the only modern director so secure in his reputation as to have a C. B. De Mille flunky tag after him, bringing hot coffee and telephone messages. "I have a very important meeting I *must* get to." For five hours, various people told us how valuable their time is. Arlo Guthrie sat in the back looking stunned, a child who could not understand why grown-ups shout so at each other. People went on egotistical rhetoric jags to explain why getting jailed is just an ego trip. Even jail, I thought, could not be this boring (but I was wrong), so I left.

What good, after all, would it do to go to jail? We were just chic if we did and chicken if we didn't. The whole thing was hopeless—like everything else that had been done against the war. The big bad Wolfe would get us for our empty gesture—or, more likely, would scare us off from making a gesture. The Beautiful People were either too good for the lockup or not good enough. Still, I wished some had come to feel that even important people can go to jail.

The next morning, there was more jawing and a jaded aftertaste in the mind, the moseying about of men and women who had come to do something noble and now felt just silly. Still, we were asked to eat a good lunch, just in case. I found myself talking too much to take the advice—trying to talk myself into something, I guess. I was tired of hearing so many people try to talk themselves out of things.

The march began like most, with mating initiatives between "spokesmen" and the cameras, Gloria Steinem and Marlo Thomas photogenic up front. We had a big petition done in fancy calligraphy; "Petition of Redress," it said, with a pious nod to the First Amendment. Marcus Raskin had written it—"We are citizens and not hostages." There was a lot of scribbling being done in this compulsively literate crowd. "Where are you going to write it up?" was a commonplace of conversation. It confirmed me in my own first resolution, the only one I was still sure of, after the disillusionment of the night before: I wouldn't write anything about this. It wasn't worth it. Besides, this was the first protest of any sort I had participated in; I had kept my coverage of such affairs hygienically separate from active involvement, at once enjoying and earning a spectator's immunity. To be on the other side, I thought, demanded a symmetrical abstinence, and I'd be damned if I'd take any notes (and would damn myself afterward because I didn't take them). When, the night before, Howard Zinn read the long press release meant to explain our long petition, I just listened, bored, wondering how some of the journalists involved could think so unquotable a document was newsworthy. I confess I did slip around to Howard after the reading and ask him to trim back a tautology—"can possibly" to "can." Shame knows some limits, and we were supposed to be literary heavies.

We waited for the cameras across from the Supreme Court, the explainers explaining, all very glib—and right, of course. But we had all written against the war. We didn't have to come here to do that. Why, despite our glibness, were we here? I still didn't know—the explainers did not explain it to me. March, petition, protest—the same old things. We didn't have much imagination. Admittedly, we were dealing with the same old thing—TV announcement, victory in the offing, bombs, incursion, bombs, mining, bombs, honor, bombs. But was lack of imagination in our adversaries any excuse for us? Joe Papp had his justifiable point: Couldn't we at least do something that was good theater? This was far from that—which made us chary of the whole thing while wanting to support it, hot and cold by flashes, feeling we stood where prominence was betrayal and the cameras our enemy. While the core of our group orated, the fringes were muttering. Still, habit lured me over in time to the journalists, at home among those taking notes even while I could not, and I had to wonder what I would say about me if I were watching me—that I looked out of place, I guess. The thing could not even be treated as a lark. Momentary discomfort is no martyrdom, and publicized discomfort is a mockery of real pain. We were a bunch of pretenders. Even if we negotiated our problematical trip to jail, it would be nothing but make-believe—or so we made ourselves believe, not really wanting to bring it off. Who knows whether any of us might be a Sydney Carton, given the secret choice? But without the secret choice, there is no sense to a Carton's sacrifice.

It was bright and hot when we reached the pillared Capitol, where Bella Abzug waited for us on her canted-pillar legs, her thick slant of hat obvious from a distance. We were kept there for more oratory in the sun. Congressman John Conyers did his impeaching bit, and another in our congressional greeting party dissociated himself from it. Then we were admitted to the House wing in driblets, guards checking all purses and packages. Carl Albert is too shrewd to say a simple yes or no to anything; he took our petition, thanked us politely, and said he would remand it to the appropriate committee. Committees are all so many plots in our country's legislative graveyard, and we were here to talk about graves of a different kind. Robert Lifton, the Yale psychiatrist, had become our spokesman and spoke well—stuff about national emergency, inadequate response, the dereliction of Congress, criminal to recess until action had been taken, etc., etc. Good stuff, all of it, as good as it had been two years ago, or four, or five. Bella, meanwhile, was trying to get a recess, for reception on the House floor of some spokesmen from our group. She used the precedent of interruption in House business to applaud and listen to an astronaut. Hale Boggs thought she was out of order, and so did Albert. He had already taken care of us.

Some congressmen came out to talk with our group, Don Riegel plugging his new book, Gerald Ford trying to act like a host whose hospitality not even lepers could ruffle. Karl Hess, the Pat Buchanan of 1964 campaign oratory, pushed forward to remind "Jerry" of their speechwriting days together in the assault on "uncrowned monarch Lyndon." Ford answered stiffly that he always supports a President when the nation is at war. Hess looked deflated, and a hippie type we had picked up somewhere shouted obscenities. Ben Spock used his best (unavailing) pediatric techniques to quiet him. Closing time came and went, more cops sifting toward us, though some congressmen were still mixed in our number. Capitol police now bluntly invited us out over bullhorns. Some of us claimed we were exercising our First Amendment "right to petition." But the guards answered that we could talk to congressmen on the steps outside; that congressmen have immunity, but we did not; that the building was closed, but we'd be given fifteen minutes to leave it. I remembered the late Duke of Windsor's advice—on a tour of honor, use the john when occasion offers; it may not come your way again—and crossed the foyer, descended stairs, asking cops the way, all the way, to identify myself as coming from the inside; but, sure enough, when I tried to climb the stairs again, a new cop shouted from an upper level that the building was closed. "I know, but we were given fifteen minutes to leave, and that's not up yet."

"That was only for those who want to get arrested. Do you?"

"I don't know. I want to rejoin my friends during the waiting period."

"All right," he said, in a you-asked-for-it voice, and accompanied me across the foyer, down the hall, telling each cop we passed, emphatically, "This one wants to get arrested." I didn't, really—or hadn't, up to then. Something in the way he said it made it suddenly plausible, put a halo of his hatred all around the idea. I guess I did, in fact, want to go to jail. If he thought I belonged there, then I must.

It came with the tense slow-motion procedures I had watched before. Women first, they would go to a precinct station—there were only twenty-seven of them. The men, sixty-seven in number, would have to go to "central lockup." I was, of course, interested in other chroniclers of arrest who had not undergone it—like Francine Gray, the Berrigans' lanky blond Boswell, who was here with her painter husband, Cleve. I was also interested in how the group solidified itself, at last, sitting on the marble floor and singing, Judy Collins amazing us with "Grace"— amazing, at least, all those who had heard her argue against arrest the night before. She had often protested in song, but never from behind bars. Those who had spoken against arrest must have got as tired as we did of hearing how important they were: They talked themselves out of it so long that they talked themselves into it. As the Judy Collinses and Felicia Bernsteins were collared, I began to wonder—if prison has such snob appeal, why is middle-class poet Judy Viorst being led off to a van while radical mannequin Steinem tactfully disappeared before arrests began?

Not that it was not chic, our advertised (and mercifully brief) incarceration. I have never been so applauded in my life—at arrest, going down the Capitol steps, getting into the van, getting out of it. When there were no bystanders, we clapped for one another. Raskin, whose genius is for affection, watched with a teary kind of pride as they took away his writer-wife, Barbara. Even Marc, who had stood trial with Dr. Spock in Boston for encouraging draft evaders, had never technically been arrested. His codirector of the Institute for Policy Studies, Richard Barnet, sitting next to me, had not even marched or demonstrated, much less been jailed. He thought the pen mightier than the sword, as had we all; and we were here in penitence.

The young officer who took me was a fumbling sort, for some reason more nervous than I was, unable for a while to spell Baltimore when I told him where I live. We posed for two Polaroid shots together—Francine Gray said she and her policeman wore self-conscious smiles, like newlyweds at a Coney Island concession. The tin insides of the paddy wagon were sizzling—it had been parked in the sun—and it was a relief to clump out of it, dripping wet, into a dark police basement. We were held until all sixty-seven were delivered, so the sergeant would not have to repeat his pitch: "I run a tight ship here—just ask the Doc." Spock had been arrested the week before in the Capitol rotunda. There was a cheer for Joe Papp as he emerged from one wagon: "I thought you had an important meeting tonight!" He smiled ruefully, looking embarrassed and proud, and embarrassed to be proud. Keys were the only thing they took from us at that point—they, too, did not think much of the pen as a weapon; though, at this literal level, I would rather be stabbed with my hotel key than with a ballpoint. We were counted off, four to a two-man cell, to wait—interminably, as it turned out—to be booked. And no phone calls till we had been processed.

The prospect of death, Dr. Johnson said, wonderfully concentrates the mind. One might expect the same thing from jail—or from the prospect of jail, which was about all I would get from this brief stay inside. Pent up physically, mentally thrown back upon yourself, you begin to hope this is a "concentration" camp, at least in the sense of focusing the mind. But no: What life in a cell actually does is fragment and fray out one's sensations. We were four men in a lightless five-by-seven-foot cell —small open stool and basin, iron-ledge bunks with no mattresses or covers. We were stilled inside a vast hive of activity and din, only scattered parts of things decipherable. One drifts automatically to the bars because, from the inner part of the cell, people pass by this brief lens opening on the world, this little cell eye, too rapidly to be recognized. One sends out antennae through all the near yet unknown regions—who is in the next cell? what is happening to us? when will we see a lawyer? what did the new arrivals say? what order are they taking us out in? Prison is supposed to have a lively grapevine, and certainly rumors spread among us in no time; but they were often uncheckable, contradictory, or merely wishful. We heard from 8 P.M. to 3 A.M. that bond was being posted for us, that we would be out any minute now.

It was good company. Karl Hess, abashed that he had been intimidated by his old buddy Ford—as if he were a schoolkid trying to get the teacher's attention. "There is something so debilitating when a man stands there and tells you a defiant lie." Ken Iverson, mathematician from Philadelphia, accountable to the great computer in the sky while most of us were self-employed or had sympathetic employers. The fourth man was young Tom Hirsch, who had been to the Paris peace talks and worked for *Liberation* magazine. Since Tom is a vegetarian, he was forced to keep fast that night: Our dinner, which came promptly to the dark cell, was stew in a Styrofoam cup—not bad, despite a natural suspicion of any food one cannot see. Its very heft and beefiness were, for us nonvegetarians, ironic obstacles to eating it, since we had to use ridiculously nonlethal paper spoons, which went soggy in no time.

Booking was an almost comic takeoff on bureaucracy. Endless forms filled and refilled by polite, bored desk men. And just when I think I've put my prints on every paper in sight, the process starts all over again. The man handling the prints has a remote air and undeflectable skills, rolling my fingers and thumbs this way and that, like things detached from me, stamping the thumb quick-quick-quick in the lower corner of eight carbon copies expertly fanned out to leave just that key spot bared. The ink is more like the grease that frames a mechanic's nails, making them look naked. It is spread in a thick film over a little cookie tin; my fingers are made to pick up exactly half the film in a set of expert rolls and then centered in ten little windows of a metal card holder; then, after some stamping here and there of individual fingers and (mainly) thumb, the tin is turned around and I pick up the other half of the grease. Even so, by the end, my thumb is made to dart back to the few remaining smudges on the tin, so that my prints stare back at me like clear little eyes from even the dimmest of carbons. It was odd to watch all this going on, as if from the side, my digits turned into stamps in a semiautomatic process, feeling like a puppet, not wanting to feel that, yet knowing things would go better if I went along with the situation, consenting, to accomplish two interdependent things—make the process easier on me, because I was making it easier for the worker on this human assembly line. One doesn't, of course, want to make his life more troublesome—the war is not his fault: but I sense why a prisoner must feel "co-opted" when he cooperates even in little things—each act is an admission of sorts, a recognition of one's total disposability at another's word. No man in prison is ever on his own. He has no choice over when or what he will eat, how he will divide his time, move, act, exercise control over himself. Authority is omnipresent, intrusive, abrasive in its hourly effects—and what a goad that must be to rebelliousness.

From fingerprinting to the mug shot, this time no Polaroid arrest shot, but with a number pinned to one's suit, as in WANTED signs—a new way to play post office. I started to pull up my tie, left loose as a hangman's knot since that ride in the paddy wagon oven, but then I remembered my greasy fingers. In honor of such chic imprisonment, I had worn my expensive tie—I don't know the designer's name, but Tom Wolfe could tell at a glance (come to think of it, I did not even notice what Felicia Bernstein was *wearing* when nabbed). Anyway, rather than stain the tie, I let it dangle in an appropriately criminal way. No wonder everyone (except, no doubt, Richard Avedon) looks like a mug in his mug shot.

Odd, how good it seems to greet one's cellmates after a mere half hour or so away from them. Out of the impersonal back to people. We are cramped, but we defer to one another somewhat elaborately, glad to have this jostle of too much humanity in too little space, aware that the opposite could well be our plight inside these walls. If ours were a "real" (i.e., nonchic) arrest, we would be coping with strange and probably antipathetic cellmates—a fact that was cruelly forced on the women when they were processed: A guard growled at them that he'd like to put them in with the lesbian regulars. We might be caged for a while, but we were there in company, as a company. What a difference that made. Even we had felt slight hints of the helplessness that comes with surrender into another's control. What would it be like to face this prospect alone? Would *that* wonderfully concentrate the mind? Well, it would do something to it, we felt sure. We talked at a forced rate, keeping off such specters.

And that was the weird thing. The very protectedness of our own state made us dwell on all the "as ifs" of a prison cell. We kept assuring ourselves that we had it easy, we were not real victims—which, given the close malodorous aids hourly supplied to our imagination, brought the alternatives vividly to life for us. Hence our imperative laughter, our joint exorcisms of the *real* plight, which we were *not* undergoing. In order to dispel the specters, we unconsciously summoned them. The bare spotted stool was repulsive, but no real threat to us. Still, one wondered—what if your cellmate were a vomiting drunk or a diarrhetic? We did Alphonse-Gaston acts of courtesy, each of us proffering the others room on the iron bunks—but what if one were caged with a surliness denying even the few possible amenities, or making the achievement of them prohibitive on any but dehumanizing terms? Every slight comfort was humbling, one's few joys impressed by their fragility. We served as each other's crutches—and were made to realize how, in all things, others get us through. We sought out reasons for congratulating ourselves on our surroundings and almost threw a party when we found that none of us smoked—why, we had been worse off in the first-class section of an airliner when the smokers lit up!

But our abject dependence was symbolized in little ways—like having to ask for toilet paper each time it was needed. Some had to request, and wait a long time for, every paper cup of water, since their basin's faucet did not work. In the cell next to us, someone complained after a while that their stool did not work. Spock, directly across from us, said, "The little button in the wall, you have to kick it."

"We did. Nothing happened."

Spock, tall and strenuous, a sailor on the verge of his seventieth birthday, said, "I don't mean kick it. I mean *kick* it."

There were noisy bangings on the iron wall, reverberating our own walls, then sudden reluctant thunder of the plumbing (everything is loud here, especially the clang of remote-controlled releasings and sealings of the cells). "You were right."

"Sure," Spock smiles, "the good old john in cell thirty-eight; I know it well."

As the hours went by, our mutually cheering conversations lagged a bit. I felt dehydrated and our little paper cup was disintegrating—one cannot talk much when drinking semiliquid paper. Young John Steinbeck, across the way, tried several times to entertain us with his flute; but each time Howard Da Silva, in the same cell with him, roared, "Put away that instrument of torture! Or give it to someone who can make *music* on it. David? Where's David?"

David Amram, the composer, answered from somewhere down the line and Da Silva persuaded Steinbeck to baton his flute along, hand over hand around the screened parts of our cells, till it reached Amram, who made tentative attempts at fingering.

"David!" Da Silva again, in his booming actor's voice. *"Give that back to Steinbeck."*

Da Silva had last been to Washington as part of the cast of the musical *1776,* playing Ben Franklin in the White House itself, at Poor Richard Nixon's request. He seemed the obvious choice to read our petition to Carl Albert. But the idea was rejected as too obvious. The theatrical people were being careful not to show off. Despite Papp's complaint, most of us had shied away from theater, adopted a deflationary style, underplaying each situation. Da Silva was especially self-effacing, despite that voice, which could now be heard in every last cell as in the last row: "With all the effort to fill these cells with prominent prisoners," he muttered, "why didn't someone try to find a prominent bail bondsman?"

It was ten o'clock, five hours since our arrest; most of us hadn't been given our one phone call, and the only lawyer to show up in the early hours was Mark Lane, a signer of our petition, who explained in a loud voice how he tried to get arrested but just couldn't manage it; how the government didn't have a leg to stand on, so we should all plead "not guilty" in the morning; and how our subsequent well-publicized trial would vindicate the antiwar movement. It was just what I expected of him, from his books.

Sitting on the bunk was not easy; it had a sharp little retaining ridge, to hold our nonexistent mattresses, that bit into the backs of our knees—until I padded it with my suit jacket. Tom Hirsch had jackknifed himself into a corner of one bunk and gone to sleep. How nice to be young, I thought; he must be the only one able to sleep here—till I looked across the aisle. Spock's white hair and tanned arm, dim under the bottom bunk, showed he had dozed off on the floor (too tall to use a bunk, even if he had wanted to claim a whole one from the three other men with him). And I realized that my first reaction had been right—how nice to be young.

At last I was taken out to the one pay phone—since some would never get to it, I could hardly complain. We had all urged one another to be brief, and I was. I called my wife, who was matter-of-fact about things; she had been sure I would get arrested, surer than I had been—which obscurely pleased me (prison, no doubt, brings one's idiocies to the surface). But I was vaguely saddened once the call was made—it had been the one definite thing I could look forward to, and now that was gone.

Finally, a young lawyer, Susan she called herself, worked her way around to our cell, asking if anyone wanted bail—it would cost $40 merely to get us out from some time in the early morning until 9 A.M., when we must show up for arraignment. I was the only softy in our cell. All I wanted by then was a stiff drink and a shower and different walls around me—or, for preference, no walls. I said sure to her question and gave her my name. She recognized it, said she had heard about my Sunday commencement address. Though she looked like a teenager, she had worked, three years ago, on the DC Nine case of priests and seminarians who had destroyed Dow Chemical files—that explained her knowledge of Baltimore clergymen and of the fact that I had addressed the graduates of their seminary three days before the arrest. The ceremony took place in church—some parents seemed disconcerted at talk of the war as their sons were graduated. Part of that sermon went like this:

> We gather here on Pentecost, the feast of gathering ("all were together in one place" goes Luke's text), of the people sharing a covenant. Pentecost is the New Testament's answer to that Old Testament symbol, the Tower of Babel—articulated fiery tongues that speak to difference, over against Babylon's attempt to forge unity out of sheer power. The tower is meant to intimidate, to give clout—"to make a name for us," as the text of Genesis puts it; "to maintain our credibility as a great nation," in the words of Nixon and Kissinger. And God confounds the effort of Babel, divides men one from another when they try to impose a brute unity of power.
>
> Other men, this very day, are trying to use the tongue against physical terror, to speak on a human scale at the mere bulk and huge sprawl of the Pentagon. They are demonstrating not far from here, and I think it is fair to ask why we are not there. Was it not the task of those God threatened at Babel to oppose the tower, speak out against it, tear the bricks down one by one if they could, or at least make the effort? The Old Testament God pardons other cities of sin for a few just men within their walls, for even one voice lifted against injustice. . . .

Was that what I was doing here—tearing bricks out of the tower? It seemed unlikely as our midnight ache and sleeplessness settled in. More likely, making an ass of myself. Being chic. I wondered how Mrs. Bernstein felt—her Panther party had ended in time to be reported in the 10:30 P.M. edition of the *Times*. Jail is, altogether, a longer kind of party—and the longer it became, the more dubious its purpose seemed. Rapid chatter still broke into the uneasy silences, but at longer intervals. An extraordinary oscillation of mood was signaled in the brightening and dimming of voices all around. I had been told that men in prison suffer this openness to incursive unforeseen moods, helpless in their ebb and flow, victims of prison's true powerlessness. Now I believed it, having seen it, if only from afar. We bobbed on a comfortable raft, in no danger of drifting off from shore; but we *were* a little way into the night surf and could look out from land with new feeling for those lost out there, for the truly uncharted experience imprisonment is.

It was doubting time, all down our cellblock. The first team-cheering spirit had evaporated. I had my own futile catechism to run through: why had I done this—to end the war? It couldn't. To feel noble? I didn't. To accomplish something? What? I was less sure the longer I stayed; and my mood was obviously shared by others, since the laughter was rueful when Marc Raskin mused humorously in one of our longer pauses, "Did you ever wish you were for the war?"

Never till now, was the first answer of instinct in me. Then: Never less than now. For what jail was giving me was just that possibility of regret. Opposition to the war had never cost me anything. I did not have to defy the draft, disobey an order, displease a superior. I had just talked a lot in favor of people taking those risks. Tonight was my chance to join them in some minimal gesture, as a first motion toward them.

But this was not *really* joining them. True—yet most of them, I hoped, would understand the gesture. Besides, to say it is "only" a gesture is to move it into the realm of symbol, make it a thing freighted with meaning beyond itself, if for no one else, then certainly for me. And I had to puzzle out that meaning.

Those all around me tonight were the honored ones of our society—a Nobel winner, the recipients of many other prizes, people whose names graced marquees, drew crowds. Not that we were all celebrities, by any means; but we were middle class conformers, not "the demonstrating type." Why had we chosen, if only in symbol, this dishonor, most of us for the first time? First arrest, first night in jail, first guilty sentences, fresh criminal records—sixty or seventy new records this night, and the promise of more to follow in the future. The whole thing was a play of symbols, a reversal of the customary language of recognition in our society, of honor as we had understood it. These were people grown partly ashamed of their honors, because they feared the honoring society. Their jailing was a little thing; but so is a medal or a prize; so is an award. This jailing—the number pinned on, the mug shot taken, the grease marks paced out and out on paper (let your fingers do the walking)—all this was a cumulative *anti*award. It cost little. But it did not—at least to us; not then, at least—mean little. It was only a symbol, as was the Star of David worn by some non-Jews in the Occupation. It was a declaration of solidarity with the hunted, not the hunters. We had been drawn into a fellowship of noncomplicity with what our government was doing, our society condoning. We could no longer be sure where the guilt ended, could not satisfy ourselves that we did not stand with the war criminals until we had taken some (minimal) pains to place ourselves among the peace criminals, drawing on their merits, as it were, saved by our new fringe membership in their company, innocent by association.

And that had to be the point—not that we, in some spotlight, replaced true heroes of resistance. None of us dreamed we could do that. Brave as he is, Ben Spock is no Dan Berrigan. Visiting jails is not living there. And not one of us in that cellblock thought ourselves, then or ever, the equal of Spock. Those who wore the Star of David in Denmark did it not only to save Jews but to be saved by them—saved from murder in its most cowardly form, murder by easy societal acquiescence. Ours was, thus, a homage to the real peace criminals, application for honorary membership, an attempt at a vicarious and saving guilt. Our hint of what real prisoners undergo was a shy move on our part toward fellowship with them.

In my "sermon" of the Sunday before, I had asked why my seminarian audience and I were not at one with the demonstrators in Washington—and then tried to trace a way in which we were. On that very Sunday, Nixon was trying to speak through interpreters on his way to the Moscow summit—while his planes blustered and mines whispered in various Babel threats across the world. Yet the meaning of all Meaning, it seemed to me, lay in Luke's tale of the divided yet uniting tongues— how all that we understand in ineffably private ways stands, yet, under a common judgment; how we meet *each other* in true understanding by way of our deepest burrowings, in our most private journeys, when we think ourselves most apart (as men counted off into separate cells). Perhaps Noam Chomsky, down the cellblock from us, was getting at something like this with his concept of the inner preformed language given us in the mere act of being men, the abrupt light of intellection we discover in ourselves by discerning it in others. We speak to one another in the shared spirit, or not at all; and fiery tongues are our only true interpreters. Speaking to the class of theologians, I suggested that studies dedicated to the spirit constitute of themselves a war upon the tower and concluded:

Unless we have this prior way of using our voices "deep to deep," it is probably vain to try purifying the language of Power at the Pentagon or in Moscow. We would only be pitting new towers against the tower, adding further Babels to Babylon. . . . When I turn over these thoughts in my mind, they reconcile me to our presence here, instead of in Washington—though I hope to see some of you over there, later.

Actually, I had seen some of those seminarians at earlier demonstrations and in Harrisburg for the trial of Philip Berrigan. And, once again, I was trying to join them, for I knew many hold a concept of ministry that will take them into risks not unlike Father Berrigan's. And the reference to Washington—to this night in jail—was, in my own mind, made because I wanted to buy the right to move in their company through crises yet to come, to keep writing "mere words" against the war and against our nation's war-making readiness.

But mainly, by now, I wanted out. Around 1 A.M. D.C. residents were released without bail, "on citation," as the term goes—so we lost Karl Hess. It made more room in the cell but left an emptiness, too. We were glad to see him go free—he was wondering, by this time, about his wife, who had been arrested with the women—but it was also sad to see our company dispersed. Susan, our lawyer, had gone to bed, promising the bondsman would arrive; but we were in for more hours of indeterminacy. A nice young Roman-collared minister across the way had not understood her rather breathless questions and was left behind, unlisted with the other D.C. residents. Now he could not even get onto the list of those to be bailed—though no results had come of that to this point. Only Joe Papp had been let out, his flunky scurrying to purpose, making separate arrangements for him.

At last, at 3 A.M., word came that ten of us could leave on bond. Others now wished to join us, but Susan was not there and guards cannot solicit for a bondsman. I wrote six shouted names and gave them to our bondsman on the way out; but he just shoved them into his pocket and headed for the precinct house, where women prisoners awaited him. Five of us jammed ourselves inside a cab—it seemed spacious; it was not a cell. We were a bit punchy, slaphappy with mere ability to go where we wanted to go, eat what and when we wanted to, bathe, shift scenes, breathe different air. The cabdriver laughed with and at us. We must have seemed like a carload of kids, drunk from our long sobriety—what a sight: just childish, learned, paunchy, overaged schoolkids giddy to be out on holiday.

Our first stop had to be the railway station. Poor Martin Duberman, the critic and playwright, had sweated out the morning hours knowing he had left a manuscript he had worked on for four years in a twenty-four-hour locker there; had it been found and thrown away? We bantered on the drive about Carlyle's famous loss, but we were all writers in the cab and knew this was no laughing matter. At the station, we tumbled out, wanting food and drink—not that we were really hungry, but hours of being unable to eat, even if one does not need to, triggers some urge toward self-preservation. We descended on the milk and candy machines like calorie freaks.

The manuscript recovered, we went back to the cab, poet Kenneth Koch asking, "What was all that crap about jail as an ego trip? My ego never felt less traveled!" At the hotel we parted, five men who had never met until last night, all backslaps and comic affection like Chaplin and the drunk in *City Lights,* knowing that tomorrow our odd link would be broken and we would be strangers again.

We couldn't get much sleep. We were on the day's docket, which meant showing up at 9 A.M., though we would not be processed until after lunch. In the hotel coffee shop that morning, Da Silva read the *Times* account of our arrest and said, "I've had better reviews." At the courthouse, we met the women again and heard of their night—five in a cell, with only one low bed, not two bunks. Francine Gray said, "Dr. Spock was right. It *is* like losing your virginity—it's even more boring, and hurts more, than you had expected." She had been trying to see her husband, who had stayed in jail all night, to tell him she would plead not guilty and prosecute the State—her own kind of imprisonment cheek.

We were advised by our lawyer to plead nolo contendere and get off, most likely, with a $25 fine (or two days). While pleading nolo, we would be allowed to make a statement, and Redress organizer Fred Branfman urged us all to make an individual declaration of conscience and to subscribe to one written in the yard that morning by those who had stayed in jail. (Scribble, scribble—what a crew.) I tried to think of something pithy for the judge and stenographer but could not, though I wanted to speak now. Anything I came up with for on-the-spot elocution would be too pompous or too elliptical. The only thing I can do with words is write them—after consideration. The pen may be the weakest of weapons—it has certainly been less mighty than the sword all through this war—but the falsest thing about my night in jail, I began to realize, was my pretense that I could escape the pen and typewriter. I have to turn words over on paper, try to link them, simply to make my mind work. I had gone to jail thinking it absurd to write about a few hours' experience, when so many others have years of ordeal to record. But already my own first impressions were fading—as they would, indeed, if I stayed inside for a decade. And if I did stay longer, I would perhaps be too dulled, or not dulled enough to the pain, to get the thing down on paper. So, despite all our resolute mental drill to convince ourselves it was no big deal to go to jail, I began to wish I had taken notes. I have spent far worse nights, and in far more dubious company; yet there is, after all, something different about being in jail.

The urge to take notes would grow, and finally triumph, as I listened to the personal testimony of the others facing our judge. Their halting words were especially eloquent for me, tongue-tied as I am without a pen in my hand. And some spoke movingly by their very silence. Tom Wolfe had dwelt on Felicia Bernstein's practiced voice and dramatic reading at the Panther party. Here, she stood silent, after the clerk proclaimed the adversaries ("The United States versus Felicia Montealegre Bernstein") and then stamped GUILTY on her thick legal-length folder. Howard Da Silva muted his Ben Franklin roar, and only asked the judge if those who had stayed in jail all night could be processed quickly. Dr. Spock, unshaven but still courtly in his rumpled suit, pleaded not guilty with his unperturbable humor and politesse. Waked at 3 A.M. with word that his bond had been paid, he asked the time, said it would only disturb his Washington hosts if he dropped in on them now, and went back to sleep on the floor. Not, however, until he had grinned and mumbled an answer to a guard taking Martin Duberman out of the cellblock. "A writer, eh?" said the guard. "What do you write?" Before Duberman could answer, modestly, "This and that," Spock mischievously volunteered, "Cheap novels. Just cheap novels." In court, Spock and the judge knew each other, from his arrest in the rotunda, and Spock now assured him, "My reappearance does not represent any disrespect for the court, Your Honor." Always the gentleman.

As I would soon discover, one feels strange and exposed facing a judge, waiting for sentence, even one so slight and foreseen, and my admiration grew for those who tried to voice their most private feelings in undergoing this first symbolic break with the social system that had coddled them. The note most often heard was one of wistful love for that system and mourning for it. David Amram said: "Your Honor, in 1945, at the age of fourteen, I went with my mother to Pennsylvania Avenue, to watch Harry Truman light the Christmas tree. I was so impressed I made a watercolor of the event when I got home that night. I was brought up to respect the office of the presidency and the other branches of our federal and municipal governments. Now, in 1972, at the age of forty-one, I have committed civil disobedience and spent a night in jail for the first time in my life.

"I feel my crime of illegal entry is morally justifiable, since it was an attempt to petition Congress in protest of a totally immoral war. All of us, as extremely active leaders in our various professions, do not relish the idea of any more days or nights in the cellblock. But many of us may go to jail again, if necessary, to help bear witness to all of the young people of this country who have died, or who have been physically, psychologically, and spiritually maimed by the unjust war their elders have forced upon them.

"I am a veteran, a taxpayer, and I love this country and the people I meet as I travel all over America giving concerts. Enough of us here today love and respect what is stated in our Constitution and Bill of Rights to go to jail again, if necessary, as a part of a new consciousness that wishes to restore America's soul, by ending this war so we can bring peace, love, and justice to all of our citizens."

Cynthia MacDonald, poetess from Sarah Lawrence, said: "When I was a child, I was a baseball fan and I used to listen to the New York Giants games over the radio, which is what we had in those days. Alone in my room, I always stood up when 'The Star-Spangled Banner' was played. As I didn't have a flag, I used to look at something red, white, and blue, and think of it. I love my country, but I detest what it's doing in Vietnam. . . ."

Even recent beneficiaries of our greatness had reason for regret. Mia Adjali, of the United Methodists, was born in Algeria: "I have only been an American citizen for five years. But before I became an American citizen, I grew up in a country which fought the revolutionary war. I fought, in many words and in many letters, what the French did in Algeria, what the French did in Indochina, and when I became an American citizen, I saw my new country involved in the same struggle. I could only do one thing. I have never been arrested before. I have tried many means, perhaps not all of them, but I wish that Congress had, perhaps, shown the kind of understanding you are showing now, and perhaps had met last night, and acted in its own way to end this war."

I had moved to the front bench and my pen was out now—I would later verify what was being said from the transcript, but it had lost something by the time I read it coldly put down on paper. The words had special urgency spoken in voices furred with shyness and sleeplessness. But it was difficult to hear them in the spectators' part of the courtroom. Judy Collins, the archcommunicator, chatting with the judge, brought this up: "It's too bad there isn't a little better acoustical system, which I would always insist on if I were singing." I had learned that she sat on the precinct steps after her release, the night before, and talked about the war with policemen going off their shift. Now she asked the judge what he thought of those who acted from conscience —a question he deftly, politely, deflected.

Others asked him to sign our petition, or to think of judges who had preserved their rectitude while Germany was pursuing covert genocide. He was inflexible but genial; and he had, after all, rejected the government's request that our nolo pleas not be accepted. Judge Goodrich was a patently honorable man—and so are they all.

I scribbled as fast as I could. Maybe I could be the acoustical system Judy Collins wanted. These few tongues, in their unifying diversity, might have no effect—the tower might not fall—but the effort had to be made and recorded. Judy ended her talk with the judge this way: "I would like to offer you my thanks, first of all. And, secondly, I would like to ask you to join us, because, perhaps, if you would join us, your court system can speed up and get on with the business that you should be involved with, and we can stop murdering babies, and women, and men. Thank you."

Then there was Patricia Simon, a gold-star mother, a schoolteacher with other children to raise, giving every spare moment to the son she can no longer care for: "I have spent enormous amounts of energy, time, money, and health in protesting the death of my son, other American sons, and the suffering and death of the Indo-Chinese people, in all the legal, nonviolent ways my imagination could produce." Has any other mother of a slain U.S. soldier gone voluntarily to jail to protest the war in which he fell?

There was nothing for me to write down as Joe Papp took his sentence, untheatrical to the end—it was 2 P.M., the time of that important meeting he had to get to. Only later did I find out what he missed. He was one of the recipients of New York State's award for cultural contribution, to be given in splendor by Nelson Rockefeller—another of society's honors for our honored company. But he was not there, he was with us in the courtroom. The medal was awarded in absentia, while he received our D.C. antiaward. His wife told New York's governor she had never been prouder of him. And so, without her right to be, was I. By God, I *did* go to jail with beautiful people.

—*Playboy*, November 1972

AFTERWORD

Since the House had not cut off funds for the illegal war, the Redress group decided to petition the Senate, too. We blocked access to the Senate chamber as we had on the House side. Karl Hess said, as we walked to the Capitol, "I hope we get put in the same cell again." I wondered why. "I've been studying Greek since that night in jail." In the long hours, I had argued that a knowledge of ancient Greek was the most economical use of one's intellectual energy. Karl, who believes in individual initiative, bought a Greek grammar the day he got out of jail. As we squatted on the floor, some senators grumbled, unable to enter the chamber, and asked the guards to cart us off without bothering to let us leave peaceably. But when Barry Goldwater came on the clotted mass and heard someone shout "Your old speechwriter is in there," he stepped over the sprawling bodies, found Karl, and pulled him to his feet with a handshake. "Why don't you ever come by and see me?" Karl, booted and bearded in Ché fashion, mumbled "I guess some of your staff would get pissed off." The Senator shot back: "Well, piss on them; you're my *friend*." Karl later told me how Goldwater had quietly but firmly rejected the efforts of racists to help in his 1964 campaign. One of the advantages of going to jail that second time was the discovery of an honorable man on the other side of the war issue.

3

Dr. King on the Case

Of course, Mailer had an instinct for missing good speeches—
at the Civil Rights March in Washington in 1963 he had gone
for a stroll just a little while before Martin Luther King began,
"I have a dream," so Mailer—trusting no one else in these matters,
certainly not the columnists and the commentators—would never know
whether the Reverend King had given a great speech that day,
or revealed an inch of his hambone.
                                        —Norman Mailer, *The Armies of the Night*

"Nigger territory, eh?" He was a cabdriver, speculative; eyed the pistol
incongruous beside him on the seat, this quiet spring night; studied me,
my two small bags, my raincoat. The downtown streets were empty, but
spectrally alive. Every light in every store was on (the better to silhou-
ette looters). Even the Muzak in an arcade between stores reassured itself, at the top of its voice, with jaunty rhythms played to no audience.
Jittery neon arrows, meant to beckon people in, now tried to scare
them off. The curfew had swept pedestrians off the street, though some
cars with white men in them still cruised unchallenged.

"Well, get in." He snapped down every lock with quick slaps of his
palm; then rolled up his window; we had begun our safari into darkest
Memphis. It *was* intimidating. Nothing stirred in the crumbling blocks;
until, almost noiseless—one's windows are always up on safari—an armored personnel carrier went nibbling by on its rubber treads, ten long
guns bristling from it (longer because not measured against human
forms, the men who bore them were crouched behind the armored
walls); only mushroom helmets showed, leaning out from each other as
from a single stalk, and, under each, bits of elfin face disembodied.

At last we came to lights again: not the hot insistence of downtown; a lukewarm dinginess of light between two buildings. One was modern and well lit; a custodian sat behind the locked glass door. This is the headquarters of a new activism in Memphis, the Minimum Salary Building (designed as national headquarters for raising the pay of ministers in the African Methodist Episcopal Church and now encompassing other groups). Its director, Reverend H. Ralph Jackson, was a moderate's moderate until, in a march for the striking sanitation workers, he was Maced by police. Since then his building has been a hive of union officials, Southern Christian Leadership Conference staff, and members of various human rights organizations.

Next to it is the Clayborn Temple, a church from which marchers have issued almost daily for the past two months. Marchers fell back to this point in their retreat from the scuffle that marred Dr. King's first attempt to help the strikers. Some say tear gas was deliberately fired into the church; others that it drifted in. But the place *was* wreathéd with gas, and a feeling of violated sanctuary remains. Churches have been blacks' one bit of undisputed terrain in the South, so long as they were socially irrelevant; but this church rang, in recent weeks, with thunderous sermons on the godliness of union dues.

I pay the cabdriver, who resolutely ignores a well-dressed young couple signaling him from the corner, and make my way, with bags and coat, into the shadow of the church porch. In the vestibule, soft bass voices murmur. I stop to let my eyes, initiated into darkness, find the speakers and steer me through their scattered chairs. They are not really conversing; their meditative scraps of speech do not meet each other, but drift off, centrifugally, over each one's separate horizon of darkness. This uncommunicative, almost musical, slow rain of words goes on while I navigate my way into the interior.

About a hundred people are there, disposed in every combination: family groups; clots of men, or of women; the lean of old people toward each other, the jostle outward of teenagers from some center (the church piano, a pretty dress on a hanger); or individuals rigid in their pews as if asleep or dead. The whole gathering is muted—some young people try to pick out a hymn on the piano, but halfheartedly. There are boxes of food, and Sunday clothes draped over the backs of pews. The place has the air of a rather lugubrious picnic—broken up by rain, perhaps, with these few survivors waiting their chance to dash out through the showers to their homes. Yet there is a quiet sense of purpose, dimly focused but, finally, undiscourageable. These are garbage collectors, and they are going to King's funeral in Atlanta. It is 10 P.M.; in twelve hours the funeral will begin, 398 miles away.

They have been told different things, yesterday and today, by different leaders (some from the union, some from SCLC). They have served as marshals in the memorial march that very afternoon, and preparations for that overshadowed any planning for this trip. Some have been told to gather at ten o'clock; some at eleven. They believe there will be two buses, or three; that they will leave at eleven, or at twelve; that only the workers can go, or only they and their wives, or they and their immediate families. Yesterday, when they gathered for marshals' school, a brusque young black shouted at them to arrive sharply at ten: "We're not going by CP Time—Colored People's Time. And if you don't listen now, you won't find out how to get to Atlanta at all, 'cause *we'll* be on the *plane* tomorrow night." The speaker seemed to agree with much of white Memphis that "you have to know how to talk to these people."

And so they wait. Some came before dark, afraid to risk even a short walk or drive after curfew. Some do not realize the wait will be so long; they simply know the time they were asked to arrive. Most will have waited three hours before we start; some, four or five. I try to imagine the mutters and restlessness of a white group stranded so long. These people are the world's least likely revolutionaries. They are, in fact, the precisely *wrong* people—as the Russian field-workers were the wrong ones to accomplish Marx's revolt of the industrial proletariat.

People such as these were the first "Memphians" I had met in any number. That was four days ago. And my first impression was the same as that which nagged at me all night in the church: these Tennessee blacks are not unlikely, they are impossible. They are anachronisms. Their leaders had objected for some time to J. P. Alley's "Hambone" cartoon in the local paper; they say, rightly, that it offers an outdated depiction of the blacks. Nonetheless, these men *are* Hambones. History has passed them by.

I saw them by the hundred, that first morning, streaming past the open casket in a hugger-mugger wake conducted between the completion of the embalmer's task and the body's journey out to the Memphis airport. I had arrived in Memphis several hours after King's death; touched base at the hotel, at the police station, at the site of the murder —dawn was just disturbing the sky, flashbulbs around and under the balcony still blinked repeatedly against the room number (306) like summer lightning. As the light strengthened, I sought out the funeral home police had mentioned—R. S. Lewis and Sons.

Clarence Lewis, one of the sons, has been up all night answering the phone, but he is still polite; professionally sepulchral, calm under stress. "They brought Dr. King here because we have been connected with the Movement for a long time. We drove him in our limousines when he was here last week [for the ill-fated march]. They brought the body to us from the morgue at ten-thirty last night, and my brother has been working on it ever since. There's so much to do: this side [he pulls spread fingers down over his right cheek and neck] was all shot away, and the jawbone was just dangling. They have to reset it and then build all that up with plaster." I pass through the fine old home (abandoned to trade when the white people moved from this area) into a new addition—the chapel, all cheap religious sentiment, an orange cross in fake stained glass. There are two people already there, both journalists, listening to the sounds from the next room (Clarence calls it, with mortician's euphemism, the Operating Room), where a radio crackles excerpts from Dr. King's oratory, and men mutter their appreciation of the live voice while they work on the dead body. We comment on the ghoulishness of their task—knowing ours is no less ghoulish. We would be in there, if we could, with lights and cameras; but we must wait—wait through an extra hour of desperate cosmetic work. We do it far less patiently than Memphis garbage men wait in their church. "Hell of a place for Dr. King to end up, isn't it?" the photographer says. "And one hell of a cause—a little garbage strike."

When, at eight o'clock, the body is brought out, bright TV lights appear and pick out a glint of plaster under the cheek's powder. Several hundred people file past; they have sought the body out, in their sorrow, and will not let it leave town without some tribute. But not one white person from the town goes through that line.

Those who do come are a microcosm of the old Southern black community. Young boys doff their hats and their nylon hair caps—their "do rags"—as they go by. A black principal threatened to expel any child from a local high school who came to class with an Afro hairdo. Possessive matrons take up seats in the back, adjust their furs, cluck sympathetically to other women of their station, and keep the neighborhood record straight with bouts of teary gossip. They each make several passes at the coffin; sob uncontrollably; whip out their Polaroid cameras, and try an angle different from that shot on their last pass. One woman kisses the right cheek. Clarence Lewis was afraid of that: "It

will spoil the makeup job. We normally put a veil over the coffin opening in cases of this sort; but we knew people would just tear that off
with Dr. King. They want to see him. Why, we had one case where people lifted a body up in the coffin to see where the bullet had gone into a
man's back."

Outside, people mill around, making conversation, mixing with
stunned friendliness, readjusting constantly their air of sad respect.
Again, the scene looked like a disconsolate picnic. Some activists had
called him "De Lawd." He always had to be given either his title
(*"Doctor* King," even the *"Reverend Doctor* King") or his full historic
name ("Luther Martin King" one prim lady mourner called him in the
funeral home, understandably stumbling over the big mouthful). Even
that title "Doctor"—never omitted, punctiliously stressed when whites
referred to him, included even in King's third-person references to himself—had become almost comical. He was not only "De Lawd," but
"De Lawd High God Almighty," and his Conference was stiff with the
preacher-dignities of the South; full of Reverends This and Bishops
That and Doctors The-Other. No wonder the militants laughed at it all.
And now, damned if he hadn't ended up at a Marc Connelly *fish fry* of
a wake—right out of *Green Pastures*.

Roark Bradford, on whose stories *Green Pastures* was based, learned
to read by poring over the pages of the Memphis *Commercial Appeal*
and he learned his lesson well: he was able to create a hambone God:
"Dey's gonter to be a deluge, Noah, an' dey's goin' to be a flood. De
Levees is gonter bust an' everything dat's fastened down is comin'
loose." These are unlikely people, I thought at that sad fish fry, to ride
out the deluge whose signs had already thundered from several directions on the night King died. But then, so was Noah an unlikely candidate. Or Isaac, who asked: "Does you want de brainiest or the holiest,
Lawd?" "I want the holiest. I'll make him brainy." And there was one
note, at King's makeshift wake, not heard anywhere in Connelly's play.
As one of the mammy types waddled out the front door, she said with
matter-of-fact bitterness to everyone standing nearby: "I wish it was
Henry Loeb lying there"—handsome, lovable Henry Loeb, the city's
mayor, who would later tell me, in his office, how well he liked his
blacks; unaware, even now, that they are not his. Connelly's "darkies"
do not hate white people: "the white folk" simply do not exist in his
play, which was meant to fortify the Southern conviction that "they
have their lives and we have ours," an arrangement convenient to the
white and (so whites tell themselves) pleasant for all. The whites get

servants, and the blacks get fish fries. That whole elaborate fiction was shattered by the simple words, "I wish it was Henry Loeb." Massah's not in the cold, cold ground. She wishes he were. These people may be Hambones, but not J. P. Alley's kind. They are a paradox, a portent white Memphis still must come to grips with—hambone militants, "good darkies" on the march. When even the stones rise up and cry out, the end has come for Henry Loeb's South.

The signs of it are everywhere—at the Lorraine Motel, where King died; it is an extension of the old Lorraine Hotel, once a white whorehouse. When the neighborhood began to go black, it was thrown to a black buyer as, in the South, old clothes are given to "the help." A man named William Bailey bought it, and laboriously restored it to respectability. King stayed there often on his visits to Memphis. It is now a headquarters for the SCLC's Project Memphis, a program designed—as its assistant director says—"to make Memphis pay for the death of Dr. King." Yet the Lorraine is run by a man who could pose for "Uncle Ben" rice ads—an ex-Pullman porter who is still the captain of porters at a Holiday Inn. He works for the white man, and does it happily, while he owns and runs a black motel where activists plot their campaigns. "I'm very proud to be part of the Holiday Inn family," he told me. "Why, the owners of the whole chain call me Bill Bailey." That's the black Henry Loeb has always known. It is the other side of him— the owner of the Lorraine, the friend of Dr. King—that is the mystery.

King made the mistake of staying, on his penultimate visit to Memphis, at one of the posher new Holiday Inns—in the kind of place where Bill Bailey works, not the motel he owns. The Memphis paper gleefully pointed out that King *could* stay in the Inn because it had been integrated—"without demonstrations." But the Lorraine is not integrated (except in theory). Neither was the white flophouse in which the sniper lurked. It is good that King came back to the real world, the de facto segregated world, to die. He was in the right place, after all. Memphis indeed had taught him to "stay in his place"—a thing it will come to regret. For "his place" is now a command post, a point where marches are planned, and boycotts, and black-history classes.

These garbage men are that new thing, Hambones in rebellion—and they have strange new fish to fry. The people who filed past King's body had said no to the whole city of Memphis; said it courteously, almost deferentially (which just made it more resounding); they had marched every day under their employers' eyes; boycotted the downtown; took on, just for good measure, firms like Coca-Cola and Wonder Bread and

Sealtest Milk; and were ready, when the time came, to join with King in taking on Washington. Patience radiates from them like a reproach. Perhaps that is why the white community does not like to see them in a mass—only in the single dimension, the structured encounter that brings them singly into the home or the store for eight hours of work. They seem almost too patient—wrong people for rebels. Yet their like has already made a rebellion. A tired woman in Birmingham was the wrong sort to begin all the modern civil rights activism; but Rosa Parks did it. King was drawn into that first set of marches almost by accident—as he was involved, finally, in the garbage men's strike: "Dat's always de trouble wid miracles. When you pass one you always gotta r'ar back an' pass another."

The buses were late. They were supposed to arrive at eleven-thirty for loading baggage (each man had been told to bring toothbrush, change of underwear, change of outer clothes if he wanted it, and most wanted it). Besides, there had been talk of a bus for teenagers, who were now giggling and flirting in the dark vestibule (surrendered to them by their elders). Jerry Fanion, an officer of the Southern Regional Conference, scurried around town looking for an extra bus; like all blacks, he was stopped everywhere he went. Police recognized him, and they had been alerted about the men who would be-leaving their homes for the funeral; but they made him get out of the car anyway, and laboriously explain himself. He never did get the bus. Later in the week, the teenagers made a pilgrimage to King's grave.

Meanwhile, the wives in the Clayborn Temple still did not know whether they could go with their husbands. About eleven-thirty, T. O. Jones showed up, with P. J. Ciampa. Jones is the spheroid president of the sanitation local—a man too large in some ways and too small in others for any standard size of shirt, coat, pants. He is content with floppy big pants and a windbreaker that manages to get around him, but only by being too long in the sleeves and too wide in the shoulders. He is a quiet man in his early forties, determined but vague, who began the strike by going to the office of the Director of Public Works and—when the director told him there was an injunction against any strike by city employees—changing into his "prison clothes" on the spot.

Ciampa is the fiery Italian organizer who came into town for the union and amused people with televised arguments against Mayor Loeb (who insisted that all negotiations be carried on in public). Jones and Ciampa have lost the list of men signed up for the buses; they don't know how many buses are coming, how many can ride on each. They try to take two counts—of workers alone, and workers with their wives; but it's difficult to keep track of those who wander in and out of shadows, doors, anterooms.

After an hour of disorder, it becomes clear that everyone can fit into the three buses if folding chairs are put down the aisles. T.O. had told me to save a seat for him, but the chairs in the aisle barricade us from each other. I sit, instead, with a sleepy young man who describes the route we *have* to take, and then finds confirmation of his theory, with a kind of surprised triumph, all along the way. The route one travels through Mississippi and Alabama is a thing carefully studied by Southern blacks. After giving T.O. a check for the bus drivers, Ciampa goes back to the hotel, T.O. swings onto the bus, and we pull out.

In the seat behind me, a woman is worried over the teenagers still standing by the church, vulnerable in the night. "How they gonna get home?" she asks. "Walk, woman," her husband growls. "But what of the curfew?" "What of it?" "I don't trust those police. If I hadn't got on the bus with you, I'd have stayed all night in the church." As the bus rolls through downtown Memphis, on its way South, the woman sees cars moving. "What are they doing out during the curfew? Why aren't *they* stopped?" She knows, of course. Her husband does not bother to answer her.

In our bus, all the animation comes from one voice in the back. A tall, laughing man I had watched, in the church, as he moved from one cluster to another, mixing easily, asked to sit beside me while I was still saving a seat for T.O. I was sorry later I had not said yes. As the riders shouldered sleepily into their chair backs, he joked more softly, but showed no signs of fatigue himself—though he had been a marshal all the long afternoon of marching. And as fewer and fewer responded to him, he moved naturally from banter and affectionate insult to serious things: "That Dr. King was for us." The response is a sigh of yesses. "He didn't have to come here." A chorusing of noes. As he mused on, the crowd breathed with him in easy agreement, as if he were thinking for them. This "audience participation" is what makes the Southern preacher's sermon such an art form. I had been given a dazzling sample of it three days before in the garbage men's meeting at the United Rubber Workers Union Hall. That was the day after King's death, and a formidable lineup of preachers was there to lament it. They all shared a common language, soaked in biblical symbol: Pharaoh was Mayor Loeb, and Moses was Dr. King, and Jesus was the Vindicator who would get them their dues checkoff. But styles were different, and response had to be earned. The whole hall was made up of accompanists for the improvising soloist up front. When he had a theme that moved them, they cheered him on: "Stay there!" "Fix it." "Fix it up." "Call the roll." "*Talk* to me!" "Talk *and a half.*" The better the preacher, the surer his sense of the right time to tarry, the exact moment to move on; when to let the crowd determine his pace, when to push against them; the lingering, as at the very edge of orgasm, prolonging, prolonging; then the final emotional breakthrough when the whole audience "comes" together.

Memphis is not really the birthplace of the blues, any more than Handy was the father of them; but these are the same people who created the form—the triple repeated sighing lines, with a deep breathing space between each, space filled in with the accompanists' "break" or "jazz." That is the basic pattern for the climactic repetitions, subtle variations, and refrains of the preacher's art. That kind of sermon is essentially a musical form; and the garbage men are connoisseurs. When a white pastor from Boston got up, he gave them slogans and emotion; but without a response from the audience—he didn't know the melody.

Nor did all the black preachers succeed, or win equal acceptance. The surprise of the afternoon, at least for me, came when an SCLC delegation reached the hall, and the Reverend James Bevel got up to preach. He and his associates looked almost out of place there amid the "do rags" and scarred ebony skulls; they were immaculately dressed, with educated diction, wearing just the proper kind of "natural" and a beard.

Bevel was the fourteenth, and last, speaker of the afternoon. It seemed that earlier emotional talks would have drained these men of all response left them after the shock of the preceding night. But Jim Bevel slowly built them up, from quiet beginnings to an understanding of what it means to be "on the case." (This is a phrase he invented a year ago to describe musicians who are perfectly interacting; it is now an SCLC phrase of wide applicability.) "Dr. King died on the case. Anyone who does not help forward the sanitation workers' strike is not on the case. You getting me?" (They're getting him.) "There's a false rumor around that our leader is dead. *Our* leader is not dead." ("No!" They know King's *spirit* lives on—half the speeches have said that already.) "That's a *false* rumor!" ("Yes!" "False." "Sho' nuff." "*Tell* it!") "Martin Luther King is not—" (yes, they know, not dead; this is a form in which expectations are usually satisfied, the crowd arrives at each point *with* the speaker; he outruns them at peril of losing the intimate ties that slacken and go taut between each person in the room; but the real artist takes chances, creates suspense, breaks the rhythm deliberately; a snag that makes the resumed onward flow more satisfying)—"Martin Luther King is not our *leader!*" ("No!" The form makes them say it, but with hesitancy. They will trust him some distance; but what does he mean? The "Sho' nuff" is not declamatory now; not fully interrogatory, either; circumflexed.) *"Our* leader—("Yes?")—is the *man*—("*What* man?" "Who?" "Who?" Reverend Abernathy? Is he already trying to supplant King? The trust is almost fading)—who led *Moses* out of *Israel."* (*"Thass* the man!" Resolution; all doubt dispelled; the bridge has been negotiated, left them stunned with Bevel's virtuosity.) *"Our* leader is the man who went with Daniel into the lions' den." ("Same man!" "Talk some.") "Our leader is the man who walked out of the grave on Easter morning." ("Thass the leader!" They have not heard, here in hamboneland, that God is dead.) "Our leader never sleeps nor slumbers. He cannot be put in jail. He has never lost a war yet. *Our* leader is *still on the case."* (*"That's it!" "On* the case!") "Our leader is not dead. One of his prophets died. We will not stop because of that. Our staff is not a funeral staff. We have friends who are undertakers. We *do business.* We *stay on the case,* where our leader is."

It is the most eloquent speech I have ever heard. I was looking forward, a day later, to hearing Bevel again, before a huge audience in the Mason Temple. He was good—and gave an entirely different speech. But the magic of his talk to the sanitation workers was gone. It was not merely the size of the crowd (though that is important—the difference between an intimate combo and some big jazz band only partially rehearsed). The makeup of the crowd was also different. Those in the Union Hall were predominantly male. Men accompany; women compete—they talk over the preacher's rhythms. Their own form is not the jazz combo, but the small group of gospel singers, where each sister fights for possession of the song by claiming a larger share of the Spirit. In a large place like the Mason Temple, women set up nuclei around the hall and sang their own variations on the sermon coming out of the loudspeakers.

But that night on the bus, there was no fighting the jolly voice that mused on "Dr. King's death." Responses came, mingled but regular, like sleepy respirations, as if the bus's sides were breathing regularly in and out. This is the subsoil of King's great oratory, of the subtly varied refrains: "I have a *dream* . . . *I* have a dream today." He must have been a great preacher in his own church; he could use the style out in

the open, before immense crowds. He made the transition more skillfully than Bevel had—and far better than Abernathy does. That very
day, the Monday before King's funeral, Abernathy had paused long on
the wrong phrases: "I do not *know* . . . I do not *know.*" He had let
the crowd fool him by their sympathy; he took indulgence for a *demand*
to linger. He did not have King's sure sense of when to move.

I suppose I heard thirty or forty preachers on that long weekend of
religious eloquence; but not one of them reached King's own level of
skill in handling a crowd. That was the mystery of King. He was the
Nobel Prize winner and a Southern Baptist preacher; and, at places like
the Washington Mall in 1963, the two did not conflict but worked together. As the man in the bus kept saying, "He was for *us.*" ("Unh-
hmmn!") "He was *one* with us." ("That he was." "That he was.")

But King's rapport with his people was not the natural thing it seems
now. He had to learn it, or relearn it. The man's voice rose behind us in
the bus: "You know what Dr. King said?" ("What?") "He said not to
mention his Nobel Prize when he died." ("Thass what he *said.*") "He
said, 'That don't mean nothing.'" ("Sho' nuff.") "What matters is that
he helped *us.*" ("Thass the truth." "That *is* the truth." "*That* is.")

King was bright, a quick study. He skipped two grades to finish high
school at the age of fifteen. He was ordained at eighteen; graduated
from college at nineteen. It was a fast start, for a career that is one long
quick record of youthful accomplishment. He got his theology degree at
the age of twenty-two. While a pastor (from the age of twenty-five), he
got his Ph.D. from Boston University at twenty-six. And he went direct
from graduate school to a position of national leadership. His major
achievements were already behind him when he became the youngest
man (thirty-five) to receive the Nobel Prize. He was dead before he
reached the age of forty; and there are constant little surprises in remembering how young he was—as when Harry Belafonte, speaking in
Memphis, referred to King as his junior by a year. Was "De Lawd" really younger than that baby-faced singer? And why did we never think
of him as young?

He had the strained gravity of the boy who has moved up fast among
his elders. That unnatural dignity is in his writing, too, which labors so
for gravity that it stretches grammar: "President Kennedy was a
strongly contrasted personality . . . trying to sense the direction his
leadership could travel." His acceptance speech will not rank with the
great Nobel speeches: "transform this *pending* cosmic elegy into a cre-

ative psalm . . . unfolding events which surround . . . spiral down a militaristic stairway . . . blood-flowing streets. . . ."

The young King wanted to study medicine. He majored in sociology at Morehouse College. He thought preachers not quite intellectually respectable, though his father and grandfather and great-grandfather had all been preachers. Even when he accepted ordination, he thought he should become a theologian-minister, perhaps a professor, rather than a mere preacher. He took his first parish—in Montgomery—to get "pastoring background" before accepting a teaching post. To the end of his life he talked of turning to an academic career.

But he was never convincing as a scholar. An account of his own intellectual development reads as if it were lifted from a college catalog: "My intellectual journey carried me through new and sometimes complex doctrinal lands, but the pilgrimage was always stimulating, gave me a new appreciation for objective appraisal and critical analysis, and knocked me out of my dogmatic slumber." He was not even a very perceptive commentator on the men who created his doctrine of civil disobedience—Thoreau and Gandhi. When he began the Montgomery boycott, he liked to refer vaguely to Hegel as the prophet of "creative tensions." It was not till someone suggested more likely patrons of nonviolent rebellion that he began referring to Gandhi and Gandhi's American forerunner—*referring* to them—as saints. He never really discusses their philosophy. And his most ambitious defense of civil disobedience—the Letter from a Birmingham Jail, written eight years after the Montgomery boycott—does not even *refer* to Gandhi or Thoreau. Instead, King uses tags from Augustine and Aquinas (hardly antiauthoritarians). Nor does the Letter deserve high marks for logic. It offers as the model of civil disobedience, not Gandhi, but Socrates, the stock platonic figure suborned for all noble causes, something of an embarrassment in this context, since Plato makes him preach history's most rigorous sermon against civil disobedience in the *Crito*.

Like Moses, he was not "de brainiest." He knew only one book well —the Bible. It was enough. All the other tags and quotes are meant to give respectability to those citations that count—the phrases sludged up in his head from earliest days like a rich alluvial soil. He could not use these with the kind of dignity he aspired to unless he were more than "just a preacher." Yet the effect of that *more* was to give him authority *as* a preacher. By trying to run away from his destiny, he equipped himself for it. He became a preacher better educated than any white sheriff;

more traveled, experienced, poised. He was a Hambone who could say no and make it sound like a cannon shot.

It is interesting to contrast him with another preacher's son—James Baldwin. Baldwin became a glib boy preacher himself as a way of getting out into the secular world. King became a student as a way of getting into a larger world of *religion,* where the term "preacher" would not be a reproach. He needed a weightiness in his work which only that "Doctor" could give him. He needed it for personal reasons—yes, he had all along aspired to be "De Lawd"—and in order to make Southern religion relevant. That is why King was at the center of it all: he was after *dignity.* His talent, his abilities as a "quick study," his versatility, his years studying philosophy and theology (for which he had no real natural bent), were means of achieving power. His books and degrees were all tools, all weapons. He had to put that "Doctor" before his name in order to win a "Mister" for every Southern black. They understood that. They rejoiced in his dignities as theirs. The Nobel Prize *didn't* matter except as it helped them. As T. O. Jones put it, "There can never be another leader we'll have the feeling for that we had for him."

Our three buses had a long ride ahead of them—ten hours, an all-night run, through parts of Mississippi, Alabama, and Georgia. They were not luxury buses, with plenty of room; the Greyhound Company had run out of vehicles and leased these from a local firm. One could not even stretch one's legs in the aisle; the folding chairs prevented that. Ten hours there. Ten hours back.

Minutes after our departure, the man behind me said, "We're in Mississippi now." "Oh no!" his wife groaned. It is well to be reminded that our citizens are afraid to enter certain states. The man most frightened was T. O. Jones. He knows what risks an "uppity" black takes in the South. He does not give out his address or phone number. The phone is changed automatically every six months to avoid harassment. He has lived in a hotel room ever since the beginning of his union's strike, so his wife and two girls will not be endangered by his presence in the house. "This is risky country," he told me. "And it gets more dangerous as you go down that road. That Mississippi!" We were going down the road.

The lead bus had no toilet, and chairs in the aisle effectively barricaded people from the other two buses' toilets. The technique for "rest stops" was for all three buses to pull off into a darkened parking lot; the chairs were folded; then people lined up at the two toilets (one bus for

men, one for women). At our first stop, some men began to wander off
into the trees, but T.O., sweating in the cool night, churning all around
the buses to keep his flock together, warned them back. "Better not
leave the bus." I asked him if he expected trouble. "Well, we're in Mississippi, and folk tend to get flusterated at—" He let it hang. He meant
at the sight of a hundred and forty blacks pouring out of buses in the
middle of the night. "You didn't see that man over there, did you—in
the house by the gas station? There was a man at the door." Some had
tried to go near the dark station, to get Cokes from an outdoor vending
machine. T.O. pulled them back to the buses. He carries his responsibility very self-consciously.

Back in the bus, there was a spasm of talk and wakefulness after our
stop. The deep rumbling voice from the rear got chuckles and approval
as he mused on the chances of a strike settlement. "We got Henry Loeb
on the run now." ("Yeah!" "Sure do!") "He don't know what hit him."
Fear is not surprising in the South. This new confidence is the surprising
thing. I had talked to a watery little man, back in the church, who
seemed to swim in his loose secondhand clothes—a part-time preacher
who had been collecting Memphis's garbage for many years. What did
he think of the mayor? "Mr. Loeb doesn't seem to do much thinking.
He just doesn't *understand*. Maybe he can't. The poor man is just,
y'know—kinda—*sick*." It is King's word for our society, a word one
hears everywhere among the garbage men; a word of great power in the
black community. It is no longer a question of courage or fear, men tell
each other; of facing superior white power or brains or resources. It is
just a matter of understanding, of pity. One must be patient with the
sick.

Henry Loeb does not look sick. He is vigorous, athletic, bushy-
browed, handsome in the scowling-cowboy mold of William S. Hart and
Randolph Scott. And he has a cowboy way of framing everything as
part of his personal code: "*I* don't make deals. . . . *I* don't believe in
reprisals. . . . *I* like to conduct business in the open." There is an implicit contrast, in that repeatedly emphasized pronoun, with all the
other shifty characters in this here saloon. He even has a cowboy's
fondness for his "mount"—the PT boat he rode during the war, a loving if unskilled portrait of which hangs behind his desk. (His office biography makes the inevitable reference to John F. Kennedy.)

Loeb is an odd mixture of the local and the cosmopolitan. He comes
from a family of Memphis millionaires; he married the Cotton Carnival
Queen. Yet as a Jew he could not belong to the Memphis Country Club

(he has become an Episcopalian since his election as mayor); and he went East for his education. A newsman who knows him made a bet with me: "When he hears you are from a national magazine, he will not let five minutes go by without a reference to Andover or Brown." When I went into his office, he asked for my credentials before talking to me (he would later boast that he talks to anyone who wants to come see him). Then he asked where I live. Baltimore. "Oh, do you know so-and-so?" No. Why? "He was in my class at Andover, and came from Baltimore." That newsman could clean up if he made his bets for money.

Loeb did not mention Brown. But he did not need to. As I waited for him in his office, his secretary took the Dictaphone plug out of her ear, began flipping through her dictionary, and confided to me, as she did so, "The mayor was an English major at Brown University, and he uses words so big I can't even find them." Later, his executive assistant found occasion to let me know that his boss was "an English major at Brown University."

But the mayor also plays the role of local boy protecting his citizens from carpetbaggers out of the North. He has the disconcerting habit of leaving his telephone amplifier on, so that visitors can hear both ends of a conversation; and when a newspaperman with a pronounced Eastern accent called him for some information, he amused local journalists, who happened to be in his office, by mimicking the foreigner in his responses. When a group of white suburban wives went to his office to protest his treatment of the garbage strikers, he listened to them, then slyly asked the five who had done most of the talking where they were from; and his ear had not betrayed him—not one was a native "Memphian." He has a good ear for classes, accent, background. He wanted to know where I had gone to college. The South is very big on "society."

But Loeb has no ear at all for one accent—the thick, slow drawl of men like T. O. Jones. He knows they haven't been to college. I asked him whether he thought he could restore good relations with the black community after the sanitation workers settlement. "There is good understanding now. I have blacks come to me to firm up communications —I won't say to reestablish them, because they had not lapsed." I told him I attended a mass rally at Mason Temple, where more than five thousand blacks cheered as preacher after preacher attacked him. "Well, you just heard from a segment of the community whose personal interests were involved. Why, I have open house every Thursday, and just yesterday I had many blacks come in to see me about different things." Imagine! And Henry even talked to them! And they came right in the front door, too! It is the conviction of all Henry Loebs that the great secret of the South, carefully hidden but bound to surface in the long run, is the blacks' profound devotion to Henry Loeb. After all, look at everything he has done for them. "*I* took the responsibility of spending fifteen thousand dollars of city money—multiplied many times over by federal food stamps—to feed the strikers." Noblesse oblige.

The odd thing is that white Memphis really *does* think that—as citizen after citizen tells you—"race relations are good." Its spokesman cannot stop saying, "How much we have done for the black" (the Southern bigot is nothing but the Northern liberal caricatured—we have *all* done so much for the black). A journalist on the *Press-Scimitar,* the supposedly "liberal" paper in town, says, "We have been giving blacks the courtesy title" (that is, calling Mr. and Mrs. Jones *Mr.* and *Mrs.* Jones) "ever since the Korean War." (It embarrassed even the South to call the parents of a boy killed in action *John* and *Jane* Jones.) But the executive secretary of the local NAACP was considered a troublemaker when, arrested in a demonstration supporting the strikers, she held up the booking process time after time by refusing to answer the officer's call for "Maxine" instead of Mrs. Smith. ("Why, *isn't* your name Maxine?" one honestly befuddled cop asked her.)

Mrs. Smith is one of the many blacks who protested the morning paper's use of the "Hambone" cartoon. But she ran up against the typical, infuriating response: "Hambone" was actually the white man's way of saying how much he *loves* blacks. It was begun in 1916 by J. P. Alley, who—this is meant to settle the question once for all—won a Pulitzer Prize for attacking the Klan. It was kept up by the Alley family (one of whom is married to the morning paper's editor), and Memphis felt it would lose a precious "tradition" if their favorite darkie disappeared from their favorite newspaper—as, at last, a month after King's death, he did; with this final salute from the paper: "Hambone's nobility conferred a nobility upon all who knew him."

Nowhere is the South's sad talk of "tradition" more pitiful than in Memphis. The city was founded as part of a land deal that brought Andrew Jackson a fortune for getting Indians to give up their claims to the site. The city's great Civil War hero—to whom Forrest Park is dedicated—could not belong to the antebellum equivalent of the Memphis Country Club because he was not a "gentleman"—that is, he was not a slave *owner* but a slave *trader.* After the war, however, he took command of the Ku Klux Klan, which made him "society." The Memphis Klan no doubt boasted of all the things it did for the black, since it *was* more selective and restrained than the Irish police force, which slaughtered forty-six blacks in as many hours during 1866. Later in the century, yellow fever drove the cotton traders out of town, and Irish riffraff took over; the municipality went broke, surrendered its charter, and ceased to exist as a city for a dozen years. Then, just as Memphis regained its right of self-government, a small-town boy from Mississippi, Ed Crump, came up the pike and founded the longest-lasting city "machine" of this century. The main social event for the town's "aristocracy"—the Cotton Carnival—goes back only as far as 1931, when it was begun as a gesture of defiance to the Depression: the city, as they say, is built on a bluff, and run on the same principle.

When Dr. King's planned second march took place, four days after his death, men built the speaker's platform inconveniently high up, so Mrs. King would be standing before the city emblem, above the doors of City Hall, when she spoke. It was meant, of course, as a rebuke to the city. But her standing there, with that background, is henceforth the only tradition Memphis has worth saving.

Yet the city keeps telling itself that "relations are good." If that is so, why was Henry Loeb guarded by special detectives during and after the strike? (One sat in during my session with him; they stash their shotguns under his desk.) Why did some white ministers who supported the strike lose their jobs? Why are black preachers called Communists in anonymous circulars? But the daily papers will continue to blink innocently and boast on the editorial page: "Negro football and basketball players figure prominently in all-star high school teams selected by our sports department." What *more* do they want?

When dawn came, our buses had reached Georgia, the red clay, the sparse vegetation. By the time we entered Atlanta, it was hot; the funeral service had already begun at Ebenezer Church. The bus emptied its cramped, sleepy load of passengers onto a sidewalk opposite the courthouse (Lester Maddox is hiding in there behind *his* bodyguard, conducting the affairs of office on a desk propped up, symbolically, with shotguns). The garbage men who brought their good clothes have no opportunity to change. The women are especially disappointed; the trip has left everyone rumpled. Men begin to wander off. T.O. does not know what to do. He ends up staying where the bus stopped, to keep track of his flock. Some men get the union's wreath over to the church.

Others walk to Morehouse College. But for most, the long ride simply puts them in the crowd that watches, at the capitol, while celebrities march by.

It was a long ride for this; and the ride back will seem longer. The buses leave Atlanta at eight-thirty on the night of King's burial, and do not reach Memphis until six the next morning. But no one regretted the arduous trip. T.O. told me he *had* to go: "We were very concerned about Dr. King's coming to help us. I talked with the men, and we knew he would be in danger in Memphis. It was such a saddening thing. He was in Memphis for only one reason—the Public Works Department's work stoppage. This is something I lay down with, something I wake up with. I know it will never wear away."

A week after the funeral, Mayor Loeb finally caved in to massive pressures from the White House. The strike was settled, victoriously. At the announcement, T.O. blubbered without shame before the cameras. It was the culmination of long years—almost ten of them—he had poured into an apparently hopeless task, beginning back in 1959 when he was fired by the city for trying to organize the Public Works Department. After the victory I went with him to an NAACP meeting where he was introduced, to wild applause, by Jesse Turner, head of the local chapter: "Our city fathers tell us the union has been foisted on us by money-grubbing outsiders. Well, here's the outsider who did it all, Carpetbagger Jones." The applause almost brought him to tears again: "I was born in Memphis, and went to school here. I haven't been out of the state more than three days in the last ten years. Is that what they mean by an outsider?" A man got up in the audience and said, "When my wife saw you on television, she said, 'I feel sorry for that fat little man crying in public.' But I told her, 'Don't feel sorry for him. I've seen him for years trying to get something going here, and getting nowhere. *He* just *won*.'"

When the strike was still on, Henry Loeb, if asked anything about it, liked to whip out his wallet and produce the first telegram he got from the union's national office, listing nine demands. He would tick off what he could and couldn't do under each heading, giving them all equal weight, trying to bury in technicalities the two real issues—union recognition and dues checkoff. When I went to see him after the settlement, he brought out the tired old telegram, now spider-webbed with his arguments and distinctions. Then he searched the grievance-process agreement for one clause that says the final court of appeal is the mayor (still built on a bluff). He assured me that, no matter how things look, *he* does not make deals. They really settled on *his* terms. But isn't there a dues checkoff? No. The *city* does not subtract union dues before pay reaches the men; their credit union does (a device the union had suggested from the outset). What about recognition of the union; wasn't that guaranteed? No, it was not. There is no contract, only a memorandum signed by the City Council. Well, is that not a binding agreement— i.e., a contract? "No, it is a *memorandum*" (see how useful it is to be an English major?)—"but we have a way of honoring our commitments." The code. Well, then, didn't the union get a larger raise than the mayor said it would? Not from the *city*. Until July 1, when all city employees were scheduled for a raise, the extra demands of the union will be met by a contribution of local businessmen. Will the mayor handle promised union agitation by the hospital and school employees in a new way, after the experience of the garbage strike? "No. Nothing has changed."

Wrong again, Henry. Preachers like James Lawson, better educated than some Brown graduates, are convinced that the God of Justice is not dead, not even in Memphis. Most important, Memphis is now the place where Dr. King delivered one of his great speeches—those speeches that will outlive his labored essays. The excerpt most often published from that last speech told how King had been to the mountaintop. But those who were there at the Mason Temple to hear him, the night before he died, remember another line most vividly. He almost did not come to that meeting. He was tired; the weather was bad, he hoped not many would show up (his first march had been delayed by late spring *snows* in Tennessee); he sent Ralph Abernathy in his stead. But the same remarkable people who rode twenty hours in a bus to stand on the curb at his funeral came through storm to hear him speak on April 4. Abernathy called the Lorraine and told King he could not disappoint such a crowd. King agreed. He was on his way.

Abernathy filled in the time till he arrived with a long introduction on King's life and career. He spoke for half an hour—and set the mood for King's own reflection on the dangers he had faced. It was a long speech —almost an hour—and his followers had never heard him dwell so long on the previous assassination attempt, when a woman stabbed him near the heart. The papers quoted a doctor as saying that King would have died if he had sneezed. "If I had sneezed," he said, he would not have been in Birmingham for the marches. "If I had sneezed—" ("Tell it!" He was calling the roll now, talking "and a half," tolling the old cadences.) He could never, had he sneezed, have gone to Selma; to Washington for the great March of 1963; to Oslo. Or to Memphis.

For the trip to Memphis was an important one. He did not so much climb to the mountaintop there as go back down into the valley of his birth. Some instinct made him return to the South, breathing in strength for his assault on Washington, which he called the very last hope for nonviolence. He was learning, relearning, what had made him great—learning what motels to stay at; what style to use; what were his roots. He was learning, from that first disastrous march, that he could not come in and touch a place with one day's fervor; that he had to *work with* a community to make it respond nonviolently as Montgomery had, and Birmingham, and Selma.

It is ironic that the trouble on that first march broke out on Beale Street, where another man learned what his roots were. W. C. Handy did not come from Memphis, like Bessie Smith; he did not grow up singing the blues. He learned to play the trumpet in Alabama from a traveling bandmaster, a real Professor Harold Hill. Then he went North, to tootle transcribed Beethoven on "classical cornet" afternoons in Chicago. It was only when he came back South, and saw that local songs *worked* better with audiences, that he began to write down some of those songs and get them published.

King, after largely ineffectual days in Chicago, returned to Memphis, the deracinated black coming home. Home to die. His very oratory regained majesty as he moved South. He had to find out all over what his own movement was about—as Marc Connelly's "Lawd" learns from his own creation: "Dey cain't lick you, kin dey Hezdrel?" Bevel said the leader was not Martin King. That was true, too, in several ways. In one sense, Rosa Parks was the true leader. And T. O. Jones. All the unlickable Hezdrels. King did not sing the civil rights blues from his youth. Like Handy, he got them published. He knew what *worked*—and despite all the charges of the militants, no other leader had his record of success. He was a leader who, when he looked around, had armies behind him.

This does not mean he was not authentic as a leader. On the contrary. His genius lay in his ability to articulate what Rosa Parks and T.O. feel. Mailer asks whether he was great or was hamboning; but King's unique note was precisely his *ham* greatness. That is why men ask, now, whether *his* kind of greatness is obsolete. Even in his short life, King seemed to have outlived his era. He went North again—not to school this time, but to carry his movement out of Baptist-preacher territory—and he failed. The civil rights movement, when it left the South, turned to militancy and urban riots. Men don't sing the old songs in a new land.

Yet it may be too soon to say that the South's contribution has been made. The garbage strike opens a whole new possibility of labor-racial coalition in those jobs consigned exclusively to blacks throughout the South. And, more important, the Northern black, who has always had a love-hate memory for the South, begins to yearn for his old identity. The name for it is "soul."

The militant activists insist on tradition (Africa) and religion (Muslimism, black Messianism, etc.) and community (the brothers). Like the young King, many blacks feel the old Baptist preachers were not dignified. Better exotic headdress and long gowns from Africa than the frock coat of "De Lawd." But the gowns and headgear *are* exotic—foreign things that men wear stiffly, a public facade. There are more familiar black traditions and religion and community. Black graduate students have earned the right to go back to hominy and chitlins and mock anyone who laughs. The growth of "soul" is a spiritual return to the South—but a return with new weapons of dignity and resistance. Religion, the family, the past, can be reclaimed now without their demeaning overtones. In this respect, the modern black is simply repeating, two decades later, King's brilliant maneuver of escape and reentry. He got the best of both worlds—the dignity that could only be won "outside," and the more familiar things that such dignity can transform. King was there before them all.

He remained, always, the one convincing preacher. Other civil rights pioneers were mostly lawyers, teachers, authors. They learned the white man's language almost too well. King learned it, too; but it was always stiff. He belonged in the pulpit, not at the lectern. Bayard Rustin, with his high dry professional voice and trilled *r*'s, cannot wear the SCLC's marching coveralls with any credibility. The same is true, in varying measure, of most first-generation "respectable" leaders. Some of them would clearly get indigestion from the thinnest possible slice of watermelon. Adam Powell, of course, can ham it with the best; but his is a raffish rogue-charm, distinguished by its whiff of mischief. King, by contrast, was Uncle Ben with a degree, a Bill Bailey who came home—and turned the home upside down. That is why he infuriated Southerners more than all the Stokelys and Raps put together. In him, they saw *their* niggers turning a calm new face of power on them.

King had the self-contained dignity of the South without its passivity.

His day is not past. It is just coming. He was on his way, when he died, to a feast of "soul food"—a current fad in black circles. But King was there before them. He had always loved what his biographer calls, rather nervously, "ethnic delicacies." He never lost his "soul." He was never ashamed. His career said many things. That the South cannot be counted out of the struggle yet. That the black does not have to go elsewhere to find an identity—he can make his stand on American soil. That even the Baptist preacher's God need not yield, yet, to Allah. God is not dead—though "De Lawd" has died. One of His prophets died.

—*Esquire*, August 1968

AFTERWORD

After I heard James Bevel's great sermon in the union hall, I tried to interview him at his Memphis motel. But he had never heard of me, and would not waste his time on a nobody. After my article appeared, he tried to recruit me as his very own publicist. Then it was "Brother!" when we met, and his considerable rhetorical skills were all marshaled to prove that the rights of blacks could be protected only if Garry Wills were to write a book about James Bevel. When I refused to become his celebrant, he went back to his first attitude of dismissal. I was, for his purposes, a nobody again.

II
Spies

||||||||||||||||||||||||||||||||||||||||||||||||||||||||||||||||||||||||||||||||||||||||||||||||||||||||||||||||||||||||||||

Ours has been the age of the spy. We spied on ourselves to keep others' spies from stealing our bomb. The President was given his own private army of spies—and the FBI quickly spied on it. Doing spy work to find the source of leaks, the Nixon White House found that the Pentagon was spying on the White House. Presidential aide John Ehrlichman spied on Defense Secretary Melvin Laird to see if he was spying on the the President. John Kennedy inspired a love for spies, for the James Bond stories, for "I Spy" and "The Man from U.N.C.L.E." That fad inspired, in turn, a series of spy "takeoffs," like "Get Smart." But the comedy writers could not outdo the real spies in absurdity. E. Howard Hunt, with his red wig and voice transformer, was a real-life Maxwell Smart. Behind the glamorous pose of James Bond were the sordid schemes of Gordon Liddy. And then, irony of ironies, it was Richard Nixon who helped catch a real spy—only to let that fact haunt him and, in time, trip him up.

||||||||||||||||||||||||||||||||||||||||||||||||||||||||||||||||||||||||||||||||||||||||||||||||||||||||||||||||||||||||||||

4

"McCarthyism"

It is unfortunate that McCarthyism was named teleologically, from its most perfect product, rather than genetically—which would give us Trumanism. By studying "McCarthyism" in terms of Joseph McCarthy's own period of Red-baiting (1950–54), a number of scholars have called the disease an imbalance between Congress and the Executive (thus contributing to a pre-Nixon glorification of the imperial presidency). It is true that the Executive opposed investigative committees by McCarthy's time; but in 1947 the President not only cooperated with these committees, but gave them the means to grow powerful. Secretary of State George Marshall cooperated with Senator Styles Bridges and Congressman John Taber in purging the State Department. Attorney General Clark cooperated with the House Committee in its Eisler "investigation." J. Edgar Hoover appeared before the Committee to praise its work and get congressional support for his own vast loyalty probings. In March of 1947, when Truman issued his executive order for loyalty tests, he designated HUAC [House Committee on Un-American Activities] files as an official source of evidence on employees' ties. The Committee congratulated him for his action, and took credit for purging the Executive. The Hollywood hearings of 1947 did not threaten Truman—the farther the Committee looked from Washington, the happier he was. Only when the Committee tried to steal the spotlight from the Justice Department's own work before the grand jury in New York did Truman ease up on his cooperation with the Committee —and by then it was too late. The grand jury had heard a witness named Whittaker Chambers, and Congressman Nixon had received from that witness some documents he would not yield to the grand jury.

Yet even when Truman dismissed the Committee's hearings as a "red herring" in the fall of 1948, he was not taking the hard stand that he would in McCarthy's time. He meant that conducting any investigative work in this special session of Congress, which he had demanded between the conventions and the election, was a distraction from the job of passing his economic program. Actually, the Committee's 1948 triumph in the Hiss case had worked to Truman's advantage. Alger Hiss may have been associated with the New Deal in the past, but his present work was with John Foster Dulles at the Carnegie Endowment for International Peace. More important, the principal witness against Hiss, Whittaker Chambers, claimed a Communist group had been formed within Henry Wallace's New Deal Agricultural Adjustment Administration; and two of those named—Lee Pressman and John Abt—were working prominently in Wallace's 1948 presidential campaign. Two more of Wallace's supporters—Harry Dexter White and Victor Perlo—were called Communists by Committee witness Elizabeth Bentley. The Committee actually summoned these people to testify during the campaign—they pleaded the Fifth Amendment. Truman had feared the Wallace threat more than a Southern split, and his workers took elaborate steps to contain that threat. The Committee completed their work.

Henry Wallace had broken with the Truman administration on the aggressive new turn our foreign policy had taken by 1947. He saw the NATO alliance, in particular, as a de facto substitute for all our commitments to the United Nations, a confession that peace had given way to war. His analysis of the Dean Acheson strategy, moving from the Truman Doctrine to the Marshall Plan to the Atlantic alliance, resembles that made by today's revisionist historians—and proves that their analysis is not something available only to hindsight. What is more, the effectiveness of Wallace's first criticism proves that the Acheson world vision only later acquired its air of unquestionable rectitude. The Administration was busy draping flags all over these programs in 1947—but it feared some flags would slip. When Wallace first broke with Truman, a poll showed 24 percent of Democrats willing to vote for him against Truman. There was still some question whether Truman was a proper heir to the New Deal. But Wallace, who was Roosevelt's first wartime Vice-President, had also been one of the original New Dealers.

Clark Clifford, Truman's campaign strategist, identified Wallace as the principal threat to reelection in his famous memo of November 1947. He said Truman should head off this threat by "some top-level appointments from the ranks of the progressives," by offering a civil rights program ("the South can be . . . safely ignored"), and by "isolating" Wallace: "The Administration must persuade prominent liberals and progressives—and no one else—to move publicly into the fray. They must point out that the core of Wallace's backing is made up of Communists and fellow travelers." It would be the job of prominent liberals —which meant, principally, of Americans for Democratic Action—to do the Committee's kind of work in a more sophisticated way, and to do it on their fellows.

The ADA was ready. America's "best and brightest" had conducted America's triumphant crusade against fascism as officers, scientists, foreign experts, intelligence men. They meant to continue America's benevolent disposal of the world's freedoms, using the tools of their mind (especially the bomb) to enforce their Wilsonian vision of the world. If they had to win support in a still isolationist country by a little saber-rattling ("scare hell out of the country"), the gains were worth the price. Besides, it was easy for any liberal to have been at meetings now stigmatized by the Attorney General's list, or to have worked closely with Russians during the war. Show-business witnesses before the Committee learned what columnists to go to for reinstatement as a loyal American—Hedda Hopper on the West Coast, George Sokolsky on the East Coast. By 1947 the liberal intellectual's way to establish his anti-Communist credentials was through the ADA, developed on the base of Reinhold Niebuhr's "pragmatic" wartime Union for Democratic Action. Formed in the wake of the 1946 election, which gave Congress to the Republicans, the ADA thought it could prevent further reaction by carrying on its own purge of Communists. When the Marshall Plan was proposed, such liberals made it the touchstone of enlightened anticommunism. The ADA accepted the heritage of the New Deal along with that of the OSS. (The CIA was another of Truman's gifts to us in 1947.)

—from Introduction to Lillian Hellman's *Scoundrel Time*, 1976

AFTERWORD

In a conversation over my introduction to Lillian Hellman's book, the author and editor Penn Kimball told me how the ADA did its Commie-hunting in 1947. A graduate student working with the Joseph Alsop papers in the Library of Congress came across a memo Arthur Schlesinger had sent to various people (including Alsop). At a lunch with the editors of the *New Republic*, Schlesinger complained that the magazine sent a member of the Progressive Citizens of America to cover the founding of the ADA. (Schlesinger would later write that the Attorney General's list was incomplete because it did not include organizations like the PCA.) Then Schlesinger gave his correspondents this estimate of Penn Kimball: "Kimball strikes me as a smart and cool Party-liner, at least. . . ." Three years before McCarthy appeared on the scene, some people were distributing their own private lists of suspected Communists.

5

Alger Hiss

Suspecting Alger Hiss was somehow, on the face of it, indecent. He was almost drearily correct. He specialized in innocence. He was innocent of failure—so he could not understand his father. He was innocent of doubt—so he could not understand his brother Bosley. He was so innocent of psychic turmoil that his sister was in and out of mental institutions for several years without his being aware of her disturbance. He passed through the thirties so innocent of ideology that he could later swear he met no Communists at all, or—if he did meet any—he could not recognize them. He was innocent of friendships except with the well-placed, with patrons. He seemed to spring fully armed from the head of Justice Oliver Wendell Holmes. He became the perfect civil servant, a political Jeeves who knew his place and filled it perfectly. If anything, he was so correct as to rule out originality. Justice Frankfurter, another of his early patrons, was puzzled in 1946 to reflect that Hiss had not quite lived up to the promise of his youth, which boded something beyond mere rectitude.

Yet this man drab with the proper virtues had known something vivid in his life, "a cross between Jim Tully, the author, and Jack London" (in his own later words). The contrast between the two men—pale Alger and Technicolor Whittaker—should have made their encounters, however fleeting, things to remember. But, no, Hiss stayed innocent of interest. He took and gave gifts—a rug, a car—while barely noticing the donor/recipient. Hiss's only recollected interest was, characteristically, a distaste for the improper. He remembered no tales from this Jack London, just bad teeth. The true succeeder can spot a failer far off.

It is a wonder he did not keep him farther off. They shared holidays and trips (Alger fraternizing, apparently, in a fit of absentmindedness), hobbies and small talk—just enough for the vagabond to bring down the paragon. Why did this two-bit Jack London do it? Envy, perhaps, or rejected sexual feelings? The failer would call it "a tragedy of history." The bewildered succeeder first tried to laugh it off as a comedy of errors. The only explanation Hiss's lawyers could come up with was a clash of personalities. The obscure Chambers grudge was more a matter for psychiatrists than lawyers. It was rumored that Chambers had been under psychiatric care (enough, then, to disqualify him as a witness to anything). When the defense could not substantiate those rumors, lawyers remedied the oversight by bringing in a psychiatrist as part of their courtroom team and calling on him to describe the "mental illness" that made Chambers incapable of the truth—one of the first and least glorious uses of psychobiography at a distance. Still, what other weapon could be used to defend a man from what was so clearly a *psychic* affront? The question was not Hiss's innocence, but the craziness of anyone who would challenge it. That Hiss could err was improbable: that Chambers would imagine, invent, and exaggerate was inevitable.

But one thing puzzled even Hiss's lawyers. A gentleman does not lie to his friends; and Hiss was, beyond doubt, a gentleman. Why, then, did he tell John Foster Dulles he had not called his wife about the Chambers accusations when he had? Why would he later write that he went to the first HUAC sessions without counsel, though his lawyer John F. Davis went with him? This precise and orderly man early began telling odd little needless lies, or suffering inexplicable "blackouts" of memory. In his first appearances before the House Committee Hiss remembered the names of three former servants, but not of the one maid with whom he had maintained contact—the maid who worked for him, full time, for three years (the key years so far as Chambers was concerned), the woman whose whole family had been involved with the

Hisses then and afterward, and the maid to whose son he gave the type-writer on which damaging documents were typed.

Perhaps he just forgot Claudia Catlett's name. But by December 7 he had remembered it. He called John F. Davis, that invisible lawyer from his first appearance before the Committee, and said he had given the Woodstock to Mrs. Catlett's son. Yet three days later he suffered another spell of forgetfulness, this time under oath, and told a grand jury he had no idea where the typewriter was. He repeated this claim five days later before the same tribunal. Nor was Hiss the only one who lied about that typewriter. His brother Donald later claimed that he learned of its whereabouts only when Mrs. Catlett's son came to him—though John Davis had relayed Alger's message before that.

Of course, Whittaker Chambers lied under oath. But that was to be expected. He was as clearly disreputable as Hiss was honorable. Why did the honorable man lie? How could *he* be doing "what gentlemen just don't do"? He claimed that doubts about his loyalty "had blown over" when he left the State Department to become president of the Carnegie Endowment. But they had not. John Foster Dulles, chairman of the Endowment's board, found out after the appointment that Hiss had been suspected of Communist ties. When Dulles questioned him about this, Hiss dismissed the investigation as unimportant, not really worth mentioning. Yet it had been important enough for him to ask Acheson if it posed a threat to his advancement in the State Department (Acheson said that it did)—important enough, in other words, to motivate a decision to accept the Carnegie appointment.

Hiss's sluggish memory, his hedged denials of each Chambers story until external evidence forced him to remember or revise (e.g., his tale of the gift of his car), led certain of his own lawyers to feel he was hiding something. But they thought he was covering up some radical indiscretion on his wife's part. Priscilla had always been more ardent than her proper husband. Even his lies were noble.

Hiss's lawyers learned not to probe too far, which explains some odd reticences in his trials. Thus the government expert was allowed to identify stolen documents as having been typed on Hiss's typewriter from a study of only ten characters. This has been called inadequate by later Hiss defenders, who ask why his lawyers did not demand further evidence (which, by the way, the FBI expert possessed). But the defense did not want to dwell on this matter—one of its own experts had not only made the same identification, but said the stolen documents had most of the typist's characteristics of *Priscilla Hiss*.

Hiss said he received a rug from Chambers in 1935—which could not have been the one picked out by the art historian Meyer Schapiro in December of 1936 and delivered after January 1 of 1937. Asked what had become of the rug, Hiss admitted he still had it. Why did his lawyers not show it to Schapiro when he was on the stand? A denial that he chose this rug would have countered the second of two perjury counts in the indictment—Hiss's claim that he never even saw Chambers in 1937. But Hiss volunteered little if anything. He listened to charges, separated them, minimized them, working always by reaction. His life in the years when Chambers knew him comes out, in all his own accounts, as a curious blank, filled with nothing but official endorsements and separate denials. It was important *not* to pursue greater detail—as in the case of the car. Counsel for the known partner in that transfer, William Rosen, warned Hiss's lawyers that it was arranged by "a very high Communist," whose name "would be a sensation in this case."

Hiss's reluctance to proffer material of his own was so marked that defenders have made a great deal of one apparent exception—his lawyer found and produced the Woodstock typewriter the FBI was searching for. But: (1) The typewriter itself was irrelevant as evidence. The charge against Hiss rested both on Hiss's handwritten memos and on the identity of the type in letters admittedly done by Priscilla Hiss and in stolen documents produced by Chambers. (2) Alger Hiss knew generally where the typewriter was to be sought by December 7 and Donald Hiss had been told by mid-February that it had come into the possession of Ira Lockey. They sat on this knowledge for two months. (3) The lawyer who got the typewriter acted on his own initiative (not prompted by either of the Hiss brothers, despite their knowledge). (4) That lawyer's motive seems to have been fear that the FBI, then questioning the Catletts, would build a case that Hiss was suppressing evidence.

Allen Weinstein has many—literally thousands—of new facts to offer in his book *Perjury*. But the discovery that Donald Hiss had traced the typewriter to Lockey two months before it was produced is probably the most damaging item of all. It destroys all theories of a planted, altered, or forged typewriter. Herbert Packer had shot holes in that theory in 1962, arguing that the typewriter was not relevant, so building one would have been both wasteful and dangerous (depending on confidentiality in all the sources of parts and expertise and planting); or that, *if* a fake machine had been produced to type out the stolen documents, the first priority after it had performed that function would be the machine's destruction, for fear the real one would show up and reveal the imposture—would, that is, show up *just as the Hiss machine did*. The Hisses knew where the typewriter was, and the friendly Catletts were misleading the FBI. They almost certainly would have known if the FBI had found and fiddled with the machine, and they would have told the Hisses.

Weinstein's discovery is important not only or mainly because it destroys the most popular theories of conspiracy against Hiss, but because it shows Hiss had reason to know those theories were flawed through all the years he has been espousing and promoting them. The honorable gentleman is misleading his own friends. Still, the dogged and infectious air of innocence around Hiss will continue to give people pause. The problem is not simply how he could have spied for a certain period, but (even more) how he could have lied to his friends in the first place and maintained the lies with assurance—even with serenity—for over a quarter of a century.

Most of those who conclude, reluctantly, that Hiss lied cannot believe that he kept a foreign allegiance after America went to war, or—even more unthinkable—after America's break with Russia. But neither could they think that of Kim Philby, another impeccable civil servant with a disreputable friend. Philby was about to be knighted, and was being groomed to head the British secret service, when the first doubts about him forced him from the service. Hugh Trevor-Roper notes that Philby had a vague reputation as an intellectual though he showed no real interest in ideas. The same is true of Hiss, whose intellectual reputation rests almost entirely on a book Justice Holmes inscribed for him.

Philby, it should be remembered, did not flee to Russia after his first uncovering. He stayed on, maintained his innocence, and worked his way back into the service. He even profited from doubts expressed about his loyalty. Harold Macmillan, lacking legally actionable proof of Philby's Russian ties, had to exonerate him in public. "He really could not have done me a better turn if he had wanted to. By naming me, he [his accuser, Colonel Lipton] virtually forced Harold Macmillan to clear me." Weinstein shows that Dean Acheson's defense of Alger Hiss, at his own confirmation hearings, was dutiful, lukewarm, and motivated by Acheson's much closer ties to Donald Hiss. Acheson refused to testify for Hiss at his trial.

It was only after clear legal evidence was found against him that Philby skipped to Russia. He was facing a death penalty. Hiss, by contrast, faced much less serious charges, and had a chance to ride them out. Chambers might crack—tie himself up in imagined self-dramatizings, go crazy, or commit suicide (as he almost did). Hiss had many defenders, and his first trial ended in a hung jury. His conviction for perjury was a comparatively minor affair. He could dream of rehabilitation. Perhaps he was urged to undertake the line he did. The Russians' interest in Hiss is indicated, according to Weinstein's research, by the kinds of FBI files Judith Coplon was stealing, and by Gromyko's suggestion in 1945 that Alger Hiss be the first UN secretary general.

Weinstein does not conclude that Hiss was still engaged in espionage after the Chambers incidents. But he shows that during and after the war Hiss did ask for highly classified material outside his area of specialization. I prefer to think Hiss was a principled believer throughout this time; that his early radical sympathy did not disappear into the organization man he later made of himself; that his energetic protestations of literal innocence drew on deep conviction that he was working for the country's (and the world's) ultimate good; that he was not summoning a mistaken heroism of denial to cover up peccadilloes; that friends were deceived to advance a cause, not to salvage a career.

Hiss's strategy of total and universal denial and forgetfulness from the very outset, his refusal to volunteer autobiography, make sense if Hiss had more to hide than the Chambers documents. That is why I would not pay him the insult of believing his denials. I would rather think he has been serving his own gods with hidden gallantry. It is only as a secret foe that he regains the integrity people have always sensed in him. Chambers, confirmed in so many other ways, would thus be confirmed in his belief that Hiss was still a Communist in 1949, an enemy he could salute even as he tried to destroy him. The drama of the courtroom was not overstated. It was a battlefield transaction.

—*New York Review of Books*, April 20, 1978

AFTERWORD

For some, the crude antics of our spy-hunters have constituted an illogical guarantee that there are never any spies to be found. One famous journalist stopped speaking to me after this article appeared. But Maurice Braverman, the Baltimore civil rights lawyer, said he had information—only some of which he could give me—that confirmed what I had written. Braverman was an avowed Communist, who went to jail under the Smith Act—to Lewisburg Prison, in Pennsylvania, where Hiss was. When they met, Braverman told Hiss that he had been a lawyer for William Rosen (who had transferred ownership of Hiss's car, according to Chambers, to a Communist). "Yes, I know," Hiss answered—and never brought the subject up again, though the two men served together on breakfast shifts, and shared letters and gossip of the outside world. Most prisoners who claim they have been framed become jailhouse lawyers, forever retrying their case with anyone who will listen. Yet here was a man privy to a key transaction in Hiss's case, and Hiss never tried to question Braverman. Moreover, when it came time for Hiss to be released, Braverman's lawyer said that the Party hoped he would have an eloquent statement to read when he came out the gate. Television cameras were bound to be there. Braverman asked Hiss what he meant to say; Hiss said he had not thought of making a statement. Braverman said he thought it would be a good idea. "What should I say?" Hiss asked, and Braverman composed his first draft for him. He behaved like one still serving the Party.

6

Hiss and Nixon

It was not a pleasant picture to see a whole brilliant career [Alger Hiss's] destroyed before your eyes.
—Richard M. Nixon, *Six Crises*

He went as abruptly as Spiro Agnew did, but he leaves a bigger space behind. He was always a kind of obtrusive blank space in our consciousness. Though he was a remote and private man, we had all been drawn into his life story. Decade by decade, crisis by crisis, we were unwilling intruders on his most intimate moments—we saw him cry, sweat, tremble; saw him angry, hurt, vindictive. The tapes even let us eavesdrop on those embarrassing conversations. Although no one really knew him, we all knew too much about him. He was too vividly present, and yet not present at all—a collection of quirks, and not a person; a conspicuous absence. The incomprehensible had become inescapable —every family's family joke, or boast, or embarrassment. We were not the Pepsi Generation, but—as Meg Greenfield put it, years ago—the Nixon Generation. He kept popping up, indestructible, no matter what happened to him or to us; and even those most opposed to him grew resigned, in time, to the fact that "Tricky Dick" was actually "Mr. President."

Perhaps it was because he had no real privacy that we always seemed to be invading it. His private life finally became our public life. He was the parasite on us—not, as we feared at times, the other way around. The last violation—the tapes—fit a pattern of private anguish as "historical record." There are tapes of him listening to his tapes, and tapes of him replaying his favorite crises. "The New Nixon" was forever brooding on the old one, bringing himself back to life just when his admirers hoped he had shed past humiliations. Even his critics were puzzled that he dwelt on equivocal past deeds after the large admitted accomplishments of his presidency. He embarrassed us and himself by endlessly resurrecting past embarrassments—as if he did not exist but for his setbacks. He kept displaying each albatross as a trophy, each pettiness as a triumph. Humiliation, it turns out, was his weapon—what he could blame on others, feeling the resentment that gave him his greatest sense of identity. His power came from a shared sense of powerlessness with the "silent majority," the silenced and oppressed victims of an undefined but omnipresent "they." When he summoned up the Hiss case, it was to remember how "they" destroyed Chambers. And he was always summoning up the Hiss case.

He was almost flirtatious about getting others to read his life story— even with Bob Haldeman, who assured him he had read it. "Warm up to it, and it makes fascinating reading. . . . I want you to reread it." He wanted to call up his ghostwriter on the book, Al Moscow, for a ten-years-later scene in his endless private showing of "This Is Your Life." Any occasion would serve to send people back and back to the same six crises. Charles Colson, ingratiating himself, read the book fourteen times. And always the main thing was Hiss.

The Hiss investigation was a stunning achievement for a freshman congressman, back in 1948. Even the hottest Commie-hunters in the House Committee on Un-American Activities were unwilling to believe the apparently wild assertion of Whittaker Chambers, a *Time* editor, that he and former State Department employee Alger Hiss had been spies for the Russians back in the thirties, giving them classified U.S. documents. Nixon not only pushed on the investigation, but turned up material—microfilm of stolen government documents hidden in a pumpkin on the Chambers farm—that led to Hiss's later conviction as a perjurer for denying the Chambers story. Hiss claimed he was being attacked for his minor connection with the New Deal; but the investigation was riskier than that for young Mr. Nixon. In the summer of 1948, Truman's defeat was assumed, and even Dean Acheson was treating John Foster Dulles as the next Secretary of State. Yet Dulles was chairman of the board at the Carnegie Endowment, and he had appointed Hiss to the Endowment's presidency. Nixon was taking a big chance within his own party when he went after a Dulles protégé. But one would think that, a quarter of a century later, after all the more important acts of his presidency, Nixon would be glad to forget a case that brought him at least as much criticism for his tactics as credit for the end result.

Yet in a single conversation with John Dean (February 28, 1973), Nixon brought up the Hiss case three times in differing contexts:

1. "When you talk to Kleindienst—because I have raised this in previous things with him on the Hiss case—he got, he'd forgotten, and I said, 'Well, go back and read the first chapter of *Six Crises.*'"

2. "These guys, you know—the informers, look what it did to Chambers. Chambers informed because he didn't give a goddamn. . . . But then, one of the most brilliant writers according to Jim Shepley we've ever seen in this country—and I am not referring to the Communist issue—this greatest single guy in the time of twenty-five or thirty years ago, probably, probably the best writer [unintelligible] this century. They finished him."

3. "Tell Kleindienst that Kleindienst in talking to Baker and Ervin should emphasize that the way to have a successful hearing and a fair one is to run it like a court: no hearsay, no innuendo . . . Now you know goddamned well they aren't going to . . . Tell them that is the way Nixon ran the Hiss case. Now, as a matter of fact, some innuendo came out, but there was goddamned little hearsay. We really—we, we just got them on facts and just tore them to pieces. Say 'No hearsay; no innuendo.' And that, that he, Ervin, should sit like a court there."

Each of the parallels suggested is farfetched or self-destructive. In the first reference, Nixon is telling Kleindienst to investigate without executive cooperation—but if Kleindienst looked at the book, he would see that Nixon condemned Truman for noncooperation. In the second reference, Nixon is reflecting on the difficulty of getting an FBI agent to inform against Democratic administrations—he likes the idea, but his mind veers off toward Chambers's destruction (by those who also tried to destroy *him*). In the third reference, he points up the weakest aspect of the whole Hiss case—its injudicious procedure. None of these ironies can reach him. The case has a totally personal meaning: the moods of determination and grievance it planted in him. He cannot see that it means quite different things to others. No wonder he could not see what damage the tapes would do him when released.

There was no end to the uses Nixon made of Hiss. On March 22, 1973, Dean is told how to flatter Senator Ervin "on the mountaintop" into granting "executive privilege" to White House aides: "We're making a lot of history. And that's it—we're setting a historic precedent. The President, after all, let's point out that the President, uh, how he bitched about the Hiss case. Which is true, I raised holy hell about it." The Hiss case was glory enough to turn any man's head—to make him enter history. Richard Nixon, Hiss is your life.

The fate of Chambers was on Nixon's mind again when he talked to Haldeman on March 27, 1973, about keeping Magruder from copping a plea: "Chambers is a case in point. Chambers told the truth, but he was an informer, obviously, because he informed against Hiss. First of all, it wouldn't have made any difference whether the informer [unintelligible]. First of all he was an [unintelligible]. Hiss was destroyed because he lied—perjury. Chambers was destroyed because he was an informer, but Chambers knew he was going to be destroyed." Even in the case that made his name, that brought him into history, Nixon dwells on the defeat of Chambers rather than on his own triumph.

In the April 16, 1973, charade, when Nixon tells Dean (for the benefit of the tapes) to tell the truth, we are given this cautionary tale: "That son-of-a-bitch Hiss would be free today if he hadn't lied [of course Hiss was free by then]. . . . But the son-of-a-bitch lied, and he goes to jail for the lie rather than the crime. . . . So, believe me, don't ever lie with these bastards." When Henry Petersen came into the Oval Office, Nixon found still another use for The Case. Trying to plant the idea that White House money for the Watergate prisoners was merely a defense fund, not hush money, the President airily said: "They helped the Scottsboro people. They helped the Berrigans; you remember the Alger Hiss defense fund?" Innocent of irony as ever, he names the kind of left-wing funds he had denounced in the past as Communist fronts.

There is no consistent thread of logic in these random musings. The case rearises like a private music or set of moods connected with past defeat-and-triumph, a sour memory of grievance and achievement. Only this private meaning could blind Nixon to all the self-destructive elements in the comparisons he invited men to entertain. The deepest irony of all would arise from this constant urging of his own private memories on others. He would be undone as a result of the effort to repeat the undoing of Hiss. It was the Ellsberg operation that had to be kept hidden when Watergate occurred; and that operation was a deliberate effort to turn the Pentagon Papers into another Pumpkin Papers story. Colson spelled this out in his 1971 phone call to E. Howard Hunt (another of those damning private tapes): "It could become another Alger Hiss case, where the guy is exposed, other people were operating with him, and this may be the way to really carry it out. We might be able to put this bastard into a helluva situation and discredit the New Left." Colson had read *Six Crises* one time too often—and the Hiss case, after destroying the careers of both Hiss and Chambers, was circling back to claim its final victim.

This was not merely Colson's private variation on the Nixon fantasy. The President himself established the Ellsberg-Hiss equation in his orders to the plumber-in-chief assigned to get Ellsberg. Egil Krogh said at his sentencing: "The President had directed that I read his book, *Six Crises,* and particularly the chapter on Alger Hiss, in preparation for this assignment." Like Colson and the President, Krogh saw Hiss in Ellsberg, and felt spurred to expose him. But if we take the President's advice, and go back to read that all-important chapter in his life, we can trace just how his last defeat was derived from his first great public triumph.

Nixon, in 1948, was fighting what he saw as an attempt at executive cover-up. He tells how FBI reports were denied his committee; how he refused to turn over evidence to the Justice Department, for fear it would be suppressed; how the pumpkin microfilms were kept from a grand jury so the committee could produce them; how he traced the cover-up into the Oval Office itself. Why, he asks, would a President like Truman shelter a Communist like Hiss? "His error was sheer stubbornness in refusing to admit a mistake. He viewed the Hiss case only in its political implications and he chose to handle the crisis which faced his Administration with an outworn political rule of thumb: Leave the political skeletons hidden in the closet and keep the door locked." The man who told others to read this chapter was clearly incapable of reading it—truly reading it—himself in 1973, when he raised the standard of "protecting our people" over the Eisenhower desire that his men be "clean."

Telling Krogh to bone up on the Hiss case is odd enough. But Nixon also recommended it to Acting Attorney General Richard Kleindienst—though the chapter portrays Nixon as refusing evidence to the Justice Department because it might be a partner in the President's cover-up: "When we arrived in New York at 7:30 that evening, we were met by representatives of the Justice Department. They went with us to the Commodore Hotel, where we were scheduled to meet Chambers at nine o'clock. There, we engaged in a violent verbal battle as to whether the committee should continue its investigation of the case or should turn over the microfilm to the Justice Department and leave the entire responsibility to them. I made it clear that we had the greatest respect for lower echelon Justice Department officials, who were just as interested in getting at the truth in this case as we were. But I also made it clear that I had no confidence in some of their superiors who were under great political pressures and who so far had made a record which, to put it politely, raised grave doubts. Did they intend to bring out any facts that might be embarrassing to the national Administration? In short, we did not trust the Justice Department to prosecute the case with the vigor we thought it deserved. The five rolls of microfilm in our possession, plus the threat of a congressional public hearing, were our only weapons to assure such a prosecution." Nixon's own attempt to use Henry Petersen would give substance to the vague suspicions he voiced about Truman back in 1948—yet he remained serenely unaware of this lesson from the Hiss case.

By the end of Nixon's regime, all the instruments being wielded against him, to his own great indignation, were weapons he had forged for use against the White House twenty-five years before. There are four such weapons described in the Hiss chapter—congressional committees, the press, government leaks, and plea bargaining.

1. *Congressional committees.* In his book, Nixon does not disguise the fact that he used congressional investigation less for legislative research than as a check on the Executive's pretensions: "The Hiss case exposed the blindness of the Truman administration and its predecessors to the problem of Communist subversion in government. It demonstrated the need for congressional investigatory bodies, like the

Committee on Un-American Activities, which could expose such laxity and, with the help of a mobilized public opinion, could force the executive branch to adopt policies adequate for dealing with the problem." Despite Nixon's weird later claim, made to John Dean, that "Nixon ran the Hiss case . . . like a court," it was actually conducted all in headlines. Even Chambers praised the histrionic tactics of committee counsel Robert Stripling for the way they kept the case alive. And when Chambers objected to public hearings, Nixon told him: "It is for your own sake that the committee is holding a public hearing. The Department of Justice is all set to move in on you in order to save Hiss. They are planning to indict you at once. The only way to head them off is to let the public judge for itself which one of you is telling the truth. That is your only chance." Nixon was using the committee to head off the courts, as a way of exposing the Executive's cover-up. When a reporter asked why the vital microfilm of stolen government documents was hidden in a pumpkin on the Chambers farm, Nixon said, "But you're a newspaperman! You must have seen the headlines we got out of it!" After subpoenaing this microfilm, Nixon left for a Caribbean cruise on the *Panama*—only to leap back, photogenically, from ship to lifeboat to military seaplane dispatched in haste; landing in Washington to squint, for the cameras, at unrolled segments of the microfilm, as if he could read them without a magnifier. From the pumpkin and the *Panama,* things converged magically, with Bob Stripling as hidden orchestrator.

This is the same Nixon whose spokesmen called the Ervin committee a circus and the Rodino committee a kangaroo court. Yet these committees had a far clearer mandate for investigating the Executive itself than did the committee inquiring into the loyalty of a private citizen (Mr. Hiss). Nixon lacked whatever excuse President Truman may have had in checking the excesses of the Commie-hunters at HUAC.

2. *The press.* Nixon once relied on publicity and the press he came so deeply to resent. He defeated Jerry Voorhis by means of public debates, and that made him too ready to accept John Kennedy's 1960 challenge. He made the first great political use of television in his Checkers speech —and that medium came back to get its revenge. And he exploited headlines in the Hiss case—long before an investigative press helped bring down his presidency. Not only was the case conducted with sensationalism and by leaks to friendly reporters, one reporter was actually conscripted into helping the Committee do its work. Bert Andrews of the New York *Herald Tribune* secretly questioned witnesses with Nixon, coordinated news breaks, and cooperated in the prosecution of

Hiss. Nixon says of one meeting with Chambers: "Andrews grilled him as only a Washington newspaperman can, and Chambers met the test to Andrews's complete satisfaction." It was Andrews who wired Nixon to return from the Caribbean. When the President's men accused Washington *Post* reporters of becoming Democratic party prosecutors of the Administration, the description did not fit Carl Bernstein or Bob Woodward—but, with a change of the party name, it could be applied to what Andrews did in 1948.

3. *Government leaks.* Nixon relied on information leaked from agencies of the Executive. He admits that the leakers were going against an express command: "A presidential order directed that no federal agency was to release information on government personnel to committees of Congress, thus blocking that avenue of investigation." But defiance of this order alerted Nixon to things going on inside the grand jury: "Hiss and his legion of supporters within the Administration still had an ace up their sleeves. They did not reckon, however, with some of the Justice Department employees in lower echelons who were so infuriated by their superiors' handling of the case that they apprised the committee of every action that was being taken." These leaks allowed Nixon to head off a grand jury indictment of Chambers by moving up his public testimony before the committee.

I learned of one such systematic leaking process from a Catholic priest, John Cronin, who was part of it. An FBI agent named Ed Hummer kept Father Cronin abreast of the FBI's leads and progress in daily reports, and Father Cronin passed these on to Nixon. Hummer's role was not mentioned in *Six Crises,* to prevent Hoover from having to discipline him—but Nixon treated as patriots the people who leaked information to him in this manner, even when it involved violating the sanctity of a grand jury inquiry. He was using the very device he denounced when Jack Anderson or Daniel Ellsberg made public the documents concerned with secret bombings or with the war rationale in Vietnam. Yet Anderson's and Ellsberg's breach of trust (if any) was less than that of Hummer or other "lower echelon" Justice Department officials, who were still part of government and were acting against an express order of the President. Mr. Anderson has never been part of the government, and Ellsberg had left its employ at the time of his leaks.

4. *Plea bargaining.* It was the tactic of Nixon's men to belittle the evidence of "stool pigeons" like John Dean, who bargained for immunity by informing. But Nixon's standards were quite different in 1948, when he quoted with approval Congressman Ed Hébert's comment to Hiss:

"Show me a good police force and I will show you the stool pigeon who turned (criminals in to) them. . . . I don't care who gives the facts to me, whether a confessed liar, thief, or murderer—if it is facts." Nixon got Chambers to testify publicly in order to prevent a grand jury indictment, and defended the de facto immunity he won for him this way: "Chambers has confessed. He is in the open. He is no longer a danger to our society."

Krogh and Colson, just like their master, were blind to this side of the Hiss case, since they made the simple equation of Ellsberg with Hiss. There were some obvious similarities between the two. Ellsberg, like Hiss, had been a trim, attractive youth who rose fast in government service, a protégé of liberals in high office. He formed, like Hiss, a striking contrast to Whittaker Chambers—Ellsberg, slight, intense, a born succeeder who began to doubt the credo of success; Chambers, podgy and diffuse, a virtuoso of dropping out. One was bright and high-strung, the other muzzily brooding; one straightforward, the other convoluted—the hunter and the hunted, the raw and the cooked.

But those were all shallow and misleading signs; Ellsberg did openly what Hiss was charged with doing secretly. Ellsberg's was an act of patriotism, a revelation of the nation's misled policy to the nation itself—and he admitted what he had done. Hiss, while denying it, was charged with stealing documents in the service of another country altogether. The real comparison, if we are to seek parallels in obedience to Mr. Nixon's direction, would be with another character in the drama of 1948. The plumbers' own CIA profile on Mr. Ellsberg should have given Krogh the proper clues:

"He absorbed the impression that he was special and destined for greatness."

"There has been a notable zealous intensity about the subject throughout his career. Apparently finding it difficult to tolerate ambiguity or ambivalence, he was either strongly for something, or strongly against it."

"He had a knack for drawing attention to himself."

"Many of the subject's own words would confirm the impression that he saw himself as having a special mission, and indeed as bearing a special responsibility."

"He also had suggested quite strongly that his action will not only alter the shape of the Vietnam war, but will materially influence the condition of our foreign policy and the relationship between the people and the government."

That does not sound like the contained and muted Alger Hiss. No, it sounds remarkably like Whittaker Chambers agonizing at stage center. He too was a zealot who doubled back on himself, a convert with incriminating papers in his hand—a Communist who would expose Communists, just as Ellsberg had helped prosecute the war he came to end. Both were, in their dramatic central act, emotional men of absolutes— self-abnegating self-dramatizers, reaching the peak of their fulfillment in a moment of confession, mixing private remorse and public charges. Each spoke with an awe at once possessive of his own achievement yet subjected to it—devout acolytes serving, respectively, the Case and the Papers. The Pentagon Papers and the Pumpkin Papers were much closer to each other than Colson or Krogh had suspected.

The points of comparison are not just superficial. Both Chambers and Ellsberg started life with driving ambitions derived from a mother— theatrical in the first case, musical in the latter. Both underwent years of artistic discipline, Chambers trying to become a poet, Ellsberg a pianist. Both gave themselves totally to the crusade they would later oppose. It was not enough for Chambers to be a Marxist literary man; he took up the demeaning tasks of espionage. Nor was Ellsberg content with computing a war—he went armed into the jungle, a foot soldier for what he believed (while he believed it). Still, despite devotion to the first cause, each man overstated his actual contribution to it—and each was attacking his own overstatements after "conversion."

The conversion was not sudden in either case, but gradual, causing confusion in the record. Neither man could handle doubt gracefully. But at last a conviction hardened, in each, that he must undo what he felt such large responsibility for—and both struck back by means of purloined documents, dramatically revealed. Ellsberg had no pumpkin, but he and Tony Russo once toyed with the idea of releasing loose sheets of the Pentagon Papers from a helicopter, snowing mystery down over metropolitan Los Angeles. Both Chambers and Ellsberg came, in time, to feel that their witness was lost on an uncomprehending world— they exaggerated the importance of their recanting, as they had that of their crusading. Each knew lapses into bitterness, despite the support of a wife involved in his mission. Such men are capable of odd, erratic conduct—the Chambers love of pseudonyms, life's underside, and suicidal gestures; Ellsberg's escape into pseudonymous orgy-touring. They were equally quick to tearfulness and denunciation; and both were extravagantly grateful to those who helped them in their time of need— Chambers to Nixon and the committee, Ellsberg to the New Left types who turned up at his rallies. Their respective trials became ideological touchstones, evoking passions that went far beyond the facts in dispute.

We seem to have wandered a fair way off from the demise of Richard Nixon—but only because we took seriously the advice he gave to so many people. He pointed us to the Hiss case as a parallel for the events of the seventies. How could he so perfectly reverse each obvious point that emerges from his own book? One way: by becoming his own victim. If the real parallel to Ellsberg is Chambers, then who is the modern Hiss? Ellsberg, remember, was exposing as a crime the war that had become Nixon's. So: In whose support were all the engines of executive cover-up mobilized? Whose policy, at the time of the leaks, was jeopardized by publication of the papers? Who, finally, felt the most personal stake in shutting Ellsberg up?

I realize that, once one takes these comparisons seriously, the whole affair becomes a little eerie—like a Gothic tale of guilt, in which one *becomes* the man he has destroyed, atoning with one's own destruction. But Nixon never felt guilty about Hiss. The strange recurrences in Nixon's life are simpler in their cause, though complex in results. By feeling the need to reenact his life's high moments, mixed as they were with bitterness, Nixon remained the underdog fighting against power— even when he had all the power available. He lived always with what he felt were prior aggressions against him, and felt a license to fight back with any means at hand. He had scarily huge means at hand as President. Everything was, for him, The Story of Nixon; and The Story of Nixon was always a tale of Nixon's mistreatment at the hands of others. He had to make that story come true in the end—and he did. He actually accomplished all the things he accused Truman of—because he thought he was still fighting Truman's kind of entrenched power. Nixon's absorption in himself became a literal act of self-consumption by the end.

We have trouble understanding that absorption precisely because we shared it for so long. Nixon has been prominent on the national scene for more than a quarter of a century. It never occurs to us to think of his political experience as brief, shallow, and one-sided—but it was. He was thirty-three years old before he engaged in any political activity worth mentioning. He had not registered to vote until he was twenty-five—and when men asked him to run for Congress in 1946, he was not even sure he was a Republican. He had then—and has to this day—no experience of state or local politics. His first campaign sent him off to

Washington at age thirty-three. He pulled off the Hiss coup as a young stranger in Washington, thirty-five years old. He was elected senator at age thirty-seven, and Vice-President at thirty-nine. Having entered his thirties as a nonpolitical man, he ended them at the top of a brand-new profession. It was a rise that not even Spiro Agnew would duplicate— Agnew dabbled in local politics before becoming county executive and then governor. Nixon, moreover, came from a state without solid party structures, and ran his first campaign as a loner. Though there had been no sign of ruthlessness in him during his obscure years as a student and lawyer, he rose through a rapid succession of envenomed struggles that destroyed the careers of his rivals—Jerry Voorhis in 1946, Alger Hiss in 1948, Helen Douglas in 1950. The central event in this dizzying six-year rise, the most important one, the one that made him Vice-President, was the Hiss case. No wonder it became for him *the* political experience—the *mano a mano* in which one is made or broken.

After his stalled eight years as Ike's Vice-President came the two defeats, of 1960 and 1962, and Nixon's exit from politics. He went as he came, quickly, a freak of national politics without any local base in normal party organization. He ran and lost as a carpetbagger in his own state in 1962—and then promptly left that state. After thirty-three years as a nonpolitical man and sixteen years of national contentiousness, it seemed that he would spend the rest of his life as a corporate lawyer in New York. But after a four-year interim, there was an even more spectacular rise, after two hard years of work, from defeated ex-politician to President. This "resurrection" has been considered a political miracle of determination and planning—and it was. Yet the rise was no more dizzying than his first had been—and in some ways it repeated that odd rocketing to fame. Once again Nixon came from nowhere, without any local political base, with his own personality as a major issue, drawback, and resource. This man, who has been called the total politician, remained in some ways an overnight sensation and exception. His strength was not in the California party, or the New York one; not in the Eisenhower camp (where he was always an intruder of sorts), nor even in the right wing of the party (which considered him a traitor to Taft in 1952). He forged a unique personal constituency of grudging indebtedness. The Republican party's heart was with Taft, Goldwater, or Reagan. Its head was with Dewey, Ike, or Rockefeller. Nixon, the first choice of none, a reluctant second choice with many, slipped past the others' puzzlement—and remained apart. His politics was his own career; and he gathered around him those with no earlier or deeper loy-

alties than to one of his campaigns. He created a new political specimen, the young crony, of which Ron Ziegler was the perfected type. He
achieved at last a mastery of politics founded on a contempt for it—give
him enough admen and he could lick all the pols in town. But the
admen must learn his lesson; read his book; trust no one; destroy or be
destroyed. They must each have their Hiss case; bag a foe; be "blooded"
politically as he was.

Fawn Brodie has turned her psychohistorical blunderbuss on Richard
Nixon's childhood and decided that he was always a liar and a cheat. I
find no evidence for that. I think he was a fundamentally decent man
until he took up politics and learned too well the lessons of a Murray
Chotiner. He learned politics as a craft—and a duty—of destruction; and
did violence to his better nature in the process. Right to the end, he
dwelt upon the Hiss case in order to nerve himself to a necessary
hatred. He could never forget that the foe was ready to do to him what
he had done to Voorhis and Hiss and Douglas—what he was scheming
to do to Daniel Ellsberg, Edward Bennett Williams, and the Washington
Post. He honestly thought all other politicians lived by this killer's code,
and therefore he remained a lone man who felt his very survival was
threatened by any signs of opposition. Even power was an enemy he
had conquered, and could never feel at home with. He had once, perhaps, been too good to be a politician. In the process of trying to become one, he became too bad. A new, an evil Nixon was born in the
Hiss case, and he felt a duty to nurture that Nixon even as it was consuming him. His "long career" was really the story of two rocket rises
followed by abrupt, self-destructive falls. It was all a matter of destroying and being destroyed—not of politics as others understand the term.
He will spend the rest of his life brooding on the Hiss case—and on all
the other victories that were in fact defeats—and he will never understand them.

—New York *Times,* August 25, 1974

AFTERWORD

When Nixon released his first huge batch of (doctored) tapes, I went to
the Government Printing Office to get a set of them, and stayed up all
night reading and taking notes for an article. By a happy accident,
Alger Hiss was visiting Johns Hopkins, his old campus, that week, and
had been invited to address my class on postwar America. I made a list

of all the references to Hiss in the tapes (which he had not yet heard of), and I asked him if he would comment on each as I read it aloud to him for the first time. I was hoping for something revealing to come of the opportunity, but Hiss was totally noncommittal. As always with the press, he volunteers nothing but trite observations on Nixon's mendacity. He lives by a severe economy, as reviewers of his book noticed, and gives little or nothing of himself to the outside world. He keeps his mystery.

III
Watergate

<hr>

In many ways, Watergate was what press secretary Ron Ziegler called it, "a third-rate burglary attempt." Even when investigation of this minor crime led to the discovery of more serious things, those greater horrors could not—many of them —be included in the planned articles of impeachment, since the American public supported Nixon's secret bombing of Cambodia or violation of civil liberties in the mass arrests of "Mayday" 1971. Then what made a comparatively trivial thing trip Nixon up? The American emphasis on electoral procedure. We might have a crook in office; but we want to be sure he gets put in office fair and square. The Watergate break-in was an attempt to rig an election. Bombing countries is one thing. But cheating on the Democrats can be dangerous.

<hr>

7

Summer of '74

July 8: History was crowding in, but the heat numbed one to it. Pulsing under a debilitating sun, the Washington Monument seemed to detach itself gently from its base and hang there, waiting. Despite ninety-degree temperature at 8 A.M., hundreds of young people were waiting at the Supreme Court, veterans of this long, drenched weekend of the Fourth. They had patiently sorted out questions of precedence, seniority, and merit; turning in their own report card for admission, working out their tables of rotation. It was participatory democracy come to witness the working of judicial authority, calling the big marble temple to account, hoping it could call the President back to accountability—an iffy bet; the hard money held off, waiting.

Still other people, also mainly young, were waiting Monday morning, just across Capitol Hill, for the first appearance of John Ehrlichman in his own defense at the "Ellsberg break-in trial." But the marathon waiters were those journalists who stood, as they have for weeks, outside the House Judiciary Committee's chamber, begging like puppy dogs for scraps of testimony to print, so they could be excoriated by the White House for "selective" presentation of the evidence—as if they had a wide range of material from which they could select.

This morning they must piece together the testimony of Fred LaRue. Mr. St. Clair hopes to prove that hush money to E. Howard Hunt was authorized by phone *before* John Dean met the President on March 21. At best that would indicate when Mr. Nixon joined the cover-up, not that he did anything to prevent it. But LaRue seems not to have contributed much, either way. The White House scored points last week, in the euphoric afterglow of Nixon's visit to Moscow, when some of St. Clair's witnesses were disallowed. It looked as if the committee was trying to head off a St. Clair triumph. At last the chairman caught on, and gave St. Clair his witnesses—like giving him enough rope.

The journalists' rut, worn outside the committee room, is a dreary beat. As one says: "I've got so used to it now, I go home and stand around in the middle of the living room. Then my wife comes in and speaks to me—but she sounds like Peter Rodino, so I go to sleep." Another draws a Snoopy hanging over his doghouse, with the bubbles-of-thought up to this balloon: "Being a witness to history gets to be a bore."

Washington is currently bored with great events, and narcotized by stimulants. Estragon doubts and Vladimir hopings reduce everyone to passive expectation. The capital is waiting, but afraid of what it waits for—which should be enough to make that "something" never come. But already unwilled things have happened, large and barely discernible, but there. The proof of that lies in the mere concentration of business around the Capitol—in both House and Senate, and in courtrooms that form the other axis of the Hill; even in outposts like John Doar's group camped over in the old Congressional Hotel. St. Clair is running from one court to another, one committee to another; and the lawyers and spectators stand in line here, back where our government was supposed to occur. That is already a momentous shift of emphasis.

The layout of power in Washington resembles a dumbbell—no pun intended—with Pennsylvania Avenue as the bar between weights. But in recent years one end, the west one, has become so overweighted that the thing could not be lifted in a balanced way. The White House, with its clusters of agencies and departments, was the real center of government. Money for projects might be authorized on the Hill, but it was spent down the street. The placement of the Treasury was a terrible first omen—then the bloating of agencies, and the spread across the Potomac of defense and intelligence additions.

It was not meant to be that way. In a swampy city, with no good landing below old Georgetown's, the government had been perched on the highest point—Jenkin's Hill. From that spot L'Enfant made the principal streets radiate—to it they must converge. Here the President was expected to come, in performance of his mandated duties—reports on the state of the Union, grants of war power, advice and consent on treaties. "The President's House," as it was first known, was just that—a private residence, like the governor's mansion in a state capital, or the president's house on a campus; removed somewhat from the man's official duties to give his family privacy. Not, certainly, a seat of government. The President's House was put on a comparative rise in the swampy area because of a Southern tradition that a mansion must have a river landing. (The President's landing was not on the Potomac itself, but on Tiber Creek, now dried and drained off but running at first along Constitution Avenue.) Around this house, land ran quickly down to various foggy bottoms—a concentration of power there defied nature as well as the Constitution.

Of course, when this private residence became the de facto seat of government, it was necessary to escape from *it* for privacy—to Camp David, in the first place; then, as air travel put presidential homes within reach, to Gettysburg, to Hyannis Port, to the Pedernales. Nixon finds it necessary to vacation in a range of homes-away-from-home: Camp David, San Clemente, Key Biscayne, Grand Cay, and Moscow. He is briefly back from one, and on his way to another, this week. But no one cares. Even Henry Kissinger will bustle down to the Hill when he comes back from vital meetings with Princess Grace and the pope. The astonishing thing is not that Mr. St. Clair is doing such a good job delaying, confusing, and counterattacking all over the Hill and its environs, but that he has to give any accounting at all—to the courts, or to Congress. This is a change in the very order of things, *novus ordo seclorum.* The Congress is so surprised, it cannot make sense of the power thrust upon it; and even the Court does not seem comfortable. If they could avoid this situation, they would. It is unavoidable.

A sense of where the power was made the line at the Supreme Court even longer than those Sam Ervin's committee had occasioned last summer. People were looking to the Court to get Congress off the hook, by placing Nixon squarely on it. Various interested parties were let in—Mr. Haldeman to sit uneasily by Mrs. Jaworski. Journalists perched on folding chairs behind the pillars—only Court regulars could tell, from their voices, which of the judges was asking a question. The crowd outside had to check its boos for the villainous brilliance of St. Clair, its cheers for Jaworski's dull honesty, as they came and went. St. Clair had argued well, for a man with his hands tied behind him by his client—forced to affront the Court's sense of its own importance by dodging any commitment to abide by its ruling, something that made his very pleas before this bar a mockery.

While St. Clair implied at the Court that Presidents are above the law, John Ehrlichman was arguing at his own trial that the President's men are above it too. He came into the courtroom cocky with an assurance that had grown from the trial's first stage, when David Young proved to be an unimpressive witness for the prosecution. On the stand, Ehrlichman suffered a momentary catch of nervousness in his throat—and put that rebellious member in its place. Ostentatiously, he poured water and drank, ran it around his mouth, swallowed, drew his top lip up, ran tongue across teeth, adjusted his jawline back and forth to the proper firmness, then raised his eyebrows at the jurors—a whole toilette in ten seconds. Here, clearly, was a man in charge.

After trying to get the trial moved out of the District because any representative jury here would be mainly black, Ehrlichman made two of his four lawyers blacks. William Merrill, for the prosecution, did all the questioning and arguing; but the blacks on the defense team would be all-too-obvious tokens if at least one was not given a real assignment. William Frates, the main defense lawyer, saved tricky cross-examination for himself, and let one of the blacks, Henry Jones, take Ehrlichman through his direct testimony. But Jones did such a poor job of it that the judge had to send the jury out and lecture him on how to ask a question. Jones tried to bull his way through, just angering Judge Gesell further. Frates, Bebe Rebozo's Florida lawyer, tried to smooth things over, but Gesell had told him beforehand he would deal with only one lawyer from each team at any one time. In desperation, Frates finally tried to get standing as an *amicus curiae!*

Jones wanted to put before the jury a series of instructions from the President to Ehrlichman. Gesell called them irrelevant, since the prosecution was not claiming (here) that the President ordered the break-in of Dr. Fielding's office. The judge said he realized that Mr. Jones wanted to impress the jury with his client's importance. "That was not my purpose, Your Honor." "Of *course* it is your purpose—part of it anyway." His expression said the man would be a jackass not to work that angle—but that he better find an admissible way of doing it. He never did.

After the lecture, Jones gave up; and cross-examination began. It revealed the picture of a man too efficient to be burdened with a good memory. In order to function, Ehrlichman explained to the jurors, he had to develop his capacity to forget nonessentials. He had a burn-bag in his brain. The President had called this man one of the finest public servants he knew; the sequestered jury is spared the revelation, on Monday, of Mr. Nixon's standards in this matter.

The House Judiciary Committee just released its undoctored version of eight tapes included in Mr. Nixon's huge release of transcript material last April. In a section Nixon himself had excised from the vital March 22 talk, he says: "That's what Eisenhower—that's all he cared about. He only cared about—Christ, 'Be sure he was clean.' Both in the fund thing and the Adams thing. But I don't look at it that way. And I just—that's the thing I am really concerned with. We're going to protect our people, if we can." John Ehrlichman's presence in this courtroom shows he could not protect his people—and maybe not even himself. What Ehrlichman is saying shows how little concern Nixon had that his people be "clean."

The released tapes record Nixon saying, "I don't give a shit what happens. I want you all to stonewall it, let them plead the Fifth Amendment, cover-up or anything else, if it'll save it—save the plan." The plan referred to is clearly that mentioned earlier in this conversation (pp. 84–85 of the committee's 131-page comparison of transcripts): "All that John Mitchell is arguing, then, is that now we, we use flexibility . . . in order to get on with the cover-up plan." In the White House, conversations of that sort are called "exploring the options"—as in the March 21 order: "Well, for Christ's sake, get it [the hush money]." Included in the committee's version of the transcripts is a fascinating question that might have eased the jurors' task of weighing Ehrlichman's guilt (not that they seemed to have much trouble taking the first unanimous vote on that matter): John Dean and the President are talking about the things Howard Hunt might say if he were not paid off. The President asks: "Including Ehrlichman's use of Hunt on the other deal?" The other deal is clearly the Fielding break-in; so the President is casually acknowledging, about his favored public servant, just what the prosecution has charged in this trial.

Each day of the week ends with long reading of transcripts twelve volumes in all will be issued, in GPO gray or green, by Friday, along with supplementary analyses. And each morning brings an expectable yapping from the President's toy terrier, Ron Ziegler. Apparently the only critical word he knows, at the moment, is "selective"—and he ludicrously applies it to the Monday release of the eight transcripts. But what is selective about it? The committee released all the tape discrepancies it could verify—i.e., all those conversations for which it had both the President's confections and its own tapes for comparison. Besides, when the President released his transcripts, he said "the whole story" was there. When the committee adds the omitted parts to this "complete" story, correcting a *first* selection, how can this be a further act of selection? The pea-shuffler makes the shells fly, and you lift the "wrong" one. When you pry open his hand and find the original pea there, it is no defense to say that you are only considering part of the evidence (look at those three *shells*).

Tuesday, July 9: William Merrill continues his patient cross-examination of John Ehrlichman. The defendant's manner suggests that *he* is quizzing the slight prosecutor, whose clipped wavy hair seems to continue his brow's permanent wrinkles up across his head. Ehrlichman's face, fixed yet mobile, is a living contradiction—a rubbery rock, a blunt rubber head always at work, butting away, erasing the past. Even the language must be mangled to suit his purpose. Yesterday Mr. Merrill asked Ehrlichman just how sure he was, now, of an answer he gave: "Are you morally certain?" Ehrlichman modestly admitted he couldn't be sure what that term meant, for a starter. "Well, what does it mean to you?" "It's a figure of speech." A figure with any meaning? None that Merrill could extract from the India-rubber Iron Man.

Today Merrill quotes from Ehrlichman's own grand jury testimony, in which he volunteered that he was "morally certain" he had not called General Cushman at the CIA to request "carte blanche" for E. Howard Hunt. In the light of the notes General Cushman's secretary took at the time, does he remain morally certain that the call did not take place? Ehrlichman concedes the authenticity of the stenographer's record. What, then, happened to his moral certitude? "That's a phrase." Obviously; but what does it mean to you? "It means *less* than certain." Spoken like the Ehrlichman who, in yesterday's transcript-release, advocated "a modified limited hangout."

He likes, he has said, to "parse" a situation. Now he will parse a phrase—parse it to death. "Morally," since it is a "qualifier," reduces the certitude of certain. But since his recollection was later improved, "certain" was too strong a term even after it had been qualified by the "weakening" adjective. So he should have used a weaker noun, with a stronger modifier—though a stronger modifier would just modify *more* in Ehrlichman logic, weakening the noun further. It is hopeless to make sense of this long excursus, delivered condescendingly to the jurors. We must settle for immoral uncertainty where Ehrlichman is concerned.

What, Mr. Merrill asks, about "carte blanche"? What does that mean? Not, of course, what you would think. "It means something less than an absolute run of the place." How much less? "It meant freely giving assistance." One would think that is the norm for civilized conduct among men authorized to do the public business. But few would think such civility a grant, to each other, of "carte blanche." The judge tires of this game when he asks Ehrlichman whether a search was not implied in his approval of the "covert operation" to get information from Dr. Ellsberg's psychiatrist. Mr. Ehrlichman doubted that he could answer that question. Why? "Search may be a term of art." Judge Gesell looked like Leon Errol on the bench, ready to do his face-smush. Merrill asked if Ehrlichman had heard of the doctor-patient privilege in law. Yes, he has. Doesn't he think that would apply to a psychiatrist and his patient? "I don't know, I haven't briefed that." Rub, goes the eraser; rub, rub, rub.

Now, the defense calls a surprise witness, a young swaggering man called William Treadwell, Egil Krogh's first lawyer after the Fielding break-in became public knowledge. He is here to say that Krogh denied, back then, any knowledge of the raid on Ehrlichman's part. The judge knows all about Mr. Treadwell, and dismisses the jury to say so: "This is the man who was calling all around, and tried to get in touch with me after the trial started. I wasn't going to talk to him. I said, 'Act like a lawyer.'" Now Gesell wants to know if the prosecution has been given the witness's dossier and all pretrial discovery material. The defense says Treadwell would not give them his notes. "Oh, c'mon, Treadwell," Gesell scowls, "did you say that?" Long pause. "I might have." Crisp and bitter: "Turn 'em over." After lunch, it will be found that his notes on one interview with Krogh contain the words: "Firm belief E. approved everything." Mr. Treadwell was first recommended to Krogh by none other than John Ehrlichman. The young lawyer left with his pompadour sadly deflated.

As Ehrlichman strode down the hall toward lunch, a young man came up to shake his hand, and asked: "Isn't it frustrating for you to be standing trial while Ellsberg is free and a hero?" The response, of course, was in Newspeak: "Only in the small rung"—accompanied by that rubber smile that is an insult. The man who asked the question had just come down two floors from another trial being held today—that of four war protesters who poured blood on the files of the South Vietnamese Overseas Procurement Office in Washington. The government was throwing the book at these dreadful reminders that the peace Henry Kissinger won a Nobel Prize for has not, yet, occurred—eleven counts in the indictment, with every offense you could dream of (insulting the dignity of foreign emissaries, no less). When I went up to the courtroom, Judge John Pratt, an ex-marine who lost his arm in World War II and presided with righteous ferocity at the 1970 trial of the DC Nine (destroyers of Dow Chemical files), was yawning over the prosecution's production of overkill evidence. The four people were putting on no defense except a statement of conscience.

When the trial was adjourned for lunch, Philip Berrigan, the priest once tried for a plot to kidnap Henry Kissinger, mustered his friends for a march to the Capitol. TV cameras on hand for Ehrlichman trained an idle camera or two on the "tiger cage" that had been brought to the courthouse as a "physical witness" to what South Vietnamese prisoners still suffer. The tiger cage has become a symbol of the largely invisible protest that Quakers and others still mount in Washington. Several weeks before this, some protesters had driven around the White House with the cage on an open truck. When they slowed to a stop behind a line of cars, park police told them to move on. Then, when they tried to, they were told to stop and wait there. They waited, for an hour and a half. At which point, without any reading of their rights, they were all arrested for protesting without a permit—including Father Berrigan, who was chained in the cage, and whose parole was thus endangered.

Today they plan to go to the Capitol with their cage—the four defendants marching with them on their lunch hour—and some expect to be arrested again. In the brain-damaging heat, a young man, Rhett Ridnour, twenty-seven years old, is handcuffed inside the cage, and ten people lift it to their shoulders (eight men, two women). Sixty people trail along with them to the Capitol and toil up the long stairs, which seem to liquefy under them in the sun. The Capitol police tell them to move, but they refuse. The negotiations go on for several hours (the defendants have to rush back to their trial), and various spots are suggested where the cage might be displayed. At last, the demonstrators win: they move it to one side, where it will not obstruct traffic in and out of the Capitol, and they are allowed to stay. Now that spot belongs to them. They will be back every day this week; and after them the Quakers take up the vigil. Their aim is to remain an accusing presence until Hiroshima Day (August 6).

While the negotiations are going on, some marchers come in from Lincoln Park—from that southeast section of Washington that has been forgotten in the concentration of power "NW" around the White House. Vice-President Ford has been over in Lincoln Park helping to unveil a statue to Mary McLeod Bethune, the great black educator. For thirteen years members of the National Council of Negro Women have been collecting money for this tribute, and Ford's presence at the unveiling is the symbol of a fugitive legitimacy still present in this town, if only as a hope.

Trying to take a shortcut back to the courthouse, I walk through the Capitol—its statues are wrapped in something that looks like dark cellophane, while restorers work on the walls—a secular Lent—and I find the doors facing west are all locked. The tripartite Capitol might as well be a sealed tomb on the side that faces the White House. It is not a natural path of communication. One looks down the Mall—the great view planned for the place—through barred windows.

So I cut around by way of the Rayburn Building. John Mitchell is appearing before the Rodino committee, "stonewalling" in the compelled service of Mr. St. Clair. When I arrive, Congressman Sandman is calling him "convincing" to the cameras—Stone Mountain is convincing, I guess. Congressman Wiggins, who runs St. Clair a close race as the President's lawyer before the committee, is giving one of his weird lectures on the Constitution to journalists in the hallway.

His theme today is the "take care" clause: the President "shall take care that the laws be faithfully executed." Wiggins says that his research indicates that this clause looks merely to *administering* laws passed by Congress, not to *enforcing* them. Journalists ask: (1) then why is the Justice Department put under the President's care, and (2) how can one administer without any power of enforcement, and (3) weren't the problems in our early days of the Constitution the power to enforce levies, etc., on the various new states, and (4) hasn't St. Clair argued that the President cannot have committed misprision by not reporting crimes to law enforcement authorities because *he* is the highest law enforcer? Wiggins shrugs it all away: "I'm just telling you what my research indicates. It's a complicated question." And back into the committee room.

By the time I reach the courthouse, Judge Pratt has already rendered his decision on the four blood-pourers—guilty on one of the eleven counts (destroying foreign property). He calls the other government charges excessive, and congratulates the defendants on their sincerity and decorum. He hardly seems the same man who sat on the DC Nine case. It is a sweltering, bewildered city, but odd hopeful things still happen in it.

Wednesday, July 10: The line before the Supreme Court on Monday seems to have moved over bodily to the federal courthouse—this is the day that Kissinger appears. Everyone is keyed up. When Gordon Liddy arrived from the jail, he theatrically threw up his hands to be searched. "You have any dangerous weapons on you, Gordon?" "Only my cock," he tittered back.

Journalists inside are pestering Dennis Bracy, the GSA man in charge of seating. I can only get a pass that lets me in for half-hour intervals—and minutes of laconic Kissinger apotheosis. But the morning's business is interesting, nonetheless. The lawyer for Messrs. Barker and Martinez is trying to enter orders and memoranda from their earlier CIA assignments. The judge asks why. To show that they had broken the law before, and it was authorized—they smuggled aliens into America, transported guns, forged documents; and received commendation from the government. The judge still asks what relevance these things have to the present case. "To show that they had reason to think they were empowered to break the law."

The argument has no legal bearing, but great psychological force. It is, unwittingly, the greatest argument against a clandestine federal police force that moves outside the law. It makes one pity the dupes of that secret agency when old comrades call them to just one more patriotic endeavor at outlawry. But Judge Gesell persists, inflexibly telling the lawyer to show him some place in the Constitution, or some law passed by Congress, or some statute in the criminal code, that gives the CIA a right to grant men immunity from the laws of the United States. It is perhaps the single most important statement of principle in this trial—but it is swallowed up in the general anticipation of Kissinger's arrival.

Henry has been called for three reasons: (1) to cause the Court trouble (this is only one of a series of apparently extravagant requests the defense made, hoping to say it could not go to trial because such requests had been denied, depriving them of necessary evidence); (2) to impress the jurors further with Mr. Ehrlichman's importance; and (3) to cast doubt on David Young's testimony. The final point is the only one a jury should consider—but, as Mr. Merrill will point out in his final argument, it has a flaw that cancels all its evidentiary value. Dr. Kissinger testifies that he did not ask for a psychological profile of Daniel Ellsberg. The CIA psychiatrist, Bernard Malloy, claims that David Young, when asking for that profile, said the request came from Kissinger. It seems to be Kissinger's word against Young's—but it isn't. The defense had neglected to ask Young for his own response on this point. There is only the second-hand report of Dr. Malloy. Kissinger's testimony could not be used to challenge Young's credibility on this issue, since Young's own testimony on this point was not in evidence.

Kissinger's appearance is just a distraction. William Frates, who called Kissinger, and would later celebrate his importance before the jury, sees Mary McGrory in the hall after Kissinger's departure, and congratulates her on a column that mocked the Secretary of State.

After a visit to the tiger cage (just to see if it is still there) I go to the House press gallery, where the Judiciary Committee is releasing its first eight volumes of evidence—the books John Doar has been explicating, with what seems to have been a boring meticulousness, behind closed doors. Naturally, the carts come over in crazy order—all Book Ones and no Book Fours. The wait extends for three hours, during which the heat is broken by a storm. Journalists stagger out into the rain with all eight volumes—some carrying two, even three sets, for their papers. Mr. Ziegler will consider the release selective; but my eyes blur before long —I fall asleep somewhere in volume three. There is, as many will notice, nothing new here—not after the leaks that let out what was new. But it is marshaled in impressive form—a series of statements, going chronologically through the years (yes, now Watergate is measured in years), with all the evidence for each point appended. It is not exciting reading, but some hearts will pound at the contents. Dull but damning evidence wondrously concentrates the minds of future defendants.

Thursday, July 11: Arguments begin at the Ehrlichman trial. William Merrill, quiet as always, spends his hour on the chronology of the case— on Ehrlichman's constant involvement with it, and his predictable failures of memory. The patterns fit each other. Then he spells out the meaning of the Fourth Amendment, "the right of the people to be secure in their persons, houses, *papers,* and effects against unreasonable *searches* and seizures." Ehrlichman admits he authorized a search of Dr. Fielding's papers, but says this was not illegal because he did not authorize a break-in.

Merrill points out that if the papers were gained some other way, there was still no right to search them, destroying their essential privacy. It was clear that the doctor would not yield them up voluntarily— why have a "covert operation" that would not be "traceable" if that were the case? In one desperate moment on the stand Ehrlichman, who had earlier said he gave no thought to how the papers might be secured (i.e., "seized"), said he *did* think they might be in the Brookings Institution. But the Brookings people were no more willing than Dr. Fielding to give them over (if anything, less); and would have no right to give them, even if they were willing; and if the papers were given, it would still be wrong to violate their privacy. It was a simple, understated lesson in the law that the judge would endorse in his instruction to the jury.

Poor Mr. Jones gets up to do the normal catechism of the defense ("burden of proof," "reasonable doubt," etc.) and to show that a black man could speak up for Ehrlichman—who certainly did not observe a fifty-fifty racial proportion when staffing the White House. Then comes Frates, trying in his overbearing way to be folksy. The effect is of the Commendatore striding into the last scene of *Don Giovanni,* slapping Leporello on the back, asking "How's tricks, pal?" and ordering a martini. He says, on two occasions, that Mr. Ehrlichman was "no big shot" (just an ordinary pal of Messrs. Nixon and Kissinger), and refers lovingly to one of his hokier touches—bringing in, as a character witness, a young navy man who served Ehrlichman his meals in the White House. He points out all the other people who had trouble remembering specifics in their testimony—but neglects the fact that their troubles did not form a complex pattern to match Ehrlichman's. He wanders, grovels grandly, flatters, and generally does a good impersonation of Ham Burger on the old Perry Mason shows. He behaves as if the jurors are cretins.

The forgotten men of the trial are given eloquent pleas for sympathy. Gordon Liddy, who sat through the trial with perky attentiveness, sometimes engaging spectators in stare-downs, his weasel head seeming to duck behind the outsized mustache, is defended by Peter Maroulis, who will not relax his distance from the Ehrlichman defense team, even to approach the jurors. Daniel Schultz portrays Martinez and Barker as the dupes of E. Howard Hunt; but has to admit that ignorance of the law is no defense. He is arguing, in effect, to the judge, not the jury—with sentencing in mind. The dupes do not even hope to get off. The big fish, apparently, still does.

Merrill's rebuttal admits the plight of the three lesser defendants, while correcting some of Mr. Frates's grosser distortions. But he goes on to say that the victim in this crime was the Constitution; that the principle of personal security from unauthorized intrusion—even from the government, and especially from a paragovernment outside its normal agencies—is tested by this case as by no other. The jurors are hurried out of the courtroom with an assurance that the turmoil outside has nothing to do with this case. The press and spectators are given time to reflect on the guarantees of the Constitution as they leave the building, in which two prisoners have seized hostages and are setting up a cell-block bastion that will hold out for days. The jurors must assemble at a different courthouse tomorrow for the judge's instruction.

Outside the cordons of police, I set off past the Capitol for the Supreme Court. Late on Tuesday, just one day after the Jaworski-St. Clair argument in the Court, the former Chief Justice, Earl Warren, died. It was like an admonition that the men called to fill his place in history should be of his caliber. The justices had done an unprecedented thing—opened the Court for Warren's lying in state, and brought out his bench-chair as the one symbol in this quiet ceremony. People have been filing past most of the day, and the big unused *cella* of the Court's central temple seems to have a purpose at last—this hall dwarfs the actual chamber that it leads to, the temple's *adyton*. L'Enfant meant Capitol Hill and its immediate ring of buildings to be our acropolis. One reason it does not shine over the city as intended is that its large Roman style was too enthusiastically adopted at the other end of the avenue, where agencies set up their own bulky temples and huge Roman boxes, complete with statues of Amazons made classically sexless with their bared square breasts. But the first instinct was right—a severe note of the Roman republic.

Even George Washington's huge obelisk comes from Egypt by way of Rome. The early imagery used of Washington (during the Revolution) was mainly biblical—he was Moses leading a people out of captivity. But very soon the images became more secular—Washington with toga or bared chest, always Cincinnatus *surrendering* power: that is the way he is remembered in the first great monument put up to him (in Baltimore), or by visitors to the Annapolis statehouse (where he resigned his commission). His best-known message came with the supreme recommendation that it was delivered as he voluntarily left office and went back to the plow. His veterans banded together as the Cincinnati. The important thing is not merely that his fellow citizens saw him in this role, but that he alone of men with his historical stature (one thinks of Caesar, Cromwell, Napoleon) lived up—or down—to the role.

There is something about national crisis that sends men's minds back to such standards. I noticed, during the week, that three columnists quoted from mottoes on the main government temples, especially from this one—the Court. The totally silent file of people circling inside reminds us of such a monument's use—to make us part of something larger and continuing, a benefaction from the past, a sign of our forefathers' faith in us—that we would be here, still honoring the republican symbols, long after the temple's designers had disappeared. Now Earl Warren is gone; what must the judges think who put on their black robes and move through this temple, adding their chapter to the building's own history? The way they have honored the late Chief Justice makes one hope they feel somewhat like Henry Adams's heroine in *Democracy*, who said: "Why do I feel unclean when I look at Mount Vernon?"

While I am this close, I make one of my favorite stops just next door to the Court, in the Library of Congress. It is the front landing that normally holds Jefferson's rough draft of the Declaration (a far more touching document than the formally engrossed and signed copy in the National Archives—though now it is temporarily removed). Here, too, is a lock of Jefferson's red hair turning gray—and the original manuscript of Washington's first annual address, beginning "My fellow *citizens* of the Senate and the House of Representatives": Cincinnatus rendering his account to the citizens empowered to demand it of him. Washington knew where the center of government was: here, on a public Hill, open to republican scrutiny. What a comedy that Congress should be frightened to see it returning here.

News has leaked, today, of new findings by the Senate Select Committee, on its last day of formal existence; so I go to the Dirksen Building, hoping to get a copy of the report that will be handed out tomorrow. Sure enough, the "dirty tricks" investigators have found something new in their last weeks of work on the Rebozo funds—it came in too late for the detailed letter sent to James St. Clair on June 6, asking for the President's response to evidence that Rebozo used campaign funds for the President's personal expenses.

The new discovery is a $5,000 set of diamond earrings purchased for Pat Nixon's sixtieth birthday, with funds that came from Rebozo through a circuitous "laundering" operation. Journalists are phoning in for a description of the earrings, and Terry Lenzner checks the details against an insurance photo on his desk. Terrier Ron Ziegler has already called this new finding "warmed-over baloney"; so the staff has bought (with its *own* funds) a huge eleven-pound baloney sausage. At first, they planned to send it over to Ron with instructions on what to do with it—some felt that he would need numbered directions and a diagram in order to understand the instructions—but prudence overtakes them in their giddy moment of exhaustion, and they will take the baloney to Senator Ervin's last Watergate Committee press conference tomorrow.

A check at the Judiciary Committee seems to indicate that St. Clair's long-awaited grilling of John Dean has not broken him down, as some Republicans hoped. Dean goes back before the committee tomorrow.

Friday, July 12: The press comes early to the Senate Caucus Room, wanting a prior look at the three-volume report whose formal release date will be Sunday. Although the Senate committee has faded from the public's mind since its televised hearings of a year ago, its teams of investigators have been using these twelve months of subpoena power with great diligence. One team, under David Dorsen, did an extensive study of the dairy contributions, and their nexus with milk price supports—it has turned over its findings to the special prosecutor, which may be bad news for John Connally. James Hamilton led the study of the Watergate cover-up—with results that show on every page of John Doar's first eight volumes of evidence presented to the Judiciary Committee.

Terry Lenzner, aided by six very bright young investigators, handled "dirty tricks" and then branched off into the Rebozo-Hughes connection. Eleven subpoenas were issued by the committee to Rebozo, who defied them all. But now this material, too, goes to Jaworski, whose enforcement powers *begin* with immediate imprisonment for contempt. The Senate investigators could never get a look inside Rebozo's own Key Biscayne bank; but they pieced together a damning picture of Rebozo's financial dealings just from his use of other banks to route campaign money around toward Nixon's private use. That picture is sketched in the twenty-two-page letter sent to Mr. St. Clair in June, and reprinted in the third volume of the report.

Some staff members regret that particular finds of theirs were omitted from the last report; but some trimming had to be done to achieve unanimous approval by committee members. One whole chapter was excised as too speculative—one tying the DNC break-in to a desire to find out what Larry O'Brien knew about Hughes money that had gone to Rebozo and other Nixon funders. Ervin buttered up the staff members whose work was excised, assuring them the volumes draw a picture of a horse—with no need for adding a caption, "This is a horse."

Ervin uses that same line in this morning's press conference, and gets a big laugh from it. Only four senators show up for the final joint statement. Senator Gurney has gone into hiding since his last indictment. I passed Senator Talmadge on my way over to the Senate building; he was posing with Georgia constituents on the Capitol steps. Sam Dash is milking his last public appearance as chief counsel for all it is worth. When the television cameras begin to dim, he brings out the eleven-pound baloney and offers newsmen slices from it.

At noon a procession forms to take Earl Warren's casket from the Court to the National Cathedral, a quizzical Gothic extravagance in this Roman town. The press is stationed in the north transept, looking head-on at the front row of mourners, with Mr. and Mrs. Nixon in the north aisle seats, Mrs. Warren in the south one. For some of us, it is the first close look at the President since reports of the phlebitis in his left leg. He does seem to favor it, to shift it as often (if unobtrusively) as possible during the long service. The first address is by Rabbi Alvin Fine, and might be considered in bad taste: "Human decency and kindness were so instinctive, honor and dignity so natural in him, that there could be no unguarded moments of indiscretion when he was anyone but the true Earl Warren." A funeral is no place for partisan expressions—yet who could hear those words, in today's Washington, and not think of all the unguarded moments of indiscretion recorded in the tapes and transcripts that fill this city's consciousness? Still, how can one criticize a man for paying tribute to Warren's unquestioned integrity?

Standing in the transept, studying Nixon's studied woodenness, the real enormity of our situation hits me. Nixon has made integrity itself a partisan issue. Any mention of honesty in his presence is an implied rebuke. The merest commonplace about human decency becomes a tacit dig at his politics, which are at one with his person. It is weirdly fitting that this insight should be enforced at the funeral of Nixon's old Republican enemy from the fifties. Warren, who tried to fight Eisenhower for the presidency, shared with Ike a concern that men be "clean," the code that Nixon expressly disowned in transcripts released this week.

Yet even the slack code of crony's loyalty has not been observed by Nixon. After telling Mitchell he would protect his own, Nixon plotted to throw Mitchell to the dogs, using hoodlum language about "the Magruder-Mitchell head-chop business" in his April 16 talk with Dean. For Nixon to sit here in public, after we have heard those cruel private schemings, lends the ceremony a surreal note, as of feigned grief at Mafia funerals. Granted, Mrs. Nixon is not wearing the "swag"—neither the $52,000 jewels from Prince Fahd nor the $5,000 earrings courtesy of Bebe. But for Nixon to sit here at the casket of Earl Warren brings back the campaign that was the foundation of their enmity, and the lachrymose but still honorable boast of those years: "Pat doesn't have a mink coat. But she does have a respectable Republican cloth coat." The days of such respect are gone, not only for Republicans but for the republic. Whatever one said about "Tricky Dick" of the fifties, at least he did not seem personally venal. Now one wonders if there is anything left for him to be charged with—any code, however low, he still claims to honor; any standard remaining beyond brute survival.

It is a country in deep moral trouble that follows the hearse of Earl Warren, and the limousine of Richard Nixon, over to Arlington National Cemetery. This was the good side of the Potomac, site of the Custis plantation, which George Washington's step-grandson turned into the principal site for honoring the first President in the early nineteenth century—far more alive with his memories than Mount Vernon itself. Here the public was received in the great battle tent of the General (now on display in the Smithsonian), and George Washington Custis lovingly showed people his grandfather's personal possessions. Even through the Civil War the mansion kept its honor bright by connection with the name of Lee. When Henry Adams sent his fictional characters to visit it, it was already the national graveyard—"the long white ranks of headstones, stretching up and down the hillsides by thousands, in order of battle; as though Cadmus had reversed his myth, and had sown living men to come up dragons' teeth." The serene squat front of the great house indicted the city then, and still does.

Adams, like Twain, saw the Reconstruction White House adrift in a muck of money and pride; not anchored in an honorable past like Arlington; unable to live up to the mottoes on its own temples. It is tempting to think the same thing of a country whose hard-core supporters of the President tell pollsters that all politicians steal, that Presidents can therefore steal more, and that the law cannot touch them—in short, that the Constitution not only gave us a monarch, but a particularly sleazy and tainted one; and that we must live with him. But all is not lost, Mr. Adams, Mr. Twain, Mr. Warren. As I turn back from Arlington, the car radio says Ehrlichman's jury has come back after a mere three hours of deliberation with a guilty verdict. And St. Clair has again failed to shake John Dean. And now that the heat has broken, one can see the Capitol very clearly from Robert Lee's mansion.

—*New York Review of Books,* August 8, 1974

8

Sideshows

AGNEW

Spiro was their hero—and such "heroism" is fallen on hard times.

He climbed dizzily to what he considered heights, the sauna bath on Frank Sinatra's estate—which should have told us something about his values and his dreams. Yet the staid Republican ladies still cheered him as he commuted from the golf course to tell them he would not resign if indicted. If they could cheer what was, in effect, a statement that he would not put national good above the safety of his skin, some will no doubt cheer when he has saved his skin while pretending he thought only of the nation's good. Some people, after all, cheered for Lieutenant Calley.

And, come to think of it, the President responded to those cheers. He had never looked Calley in the eye and said, "This guy has got it"; he had not personally chosen Calley, or promoted him to office; but a mechanism in Nixon for responding to the base appeal made him champion that tawdry hero, or at least play to his disappointed gallery.

Not that Nixon still thinks Spiro has "got it." He has known for some time now that they would get his Vice-President. Any regret he might have felt on that point is no doubt tempered by the Agnew Mafia's attempt to use Nixon's foreknowledge, implying that the President was "osmotically" conspiring to do in his own second in command.

As the plea bargaining went on (Agnew meanwhile denying it), press agent Victor Gold was everywhere on TV blaming this man and that, saying Mel Laird wanted Agnew's office or General Haig was a new Svengali. Others called for Henry Petersen's head, or thought Nixon and Connally were doing in a good man. Agnew himself went out as he came in, vilifying others; and nothing became him less in office than his taking leave of it. When the last dodge was exhausted and the last subpoenas out, after all his denials of resignation and plea bargaining, after attacks on the law-enforcement tools his own Administration had promoted (including the use of immunity), he went out calling his old friends perjurers and the President's appointees a set of legal blackmailers.

It was a sordid scene of thugs and hoods hacking at each other, weeks on end; not only undeserving of the public trust, but patently not trusting each other, trying to destroy before they were destroyed.

—Newsday, October 1973

MITCHELL

Fairly early in Nixon's first term, he held a bachelor dinner in the White House for "intellectuals" sympathetic to his reign (the gathering was small). One guest at the meal said, later, he had looked forward to hearing from the President himself, from such little-known quantities as Haldeman and Ehrlichman, or from the other guests. But John Mitchell was the man who commandeered the table, steered and owned the conversation, thumped down opinions, course by course, not pausing for a question—and, all the while, Nixon fairly hung on his friend's words, looking proud of his performance. The least interesting man present had succeeded in interesting the one man who counted.

It is difficult now, after his fall, to appreciate the magic of John Mitchell's brutishness in full blossom. Only a slight touch of that charm lingered when he appeared, rheumy-eyed and mottled, before Sam Ervin's committee—a trace of the old manner preserved long after its base had been eroded. This trapped man could muster heartier contempt for his baiters than they could bring to bear on him. Without a leg to stand on, he remained stronger than most of the preceding witnesses. A worshiper of power could still be impressed—William S. White wrote a column full of praise: "In John Mitchell the President selected a man and not some spuriously golden-haired boy." But White learned, in his LBJ days, to enjoy being bullied. Most viewers saw only a lumpish insensitivity under the loose veil of liver spots woven across Mitchell's face.

To see what Mitchell looked like by candlelight to Richard Nixon, we must recall his time of power: setting up a government, taking on a job (at Justice) without bothering to learn its requirements, casually suggesting to his predecessor (Ramsey Clark) that a friendly gesture initiated by Clark—after all Nixon's campaign attacks—would make the transition easier. Just before he was sworn in, Mitchell condescended to dispatch a young aide—Kevin Phillips—to scout out his new assignment. Phillips seemed weirdly uninquisitive to those trying to instruct him in routines of the department—as if any question would imply self-questioning, convict him of Clark's own dubiety or hesitance. But Phillips made it clear he was not worthy to unlatch his master's sandal when it came to self-assurance: "Of course, Clark came up through the ranks and is more chummy with his staff than an outsider would be. But Mr. Mitchell will be the only personality in that room when he takes over. The others will be his assistants." Later, Mitchell would put this minion in his place: When asked if he agreed with Phillips's Southern strategy, outlined in *The Emerging Republican Majority,* he said: "I don't really have a practice of subscribing to the theories of my aides. It generally works the other way."

Last spring, I asked Phillips whether he had become disillusioned with Mitchell yet, and he replied: "I saw through him eight months ago." It took him, in other words, only four years—and Phillips is a bright young man, not one to linger with a loser.

Yet Phillips, in the long run, gives us no better clue than White does to the original force of Mitchell in his friend's Administration. White liked being kicked, and Phillips was going to arrogance school—apt pupil to a master teacher. Nixon, on the other hand, prefers kicking when he can get away with it, and realizes it is too late for him to acquire an imperturbable public air of assurance. Besides, the mystery of that White House dinner is only partly tied to Nixon's own admiration for Mitchell, his reliance on him during all his early steps in office. There is even greater puzzlement in Nixon's use of the man for this unlikely assignment. The dinner was meant to woo, however discreetly, men from a sector of the population hostile to this Administration—even though Nixon has, in the past, indicated that he thinks of himself as an intellectual, too. He works like a scholar in the areas he knows well, and thinks Edward Kennedy (for instance) an empty dilettante in foreign-policy matters. As he once told Jules Witcover: "In order to make a decision, an individual should sit on his rear end and dig into the books." Nixon takes a justified pride in the number of books—and the number of countries—he has "dug into" by sheer dint of study. Yet no one, not even Nixon in his most smitten days, could think of Mitchell as a very great digger into books. The man never hid his contempt for academicians—look, for God's sake, at the salaries they get. Challenged by party regulars over his lack of political experience before the 1968 campaign, Mitchell just reminded them of the money he had made—for him, that skill was the measure of all others. Asked if he might be awed now that his boss was President of the United States, he told an interviewer: "I've made more money in the practice of the law than Nixon, brought more clients into the firm, can hold my own in argument with him and, as far as I'm concerned, I can deal with him as an equal." No mere professor can make that boast.

Mitchell was as unlikely to impress intellectuals as they were to upset his certitudes. He spelled out his contempt in a 1970 pronouncement: "[Nixon is] aware of everything that's going on. I'll tell you who's not informed, though. It's these stupid kids. . . . And the professors are just as bad, if not worse. They don't know anything. Nor do these stupid bastards who are running our educational institutions." Nixon has, indeed, certain claims to being an intellectual, which just deepens the mystery: Why has he been, from the days of Murray Chotiner to the days of Charles Colson, an intellectual who travels in the company of thugs? It would be one thing if he used them as his buffer when dealing with the harsher side of politics—to awe the businessmen or show regulars how tough he is. But he brings out his less appetizing specimens precisely when he wants to move into circles where he should feel at home himself.

But of course he is not at home among even the friendliest intellectuals. He talked once of teaching at "one of the fine schools—Oxford, for instance," if he lost his bid for the presidency. Nothing could be less likely. Cast by history on the side of pseudo-Populist antiestablishmentarians like Joe McCarthy, Nixon—who lacked their relish for assaults on intellect—was doomed to champion the "common folk" against uncommon elites. He continues to star in that comedy, even now. His is the Administration that will "play in Peoria," though Nixon has less of the common touch than any President of modern times. He hides away with millionaires, decking out his White House with trumpets and formality; he never feels more at ease than when secluded with a German professor talking about Asiatic statesmen. Nixon is a psychically displaced person—not at home with the crowds, even when cheering their attacks on Establishment citadels; not at home in the Establishment, even when its power centers have been put in his control. He is always a wary intruder—pirouetting from aggression to obsequiousness in mid-sentence, wooing and affronting at the same time. He mixes his deference with resentment, his admiration with envy, in ways that make him a man of half gestures and permanently checked impulse.

That alone can explain his dinner and the Mitchell monologue. He was saying, by their mere invitation, that he wanted these men and could use their help. But he was simultaneously anticipating rebuff, letting Mitchell signal that he did not really need them. He does not have the righteous contempt for excellence that his Wallace-like role demands—he must lean on the brutal types for that. He relies on them to be worse than he is in front of those he obscurely considers better than he is. It is a curiously self-effacing assertiveness—he "toughs it out" through his minions because he is too sensitive and intelligent to do his own contemning. He travels with thugs because he is not a thug himself —theirs are hired insensitivities. He told Theodore White, in an unusually revealing interview, "I never shoot blanks." He meant that his bullyboys don't. Just as he said he knew nothing about Watergate—"the boys," as Mitchell called them, were supposed to take care of that. The tone was set by that Attorney General whose first judgment over Watergate was that "Kay Graham has her tit caught in a wringer."

Mitchell, entering office, had boasted that Nixon's administration would be even farther to the right than people expected. He complained when Nixon took even mildly liberal stands. He hated to hear about a Bob Dole or a Bob Finch talking mushily when he was orchestrating barks and growls. After using Bob Mardian to sabotage Finch's HEW on the busing issue, Mitchell—beginning his own slow decline—still served as a bumper between Finch and Nixon in the White House. Mitchell was also upset at Ripon Society types who gravitated toward the Moynihan office. He once referred to Ripon's young members as "juvenile delinquents," and the society was a particular target for Kevin Phillips, who disliked its Establishment style. Most politicians try to reach beyond their immediate constituency; Mitchell kept expelling people from that small first circle of Republican intellectuals—whence his first-strike offensiveness at the White House dinner.

Others were learning the lesson of that dinner, along with the invited guests. If Nixon admired the boorish strength of Mitchell, a preemptive rudeness that anticipates insult, then Haldeman and Ehrlichman knew what path they must follow upward. And their righteousness had a solider base than Mitchell's mere self-satisfaction. Haldeman, lean and ascetic, with an insect's economy of feature and death's-head nose, was trimmed down to monomaniac devotion. Ehrlichman the teetotaler was meant to deal with the Hill, to indulge his contempt for drunken congressmen—he just widened the voracious smile, as his guillotine eyebrows were gleefully drawn up and dropped. These two could out-outgrowl Mitchell in distrustfulness, could believe at last that Mitchell was not single-minded enough in his loyalty.

Mitchell later presented himself as only half-observant of late campaign maneuvers in the "Watergate" year 1972. But he had initiated the kind of strategy sessions that led to Watergate. In the Massachusetts elections of 1970, he wanted a hard right-winger to take the Republican nomination for the Senate, one who could savage (though not defeat) Edward Kennedy. For Mitchell, Chappaquiddick was not crippling enough. (Charles Colson, whose own dirty tricks were just coming to the fore, recommended Al Capp for this mud-slinging role.)

Mitchell was eased from stage center after his disastrous vetting of Supreme Court candidates. The defeat of Judges Carswell and Haynsworth was blamed on the Establishment's hatred of the South, during a cruise on the presidential yacht where Mitchell encouraged Nixon to strike back at the judges' foes. But Haldeman and Ehrlichman, using Mitchell's own norms, decided Mitchell's usefulness was coming to an end. By cutting back on Mitchell's influence, they unwittingly increased Charles Colson's. The falling tough guy yielded to tougher guys, his legacy assured.

—Playboy, December 1973

WATERGATE CLERGY

ANDREW CARNEGIE: This is a Christian country.
MARK TWAIN: Why, Carnegie, so is hell. But we don't brag of this.

"This nation is God's nation. The office of the President of the United States is, therefore, sacred." That is the gospel of the Reverend Sun Myung Moon. When Nixon, playing Mr. Bones to Mr. Moon, came out of the White House last December 13 to greet his disciples of the Freedom Leadership Foundation, they knelt down to worship him. That was as good an indication as we needed of Mr. Nixon's desperation. Only the idolaters were left in his retinue. When a man has sacrificed all honor, he must settle for adoration. The poor Roman emperors had no way left to get their fellows' attention but to make themselves gods.

Nothing crazier could have been imagined than for Richard Nixon—the voice of the South and the Bible Belt, of Middle American Protestantism, of antipornography law-and-order—to be incensed by an Oriental power; unless it were for him to trot out a Jesuit casuist to justify these ceremonies. Back in 1960, the Reverend Norman Vincent Peale endorsed Richard Nixon in order to keep Mr. Kennedy and John XXIII out of the White House. And even this summer some regions of the dark outback called the attempted impeachment of Nixon a Catholic plot (just look at those weird Italian names—Rodino! Sirica!). Meanwhile, Nixon, digging in for the last assault, drew his Catholic Mafia close around him—General Haig, more fierce in piety than his Jesuit brother; Pat Buchanan and Rose Mary Woods; Robert Abplanalp (a Villanova product). The Christian Scientists banished or jailed, only the Catholics were left inside to encourage outside fanatics like Sun Moon and Rabbi Korff.

Dr. Johnson said patriotism is the last refuge of scoundrels. That's wrong. Religion is. It perfectly fit Nixon's hope that we would not notice him, so awed were we by his office. He was serving, by saving, the presidency—as his flunkies burgled and lied, in 1972, to reelect the President. There is a double abasement—to the presidency by any one particular President, and to "Mr. President" by mere Richard Nixon. Nixon was an admittedly unworthy upholder of his position's majesty—which is just the approach that allows the clergy to combine personal humility and institutional grandeur. The theological formula is *ex opere operato*—that is, sacramental miracles are wrought "by the efficiency of the ceremony itself"—not, in the first place, *ex opere operantis,* "by the efficiency of the celebrant." So we are to forget the lecher inside the confessional box, or the drunkard at the altar—forget the man, and think only of his vestments, the insignia, the paraphernalia of office. Mr. Nixon was a mere appendage to the *Spirit of '76,* and we must bend our knees to It, no matter who its occupant might be. Nixon presided over his own presidency as its liturgist. And this plainest of men made it the gaudiest of principalities. When seized with self-doubt, he solaced himself with palaces.

Yet, even as the priest-king, he had to settle for a sorry crew· of diviners and haruspices. He had a millionaire valet—but it was just Bebe. He had a slick White House chaplain—who is his own order's pariah. He was worshiped—but by a corrupt Korean regime's least honorable extension. And he had a rabbi who cast himself as a stand-in for all our dead ex-Presidents. There is something very sad about such large claims when they come marched out by such a crew of moral dwarfs.

In his demonstrations and lobbyings, the right-wing version of Maharaj Ji used the motto "God loves Richard Nixon." That could be a deep theological statement, an attempt to provoke God into theodicy. The Lord boasted to Job about Behemoth and Leviathan, like a parent proud of even ill-favored children. But he must be tempted to hide Nixon somewhere in the last row when trying to explain his universe. Yet Mr. Moon did not mean it that way—he obviously considered Mr. Nixon one of God's better strokes, a thing to boast of. Isn't he powerful? He leads the world's greatest empire. What further sign do we need of the heavenly favor?

Caliban said that his colonizers had taught him a language in which he could curse them. We taught Sun Moon a religious vocabulary in which to bless us—but his blessings are a curse. We taught him that God loves a succeeder—not like that prodigal Son of his who went off and got crucified and had to be disowned. True, the crucified later got impaled on gold instruments of torture, to bleed rubies—and then he was let back into the family. But on terms.

That is the religion we taught our subject colonies, and Moon comes innocently to remind us of it, and get paid. Make a buck for God. How many guns does the pope have? Not many—not nearly as many as Mr. Nixon had. So Nixon was our pope. The great Korean succeeder, reported to be worth $10 million, comes to his mother country (the colonizing one) to show us he has built a better churchmouse-trap. Once emperors become gods, the empire sends back from its fringes rather embarrassing forms of cult offering. The surprising thing is that mainline Protestants go along with this Oriental cult. Or perhaps it is not so surprising. All absolutisms in ruin tend to lean upon each other. God as a daddy with some power left to spank has gone into hiding, and only the nuclear hickory stick looks credible still. In that sense, Nixon did more to bail out God than vice versa.

The trading of absolutisms was something we became familar with in the fifties, when there were so many converters from communism to Christ. Even Whittaker Chambers gave up one faith to embrace Richard Nixon and Christianity. The Colsons of this world have just gone one step farther, giving up even Nixon for Christ. But far more try to cling to both at the same time—and tend to mix them up. Even a rabbi like Baruch Korff compared Nixon to a Christ betrayed by Judases of the media.

A few months ago, the front pages of our newspapers showed Mr. Nixon accepting, with a rather apprehensive look, a very slim volume of his praises from an inveterate self-praiser named Korff. The President had good reason for uneasiness—Korff had, among other things, just tried to incite the Vice-President of the United States to disobey the Constitution. But if Nixon was not going to take a book from this source, where else could he look?

You may have noticed that Mr. Nixon's grand reign produced no celebratory literary work. This was not for lack of trying. Early in Nixon's reign, a *Time* reporter sent me an interoffice memo from Washington which said there was a hunt on for an official biographer. There was reason to hope, at that point, for someone respectable, which ruled out some of the obvious hacks who had written about him before. I later asked Pat Buchanan if the hunt had been successful, and he said it had: Jeffrey Hart, a *National Review* editor, had been anointed and had gone to work. The plan was for the book to be ready by the 1972 election, but Hart told me his publishers had some objection to his first manuscript. (He was pretending that the book was not authorized.) Then James Buckley, *National Review*'s candidate for anything, came out for Mr. Nixon's resignation. The book is not expected to appear. So Nixon, who turned down the hacks, had to settle for a kook named Korff.

Rabbi Korff's religion seems to be Rabbi Korff, so maybe he does not belong in this survey. He presents himself as filling in for all the ex-Presidents who died recently and could not come to their successor's defense. When Gerald Ford did not accept his plan—which called for Ford to refuse the presidency should Nixon be impeached—Korff decided Ford was equivocal: "I am not afraid to make categorical statements, unlike the Vice-President." The rabbi not only refuses to take oaths himself; he thought the only categorical act for Ford was to break his oath to the Constitution.

When not egging on Mr. Ford, the rabbi elicited mumbles of agreement from Nixon. Isn't George Meany the "archdeacon of impeachment"? Well, sort of. Wasn't John Ehrlichman's conviction a "blot on justice"? Korff said the President did not disagree. Indeed, the President went further than his imp of suggestion on this one, adding that it would be hard for anyone to get a fair trial in Washington. When Kingman Brewster said blacks could not get a fair trial in New Haven, Spiro Agnew treated that as an attack on our whole system of justice.

Rabbi Korff thinks very well of himself and Mr. Nixon, but not so well of the press or of women. He boasts of his conquests to Sally Quinn, telling her that his wife—a quarter of a century younger than he is—has good reason to fear his "digressions." But women should be kept in their place, so men can be "aggressive and domineering." Sons, he added, should never "see their fathers yield to their mothers." His swagger is paired with a cringe, forging a link between him and the resentful client: "The entire Administration is held captive by the Washington *Post.* . . . I feel like I am in Hanoi and not in Washington." Both Korff and the President are martyrs to their religion, which is their selves. The rabbi feels Mr. Nixon would not indulge in racial epithets—not that it would make any difference if he did: "Frankly, I have referred to my friends as that *goy* or that *shiksa.*" We are not told what these acquaintances call the rabbi.

Father McLaughlin is a displaced parson. He was not part of the Catholic old guard at the White House—he began to see the ex-President about the same time, and for the same reason, that Korff saw him. McLaughlin is not even part of the new Catholic majority Nixon tried to create in response to Kevin Phillips's ethnic Republican theology. Pat Buchanan, who is at least a bright defender of his favorite crook, called Michael Novak to the White House in 1972 to fish for the new ethnics.

Father McLaughlin does not speak for this constituency. With his superficial Eastern overlay of sophistication, he is a throwback to the higher Bing Crosbyism of the fifties. The first tactic of Catholics wanting to "belong" as Americans was to croon and join the country club. But by the fifties John Courtney Murray had polished this act to a subtler shine. He presided at Henry Luce symposia as a clerical Robert Hutchins. His type largely disappeared with the fall of Camelot, but it obviously shaped Father McLaughlin's ideals. He aspired to be a with-it "media priest"—even though he still has not learned that the noun "media" is plural: he keeps informing us on TV that the media *is* evil. He also became an instant expert on impeachment law, quoting with great approval the Yale professor whose name he still mispronounces (Bick-*el*). Names trouble him—a real trial for a name-dropper.

His religious superior made an ill-advised attempt to recall Father McLaughlin after the White House priest had canonized the Nixon of the transcripts. This was taken as a political punishment, though I am told that the real objection was religious: Father McLaughlin, a man who took a vow of strict obedience to religious authority, had told an audience that he was responsible to only one man on earth—Richard

Nixon. McLaughlin weathered his recall by pointing to activist Jesuits like Robert Drinan and Daniel Berrigan. The comparison does him no honor. When Cardinal Spellman pressured the Jesuit provincial into banishing Berrigan in the sixties, Father Dan obeyed and went off to Mexico. Besides, both Drinan and Berrigan have stayed on good terms with their superiors, getting approval for their mode of life, which is not extravagant or secular. By contrast, the provincial who gave McLaughlin permission to go to the White House, and the one who later tried to recall him, have both said that he misled them. McLaughlin's sense of duty seemed engaged by nothing higher than the preservation of Richard Nixon at any cost. His theology of therapy concluded that the President should blow off steam by encouraging young aides to think through the logistics of paying hush money. Older-style moralists would call that offering "an occasion of sin." And the arguments used by McLaughlin proved too much even for William Buckley to abide, prompting one of that columnist's rare moments of outrage over Watergate: "Did he [Nixon] believe that a paid consultant wearing a Roman collar could transubstantiate the tapes from barracks-room discussions about how to lay the Statue of Liberty into sacrosanct deliberations of a man identified by Father McLaughlin as 'the greatest moral leader of the last third of this century'?"

The Buckley outburst, when it came, was a honey: "Either Mr. Nixon believes what Father McLaughlin says, in which case he has completely lost touch with reality; or else he doesn't believe it, but he thinks it will work, in which case he has completely lost touch with reality." What prompted Mr. Buckley's anger is the concept, now related to the witness of Thomas à Becket, of "the honor of God." And perhaps the last, best reason for impeaching the President would have been to save God's honor from all the godly defenders of presidential crime. As Buckley said: "This venture will bring yet further discredit to President Nixon—and discredit to Father McLaughlin. And, I might add, should do so, even as we dishonor those tame priests who have been trotted out in history by emperors and princes to baptize their grimy deeds." Even the mild-tempered G. K. Chesterton responded in this way when F. E. Smith (later Lord Birkenhead) argued in Parliament that Welsh disestablishment had "shocked the conscience of every Christian community in Europe." Chesterton's reply has become a classic:

> Are they clinging to their crosses,
> F. E. Smith,
> Where the Breton boat-fleet tosses,
> Are they, Smith?

> Do they, fasting, trembling,
>   bleeding,
> Wait the news from this our
>   city?
> Groaning 'That's the Second
>   Reading!'
> Hissing 'There is still
>   Committee!'
> If the voice of Cecil falters,
>   If McKenna's point has pith,
> Do they tremble for their altars?
>   Do they, Smith?

The poem ranges out among Russian peasants, who have to get Hansard translated to find out what Wales is—so they can learn there is a curate's pay in Cardiff saved by Smith:

> It would greatly, I must own,
>   Soothe me, Smith!
> If you left this theme alone,
>   Holy Smith!
> For your legal cause or civil
>   You fight well and get your fee;
> For your God or dream or devil
>   You will answer not to me.
> Talk about the pews and steeples
> And the Cash that goes therewith!
> But the souls of Christian peoples . . .
>   Chuck it, Smith!

It is the only answer that reaches far enough down to the current defenses of our pious men. Mr. Moon suggested that God made Nixon the center of empire on earth? Chuck it, Moon! Rabbi Korff claimed he did not "bring in religion," and then compared Nixon to King David? Chuck it, Korff! Father John claimed Nixon's temporizings would save us from situation morality? Chuck it, Mac!

—*Harper's,* October 1974

AFTERWORD

Mr. Buckley's indignation did not last long. After leaving his order and the priesthood, John McLaughlin next turned up as a regular columnist in the pages of Buckley's magazine. The defenders of Joseph McCarthy could not be fastidious about a defender of Richard Nixon.

9

Dunces

When lo! a burst of thunder shook the flood.
Slow rose a form, in majesty of mud,
Shaking the horrors of his sable brows,
And each ferocious feature grim with ooze.
Greater he looks, and more than mortal stares;
Then thus the wonders of the deep declares.
[*The Dunciad*, 2.325–330]

H. R. Haldeman's book was supposed to "tell all" about Watergate, and he tried hard—he tells us more than he knows, and hints at more than he tells. He also manages to include just about everything said by just about everyone involved in or looking at the tangle of Watergate events.

He agrees with Nixon's defenders in all their countercharges. He takes the Victor Lasky–William Safire line that "it didn't start with Watergate." He also thinks, with Raymond Price and Pat Buchanan, that press enemies and Democrats drove Nixon to it. He concludes that Watergate was a Democratic trap, sprung by Larry O'Brien's dirty tricks and closed by John Doar's dirtier ones.

Haldeman justifies the secret bombing of Cambodia on the grounds that mean kids in the street wouldn't let Nixon commit his patriotic acts in daylight. And since the kids were being manipulated by Communists, that means the President of the United States was a captive in his own house of the enemies he was trying to bomb with honor in a distant land. No wonder he had to sneak and cover his nobler moments. With William Buckley, Haldeman thinks *any* President would be forced to hire gumshoes like Howard Hunt in a nation that gives its Pulitzer Prize to Jack Anderson. (Anderson, of course, was part of the Democratic trap in Haldeman's reading of Watergate.)

But after attacking all Nixon's enemies, Haldeman turns around and embraces all their theories too. So far as conspiracies are concerned, it is "come one come all" for Haldeman.

> To him the Goddess: "Son! Thy grief lay down,
> And turn this whole illusion on the town."
>
> [2.131–132]

He holds the Terry Lenzner–J. Anthony Lukas view that the DNC break-in came from Nixon's desire to know about Larry O'Brien's ties with Howard Hughes (and to know what O'Brien knew about Hughes's ties with Nixon and Rebozo). Haldeman gives himself too much credit when he calls this "my own theory of what initiated the Watergate break-in." But he admits Daniel Schorr supplied him with a CIA theory of Watergate (one prompted, as well, by Ehrlichman's novel, *The Company*). Then he adds curlicues to that theory from Miles Copeland, Howard Baker, and Fred Thompson.

Haldeman agrees with Nixon's critics that the tapings were for "blackmail" purposes—especially useful for keeping Kissinger in line. Haldeman also welcomes all variants—those of Nicholas Von Hoffman, Renata Adler, Edward Jay Epstein—of the "bureaucracy" theory of Watergate. The more people getting in on the kill, the better for Haldeman: it shows Nixon was right to see everyone (except Haldeman) as an enemy.

Haldeman, Haldeman tells us at length, was not Nixon's enemy, just Nixon's dupe. He lives, like his master, in a Hobbesian *bellum omnium contra omnes*—and his only mistake was not to see that Nixon was his enemy, like everyone else. The Nixon White House looms darker in this book than in anything written by Woodward and Bernstein. Nixon is not drunk at night, just so tired he cannot talk or walk straight. It is a wearying task, this daytime pretense at being "nice Nixon." Haldeman tried to put him back into his coffin every night before "nasty Nixon" got a chance to burble evil plans to someone like Chuck Colson. Ehrlichman loves to be devious. Kissinger is an egomaniacal paranoid. Butterfield may be a CIA plant. Haig believes in wild Jewish plots. Buchanan overheats Agnew. Even Pat Nixon is given a slap for her taste: Nixon "warned me that . . . Pat would decorate the office for him. Eventually she did it in bright blue and yellow because she wanted it to be 'cheerful.' "

Haldeman admits to one mistake (trusting Nixon) and one fault (giving Colson access to Nixon). Not much: but you have to remember that anyone who exposes even partially a mind so fetid with distrust cannot be accused of self-flattery. Haldeman's convenient lapses of memory are by now standard in the treatment of Watergate. He presents most of the White House "horrors" (in John Mitchell's term) as reactions to outside aggression—the "kids" were tying the President's hands; people were stealing Pentagon Papers right out from under the government's nose; nobody would cooperate with Nixon (Haldeman thinks the IRS was being used by Democrats to "get" conservatives). When he sees occasional excess, it comes as a surprise to him—John Caulfield says that Colson wants the Brookings Institution firebombed, Nixon himself sends Caulfield to tap Joseph Kraft's telephone. Haldeman does not tell us why Caulfield was there in the first place. Like most people, he begins the horrors too late to explain them properly.

Who remembers Tony LaRocco? Not many. He never got around to doing much in the line of work for which he was hired—but that made him something of a symbol for me. He demonstrated the early desire to have gumshoes hanging around "just in case." He shoots down the whole idea that Nixon was responding to prior aggressions. For Tony LaRocco was hired in December of 1969, and paid from the secret White House fund, just to be ready. He was brought on before there was any actual work for him to do.

He was acquired through Caulfield, who had been hired in March of 1969—less than two months after Nixon entered the Oval Office. Ehrlichman arranged for Caulfield's payment from the Kalmbach fund—but Haldeman had interviewed him for his first job (security during the 1968 campaign) and recommended him to John Mitchell. When Mitchell did not hire him, he went on the private payroll; and though Ehrlichman put him there, William Safire remembers him as the "Haldeman surveillance man" (*After the Fall*, p. 298) and John Dean says Haldeman assigned Caulfield to his office (*Blind Ambition*, p. 35). It is clear that Haldeman and Ehrlichman wanted their own secret snoops long before Colson arrived on the scene—and before the drug team was formed from which Edward Jay Epstein dates the Watergate madness. This was, as well, before "theft" of the Pentagon Papers gave the White House an opportunity to put "plumbers" on the public payroll—an excuse for alternate financing, *not* a first occasion for political spying. The spying began in earnest after Chappaquiddick, when Caulfield sent Tony Ulasewicz to dig up dirt on Edward Kennedy. That was during Nixon's first summer in office.

Every memoir from the White House testifies to Nixon's obsession with Kennedy, and his willingness to use any weapon against him. Once again, Haldeman associates this primarily with Colson, who had his own grudges against Kennedy from days in Senator Saltonstall's office. But the Chappaquiddick capers were going on before Colson had acquired any power in the White House. Haldeman admits that Colson gained power over Nixon by boasting of "dirty tricks" performed on his behalf. Did Haldeman and Ehrlichman perform the Chappaquiddick tricks and take no credit for them where they *knew* they would be most appreciated? Haldeman and Ehrlichman knew how to please their master—who, in turn, clearly knew about this secret team. Haldeman says that Nixon himself ordered Caulfield to tap Kraft's phone. How did he know Caulfield was available for such missions? Obviously from the man who hired him in the first place, from Haldeman.

Haldeman says Nixon had to resort to private financing of secret agents because the bureaucracy had resisted his orders to find the Communists at demonstrations, plug the leaks, tap phones, etc. But there had been no time to test that resistance when Caulfield and Ulasewicz were hired—and, later, Tony LaRocco was added to the team (and, later, Hunt et al.). There was a prior determination to use gumshoes for missions so mean and personal that no one could be trusted to know about them but the little Oval Office triumvirate. The fungus, thus in place, was bound to spread.

It was a blessing to the nation that Nixon could trust no one. Those who accuse him of politicizing the IRS or of an attempted coup on the CIA and FBI do not realize that he was too frightened of those agencies to let them in on his real vindictiveness. He was an amateur at coups; Tony Ulasewicz was more his speed than Richard Helms. That is why, eventually, he got caught. The bureaucracy did not react, in fear of his reorganization plan, all of a sudden in 1973. Nixon had been laying up powder for his own destruction all through the years of secret little crimes, of buggings and break-ins and a growing band of bumbling spies tiptoeing in and out of the White House.

Nixon's distrust of others was fully shared by Henry Kissinger; and the hallmark of their diplomacy was secrecy. Not only was Congress to be kept in the dark; so were the State Department and the Pentagon. The secret bombing of Cambodia, the secret mission to China, the secret peace initiatives, the secret motive for the Christmas Bombing—all these represent attempts to run the world out of the White House, letting few others in on the act.

Here we get, I think, the explanation for Haldeman's bizarre claim that Nixon secretly averted war on two occasions—when he bluffed the Russians back from China's border, and threatened Russia out of its

Cuban submarine base. The feints and counterfeints of diplomatic intelligence gave Haldeman some slim basis for those exaggerated claims. But who did the exaggerating? Haldeman himself? I think he is relaying the fantasies of his master—which would explain the confidence with which he puts forth a thesis instantly denied by Secretaries Kissinger and Rogers, as well as by the Russians. Nixon liked to think he was saving the world all by himself, in ways that no one knew. It is important to remember that Haldeman was talking with Nixon about the President's book right up to the Frost interviews, which made him cease, but only in part, from the role of Nixon's alter ego. The bunkered Nixon still speaks through his puppet about the ways he saved the world.

All the White House memoirs—even that by the dogged loyalist, Raymond Price—talk of a good Nixon and a bad one, of a light side and a dark side. (Each author, of course, claims that *he* was there to make the light side prevail.) But the Nixon of the secret mission to China was not so distant from the Nixon who dispatched a spy team to Chappaquiddick. Price becomes an object lesson in the way *good* Nixon could use deviousness to involve others in mendacity. Price wrote the white papers and speeches Nixon used to exculpate himself in the cover-up period. Price was lied to by his boss, then helped him lie to the nation. Yet he does not resent being lied to, or repent having helped in the lies. He thinks the good things that came of Nixon's deviousness compensate for all the bad, and quotes the master as telling him: "We never could have brought off the opening to China if we hadn't lied a little, could we?"

Price was supposed to be one of those who brought out the good Nixon—the house intellectual (something he tries to vindicate in his book by pseudoscientific explanations of Nixon as a "left-brain" classicist). When Watergate revelations were breaking, Daniel Patrick Moynihan fatuously wrote from India that "strong and moral men of the present Administration" *like Raymond Price* would assure the team's integrity. Yet Price still clings to Nixon, equivocating and excusing—helping him prepare each slither of the Frost interviews. *Corruptio optimi pessima.* Those who believed in the good Nixon all ended up serving the bad Nixon.

It is distressing to see Nixon's critics acquiring the paranoid vision of politics that he had. We now get a wave of conspiracy theories, working from the *cui bono* maxim. It would make policemen's work very easy if that maxim were truly the guiding one: they would automatically send

to jail the principal beneficiary of any victim's insurance policy. The
saddest irony is that Nixon and Joe McCarthy used that maxim in their
1950s charges: *who benefited* when the Democratic administration
"lost" China or failed "the captive nations"? Foreign Communists
benefited. *Therefore* foreign Communists were running the American
government. No wonder Haldeman welcomes all such arguments. Nixon
infects men with distrust, and even those opposed to him contribute to
his effect when they become its "carriers."

Most conspiracy views begin with the fact that the DNC break-in was
undermotivated and overbungled—so it must have had some occult purpose that was deliberately sabotaged. But the Dita Beard affair was undermotivated and overbungled. So was the Ellsberg break-in. So were
the Kennedy plots. So was the "spying" on public demonstrators at the
Statue of Liberty. So were the faked cables on Diem. So, probably, were
other things we do not know about. I find Haldeman plausible when he
claims that Nixon often called for retaliatory, self-defeating acts that
Haldeman quietly ignored. Nixon's vindictiveness *was* petty and ill-
aimed, and ultimately self-defeating. He was clumsy at everything else,
why not at crime?

Years ago I marveled at the way Nixon could boast in *Six Crises* that
he got in a vicious little kick at a rioter in Latin America who was
being dragged off by the police. Haldeman claims in his book that Nixon
used to boast with similar pride of the way he told the press off in his
"last press conference." He took deep satisfaction in things others would
be (or pretend to be) ashamed of. In discussing that *Six Crises* passage
I called Nixon a "late kicker," one who discharges his resentment
fecklessly, in clumsy delayed reactions. Haldeman recounts an incident, from the 1960 campaign in Iowa, that is almost a parable: Vice-
President Nixon was on a tour that had been poorly planned, so that
hours were wasted in driving from town to town.

> Nixon seethed with anger. He was riding in an open convertible and Air
> Force Major Don Hughes, Nixon's military aide, was in the seat directly
> in front of Nixon's. Suddenly—incredibly—Nixon began to kick the back
> of Hughes's seat with both feet. And he wouldn't stop! Thump! Thump!
> Thump! The seat and the hapless Hughes jolted forward jaggedly as
> Nixon vented his rage. When the car stopped at a small town in the mid-
> dle of nowhere, Hughes, white-faced, silently got out of the car and
> started walking straight ahead, down the road and out of town. He wanted
> to get as far away as he could from the Vice-President. I believe he would
> have walked clear across the state if I hadn't set out after him and
> apologized for Nixon and finally talked him into rejoining us.

The only feelings Nixon could convey with some authenticity were resentment and self-pity—valuable items in 1968, when the reaction to war and riots and demonstrations led to a repressive mood in the electorate. The voters had their *own* enemies list that year. We are asked how the system could go so wrong in giving us Nixon. It didn't go wrong. People got what they wanted. They knew Nixon as well as any candidate has been known, his strengths and his weaknesses: but they were in a mood to accept his weaknesses as strengths. A great many of them wanted something even worse—wanted Agnew or wanted Wallace. The nation at first took a self-defeating satisfaction in its choice—just as Nixon did in all his own late kicks, and silly raids, and secret retributions that backfired.

Of course there will always be some mystery around the Watergate events—but not because of labyrinthine plots and conspiracies; not because the "inefficient" bureaucracy suddenly turned marvelously efficient in destroying a President. These events have the darkness of hatred itself, the futility of envy, the nihilism of spite. Alexander Pope knew there could be idiocy on a heroic scale. The Nixon White House has been called Kafkan or Orwellian, but it was not a place where the lone person faced an impersonal system. It was a world of little men using large powers incompetently from a combination of suspicion and panic. It was a place far closer to Pope's vision, in *The Dunciad,* of clownish bungling that verges on anarchy, and of anarchy verging on apocalypse, so that darkness covers all. Everyone was skeined in self-covering layers of suspicion about everyone else:

> So spins the silk-worm small its slender store,
> And labors till it clouds itself all o'er.

> [4.254–255]

Even when one of this crew wants to shed light on the complex scurryings in and around the White House—as Haldeman probably does—he can only throw his own darkness:

> Of darkness visible so much be lent
> As half to show, half veil the deep intent.

> [4.3–4]

These were not men familiar with the day. Darkness was their element, and it will cling to what they did forever.

—*New York Review of Books,* April 6, 1978

IV

Politicians

Politicians make good company for a while, just as children do—their self-enjoyment is contagious. But they soon exhaust their favorite subject, themselves. The idea that a politician might be interested in something else is so refreshing that some people have made the mistake of hoping to find out something new or interesting about modern art from Nelson Rockefeller, about opera from Lowell Weicker, about Teilhard de Chardin from Jerry Brown. One soon enough finds out that Nelson Rockefeller was only equipped to tell you that Nelson Rockefeller was interested in modern art, or Weicker that Weicker liked opera, and so on.

There is a kind of noble discipline in politicians, in persons prepared to devote a lifetime to discourse on a single subject, over and over, with anyone who will listen, anywhere. It inspires a goofy awe, this sight of them ringing a single bell all their lives, hammering at their own heads.

10

The Senate

Rome still survives in this assembled Senate!
Let us remember we are Cato's friends,
And act like men who claim that glorious title.
      —Joseph Addison, *Cato*

We [the Senate] looked and laughed at each other
for half an hour, and adjourned.
      —Senator William Maclay, 1790

The Senate can turn on you with anachronistic relevance, trendy yet dignified—Sam Ervin coming into his own at last. It is an embarrassment and a source of strength, like the South. The two, of course, intersect; they reinforce each other, eerily. I don't mean, simply, that Southerners have guided and stalled Senate debate—the superb courtesy of that chamber is proved by the way it still calls what it does "debate." Even non-Southerners, when they settle fully into Senate ways, often do it by getting Southernized. Who was a better master of mint-julepy eloquence than Everett Dirksen, from Illinois? The quirkier such people get, the more we see in them a type. The toga is the last refuge of ruling idiosyncrasies. Even a rebel and deserter of the Senate like Eugene McCarthy has become more whitely flown of mane, more pirouetting of stance—suggesting, in manner at least, the Bilboizing of the intellectuals. McCarthy became more senatorial by leaving the Senate—the last way to join the Club was by ostracizing himself.

A senator's importance depends largely on *where he is*. Robert Hartley describes the opposition in Illinois to Senator Percy's move from the Appropriations to the Foreign Relations Committee. This was rightly perceived as a result of inflamed presidential yearnings; so home-state voters felt he had traded ability to do real good in Illinois for increased candlepower in the national press. He had to counter this resentment with special efforts to show he could still deliver the pork. Considerations of this kind go into every decision to move or place oneself in the Senate. And they should. After all, if a general tendency to vote with the party is wrong, why are senators members of parties in the first place?

It is true that Senator James Buckley's votes seem at times to sort ill with New York's concerns. But that is not because he is really free of all ties. He is determined in most of those votes by his debts to an Issue constituency. The Senate is "issue-oriented" in ways the House cannot afford to be. Representatives are tied to one district; they must seek reelection in it every other year. Power in the House grows by intermeshings. It is true that some committee members acquire expertise in their committee's field of attention—but that field must often be one that matters to the man's district. Besides, the powerful committees of the House tend to be those concerned with organization itself—Rules, Ways and Means, Appropriations. Representatives can influence things like foreign policy only through the constitutional back door of the appropriating process.

A senator has to campaign statewide—which means largely in dense urban "media markets." Once elected, he has more time before reelection to set long-term goals, which always include the presidency. This means choosing an Issue; often "writing" a book on it; drafting a bill; specializing. So, what William Fulbright was on foreign policy, Edward Kennedy will be on health care. (Kennedy, when he still had to be careful about the Vietnam war because of his brother Robert's presidential hopes, kept up peripheral fire by specializing in *refugees* from the war.) The Issue can become a kind of theatrical routine (Proxmire on government waste) or a traveling roadshow (Hollings on hunger). Abe Ribicoff settled for car safety. Environmentalists have looked to Muskie, as labor has to Humphrey. Jackson, the senator from Boeing, is also the senator from Israel. Senator Ervin was a constitutional nitpicker on blacks' rights and a constitutional purist on the right to privacy. One reason for the South's great power in the Senate was that all its separate grandees *had* to be united on one Issue: white supremacy.

Choosing one's Issue is very important. It depends, in part, on which Issues are open. No senator wants to be overshadowed by a prior claimant, or linked too closely with a man who might embarrass his more general views. Some Issues get stolen from a man—as Dirksen fathered

Percy's All-Asian Peace Conference off on someone, anyone, *else* to keep his junior colleague from getting too much credit if the thing worked. (Percy had earlier relied on a public housing scheme for his Issue, and watched that get amended into other people's custody.) Joseph McCarthy's shopping around for an Issue is famous—Georgetown's Father Edmund Walsh served up communism over lunch at the Colony. Sometimes the Issue is a complex of ideas, like Goldwater's "conservatism," or even a network of feelings. For Robert Kennedy, being a Kennedy was the Issue. That is why Edward Kennedy cannot shed presidential solicitings even if he wants to; being a Kennedy is of itself a presidential Issue, with its own special constituency. (Even Robert Kennedy, if he had succeeded in becoming Johnson's Vice-President, as he desired, would have found his constituents balking, or prodding him to dangerous subversion of a non-Kennedy President.)

Issue constituents are at least as demanding as the money and interest ones. That is why it helps to find the right blend of Issues. Senator Jackson is hawkish on foreign policy—a stand dictated by his local constituency, and appealing to the right in general. Not only Boeing, but a general xenophobia, is part of his origins; he was one of the fiercest anti-Nisei orators of World War II. But he is also a strong supporter of Israel—which appeals to some on the left, and gives him a presidential constituency willing to fund his campaigns. The two Issues fit neatly together.

Senator Buckley, on the other hand, chose environment as his second Issue. (The first is the same as Goldwater's—is, in fact, more Goldwaterism than conservatism.) The interest in environment came from Buckley's hobby as an amateur zoologist—much as Goldwater's interest in Arizona Indians came from a hobby as his state's photographer. But when, on things like the SST, Buckley has to vote with his *first* constituency (business and defense), Sierra Clubs get angrier at him than if he never voiced environmental concerns. When Buckley is called unpredictable, it depends on what constituency you use to make your predictions. He managed to score a perfect zero in last year's ADA ratings.

One's Issues should be complementary, even compensating. Support of Israel is a presidential Issue for Jackson. It is only a local one for Senator Javits—if he still had a dream of the presidency, he would have to find a balancing national Issue. American Indian policy, a minor Issue for Goldwater, Fred Harris, and James Abourezk, was a minor nuisance to McGovern. Blacks, a national Issue for Robert Kennedy, would be for Edward Brooke a local Issue to be balanced off. The problem with such balancings is that one can reach a state of equipoise resembling paralysis. Henry Jackson, the invisible campaigner, covers so many bases that he is considered the Hill's most effective senator—yet these very achievements seem to block his upward path, as did Senator Robert Taft's.

Henry Jackson is an interesting study in energetic self-immobilization. The more he talks, the more of him fades, leaving nothing but a Cheshire frown. It would be interesting to know why he, churning away so well at lower levels, seems incapable of getting airborne. He surfaces in a presidential year much as Wilbur Mills did in 1972, a mole come up into daylight, blinking away, trying to fly with his eyelashes.

The first thing I ever heard about Chuck Percy came from a young Chicago *Tribune* journalist who went to interview him. They swam in Percy's heated pool, in the Kenilworth mansion Percy chose for his own while going by it as a paperboy. The senator, who had been a water-polo star at the University of Chicago, proposed a race; and contrived, every time, to lose by a flattering split second—and kept doing that even when the journalist experimented to see if *he* could lose. Percy is a champion deferrer. He tries to rise by bowing.

At the 1972 Miami convention, when Percy came down to the pool of the Playboy Hotel (where the Illinois delegation was staying), I saw him dodge admiring political wives for a while, then swim desperately out into the ocean to shake off the most persistent young matron. Percy, nearing sixty with a cryogenic boyishness, has a hard time keeping some people from chucking him under the chin. Yet that very charm attracted all his first patrons. He arrived on the Republican scene in the deceptive Eisenhower dawn of study clubs and seminars that were supposed to grow up and become the New Republicanism. Eisenhower adopted Percy, and told him to write the 1960 platform; but the Republican convention did not want a Great Books anthology. (The only one who has risen very far on the fifties mountain of Rockefeller Reports is Henry Kissinger.) Eisenhower, like George Washington, would have no political progeny.

Percy came to be seen by a Goldwaterized party as the client, rather, of Rockefeller, the Reports man himself—and the right wing's Father of Lies. Percy has tried to live down this damaging impression, courting Nixon as long as he could; but the right wing has consigned him to that limbo of mythology where Pocantico's aging capitalist becomes a badly disguised Marxist. Soon all his friends will say of Percy, as they have of

Rockefeller, that he could have got elected but never, in *his* party, nominated.

Yet he will continue to be "mentioned" as presidential material. Most senators are. What, after all, is the modern Senate but a presidential launching pad? Another reason the South once made such a mark on the Senate was that Southerners were automatically excluded from consideration for the presidency—they had to concentrate all their flair on Senate power. But now even an old Ku Kluxer like Robert Byrd dreams of the White House. Eight of the last twelve nominees for President or Vice-President have been senators (and three of the eight ran twice). This year seven candidates or near-candidates for the nomination are senators (Jackson, Bayh, Humphrey, Muskie, Church, Byrd, Mathias). Two others—McCarthy and Harris—retired from the Senate to campaign semipermanently for the White House. Three others have withdrawn themselves from this year's race: Kennedy, Mondale, and Bentsen. And when Vice-Presidents are being considered, a flock of senators hovers and hopes—Baker, Brock, Brooke, Stevenson. There is a club of those who have run in the past—Goldwater, Thurmond, Sparkman, Humphrey, Muskie, Eagleton, McGovern. A younger crop of men is being groomed, or is maneuvering, for later consideration—Adlai Stevenson III, Weicker, Biden, Buckley himself. We can assume that, with the exception of some old survivors from the South's proscribed time, almost all the sitting senators have been, are, or will be presidential candidates.

It was not always so. One did not, in the nineteenth century, go to the Senate to become President. One often went there to collect large bribes. Senators were chosen by state legislatures, and were supported increasingly by Gilded Age rakeoffs. It became a millionaires' club; content, in Jefferson's distinction, to represent the nation's wealth rather than its wisdom. Bright innovative men, men like Henry Clay, preferred the House of Representatives to this soiled House of American Lords. The Speaker became the most powerful man in Congress (and sometimes in the country). We cannot imagine a modern President—an Eisenhower, say, or Johnson—running for the House of Representatives after completion of his presidential term, like John Quincy Adams. Even Ford, the man of the House, cannot go back there if he loses in 1976. It would be too great a comedown. When Ford moved to the White House, Representative Michael Harrington of Massachusetts said, "They turned him loose after twenty years in the cellar."

The rebirth of the Senate came with the success of the muckrakers' drive for direct election of senators (1913). The Speaker's power had been broken under Joe Cannon; and the senators, now campaigning statewide for numbers where they were densest, began to redress somewhat the House's rural overrepresentation. The Senate took up again its most important trust, the treaty-making power. Senator Lodge was now the king of the Hill, facing Wilson down. Roosevelt, for all his imperial presidency, had a worthy adversary in Taft. Vandenberg and Fulbright partly maintained the Senate's claim in foreign affairs.

Some think we have gone too far in looking to the Senate for presidential material—that administrative talent should be called upon from elsewhere: from governors, university presidents, executive department heads. The President, we are told, *executes;* more like a governor than like legislators. But that maxim can mislead. The modern President still initiates the legislative program and forms policy goals, dealing with the biggest Issues when not dealing in large symbols. Does organizational talent equip one for the presidency? Then the best-equipped man of modern times was Herbert Hoover.

Besides, it now takes great executive skill to use the Senate for aiming at the White House. Even a representative of some seniority in the House has personal and committee staff drawing down a $150,000 payroll; and an important senator has interlocking staffs in his home state, his Hill office, and various committees, that make him command a million dollars of man-hours every year, just counting those on public pay. This, along with volunteer and privately funded campaign workers, gives such a senator an empire to govern. Kennedy, Humphrey, Jackson —each holds four chairmanships, with power to hire and fire staff. Jackson serves on twenty-four committees and subcommittees, Kennedy on twenty-two, Humphrey on seventeen. Each has a base for dealing with many agencies, Jackson mainly at Interior, Kennedy in his two Judiciary chairs, Humphrey in the Joint Economic Committee. All of them naturally want bright staff people, with good ties to the departments. An important modern senator runs a kind of minigovernment, warming up for the White House.

The imperial urge for committee posts—most senators hold at least sixteen—scatters a senator's attention. But the obvious remedy—increasing the number of senators to do the work—is unacceptable to this sacred chamber, since it would also dilute power. Turf is jealously guarded. This means jockeying among the senators themselves for committee appointments, chairmanships, staff, office space, etc. But it leads to an even more intense common front against anyone who would curtail a senator's prerogatives, the dignity of his order.

Senators take very good care of other senators. Even the exceptions prove the rule. The Senate censured Joseph McCarthy, but only after he had taken to things like calling Senator Fulbright "Half-Bright." Lyndon Johnson could vote in good conscience for censure, not on any evidence brought before the Watkins committee, but because McCarthy insulted Senator Watkins. Thomas Dodd was treated with kid gloves by the Stennis committee until he demanded the removal of the committee's vice-chairman, Senator Bennett, for prejudice. The senators' code with each other is "Mock not, that ye be not mocked." They will laugh at each other; but they get very nervous if they hear outsiders joining in. In this, at least, the Senate is still the place Mark Twain described, presenting Senator Dilworthy's ordeal:

> "Don't use such strong language; you talk like a newspaper. Congress has inflicted frightful punishments on its members—now you know that. When they tried Mr. Fairoacks, and a cloud of witnesses proved him to be—well, you know what they proved him to be—and his own testimony and his own confessions gave him the same character, what did Congress do then?"
>
> "Well, what *did* Congress do?"
>
> "You know—Congress intimated, plainly enough, that they considered him almost a stain upon their body; and without waiting ten days, hardly, to think the thing over, they rose up and hurled at him a resolution declaring that they disapproved of his conduct!"
>
> "It *was* a terrific thing—there is no denying that. If he had been proven guilty of theft, arson, licentiousness, infanticide, and defiling graves, I believe they would have *suspended* him—for two days."
>
> "You can *depend* on it. Congress is vindictive. Congress is savage, sir, when it gets waked up once. It will go to any length to vindicate its honor at such a time!"

Although few can work up quite the regard for senators that senators have, the body as a whole has never stood higher in general esteem. It almost meets the expectations of the founders. Governors move over to the Senate. But would Humphrey now trade office with Governor Wendell Anderson, or Goldwater with Governor Raul Castro? Senator Percy will stay in the Senate as long as he can, short of moving to the White House; it is even better than being president of a large corporation. But from the House he would have to look back wistfully toward Bell and Howell. Even an ex-President may now become a senator with dignity—as easily as foiled presidential candidates return to its chamber.

As a place of such prestige, as the elective national office with longest tenure, the Senate attracts about as good a brand of politician as we are likely to produce. Even inferior material can be improved there, since the elaborate courtesy of senators to senators enacts, in small, the civilizing process. Even a rough Klanner like Robert Byrd has moved up above the level of the House's vulgar pettiness (exemplified in Wayne Hays) to become a gentleman's butler to the chamber, and almost a gentleman. That would amaze Mark Twain.

With important help from Nixon, the Senate has almost nerved itself to act like the body envisaged in our Constitution. It turned President Ford down when he proposed last-gasp aid for Saigon and second-step commitment to Angola. It scared the senators to act like Cato's friends, and they are backing off somewhat on the CIA. But Madison would recognize roughly what he had in mind if he looked in on the Senate today. Maybe the wanting to be President can make senators act like men. It swells the head to be sure—and makes the Senate look like a hundred Franklin kites; each coaxed up on the wind current as high as possible, to tempt the presidential lightning. But what a gaudy sight that gives us for our money.

—*New York Review of Books,* March 4, 1976

11

The House

In 1972 an elated congressman on his way home from a party ran a stop sign in Georgetown, hit a car, hit a fence, hit two trees, hit a brick wall, hit another car. The ruckus attracted a small crowd, and the emerging driver did the natural thing: he went around shaking each person's hand. There you have the essential House of Representatives. I fancy the real explanation of Richard Nixon's pardon is that Gerald Ford, totally of the House, Housey, just wanted to be friendly with everybody after the accident.

The House of Representatives looks like history's joke on the framers. In the constitutional scheme of things, the House was to be the most "radical" wing of the most popular branch of our government. Members of the House of Representatives, all of them up for election every other year, each in his own small locality, would be the most democratic element in government, a barometer of shifting popular moods. The cooler, more remote men of the Senate must steady and delay change urged by House firebrands at the populace's beck and call.

So how did we get the House we've got, from a plan like that? It is the most conservative wing of the conservative branch. It is the land where the lobbyists play. It is stingier with public funds than the Senate wants to be. Its potentates have been men like "Judge" Howard Smith and Wilbur Mills. What went wrong?

There are two main answers to that question. One can say that the framers' preconceptions were false—that the people are *not* radical, but cautious; that elites initiate change; that the popular wing is conservative because the populace is conservative. This answer offends liberal orthodoxy and is not indulged in polite society.

The other answer is to say that the framers were right in their day, but since then the machinery has gone awry. The popular branch does not reflect the popular will because of changes over time. Whatever its other shortcomings, this answer *does* point to relevant change. The representative, it was presumed, would begin and remain a man close to his district. The Congress would have comparatively short sessions. The difficulty of travel would keep a representative home except when Congress was in session. The normal scene of his action and concern would be in his district. When he went to sessions of Congress, it would be as a stranger to the Capital, reporting on the area he knew best and lived in.

Senators, by contrast, representing the whole state, serving as links between state legislatures and the national government, would have less contact with the general run of their constituents (except their own immediate neighbors). Only Presidents were expected to live much (not all) of the time in the Capital. They were chosen by the electoral college, the elites' elite, and would stand farthest off from individual constituents, few of whom would ever see the man their electors had chosen in their wisdom.

Modern transportation and communication have turned that scheme almost exactly upside down. Few people can tell you the name of their congressman. Far more know at least one of their state's senators, by name, face, or repute. And many citizens now live vicariously with the President's family and dogs, by way of gossip magazines and TV. Congressmen are, of all these people, the ones most bound to Washington. They have fewer opportunities for travel and exposure than do senators. And now the difficulty is with travel to their districts; not, as at the outset, with travel to the Capital. They become citizens on the Potomac where senators are often just celebrities.

It is true that the representative has a comparatively small district he must service. His staff must get out his newsletter, answer mail, schedule him back before district gatherings as often as the budget will allow. But congressmen spend as much time servicing each other as in servicing their districts. They become each other's constituents. A freshman congressman barely finds his way around before he must start running for reelection. Since general recognition of congressmen is so low, he has a tremendous advantage in mere incumbency—more so than do candidates for higher office. Where the voter knows practically nothing about either candidate, the mere fact of prior election to the United States House of Representatives is itself a large recommendation. And that advantage will grow, campaign by campaign, by increments of recognition, by gradual achievement of seniority and some power in the House. If a congressman can do only one widely recognized favor for his state in his first two terms, but two in his third term, and three in his fifth or sixth, he acquires virtual immunity from challenge by a man whose prospects in the early terms are no better than the incumbent's were when he began. With a congressman, you buy 1/435 of a vote in the House. With a senator, you buy over four times that power in *his* chamber.

The individual congressman weighs little. His power is a power of combining with his fellows in the House. His influence must grow by agglutination, which takes time. Thus a congressman uses his Washington connections to commend him to his district's constituents, rather than vice versa. The man whose seat is still not safe asks colleagues to come and tell the home folks how important and respected he is in Washington. Notices in the national press are sent back home. By the time a man is established in the House—often becoming reelectable for life if he is content to rise no higher—he has incurred a number of debts of this sort; and the way to increase his power in the House is to serve other members as he was served, acquiring his own due bills.

This would be an almost inescapable task, even if the congressman wanted to escape it. But most House members like the job, or learn to like it, of doing favors for their fellows. The Senate likes to think of itself as a very exclusive club. The House is a bit embarrassed about its inclusiveness. Mutual puffery sustains the members' self-esteem, abraded elsewhere in the Capital. Most congressmen drive about Washington, shop, and go to the movies, without being recognized. At least around the House their fellow club members and their shared employees give them a respectful welcome.

Babbitt, you remember, can hardly wait to escape from the anonymity of even Zenith's streets into the world of nicknames, established jokes, and achieved little niches at the Athletic Club. The District of Columbia is Zenith on the Potomac; the House wing and office buildings make up the Athletic Club. (The Senate is the Union Club, from which Babbitt must lower his sights.) Congress *bunches*. Lifelong House men move in clusters, amoebalike, by continual adjustment. There are clubs within clubs, and jealous little calibrations of privilege. Rivalry is intense but low-key because of the embracing code. The jealousies are school-size, a world of afternoon cabals, of honor pledged, of little betrayals, as clique worries at clique. But no matter what clashes occur, everyone can be counted on to rush around shaking everyone else's hand.

—*New York Review of Books,* October 16, 1975

AFTERWORD

The 1980 election seemed to reverse things again, making the Republican Senate conservative and Tip O'Neill's chamber liberal. But it is early to say. As late as 1978, the Senate had passed the Panama treaties with much Republican help; and Reagan fans were soon deploring the "moderate" leadership of Howard Baker, Charles Percy, and (of all people) Robert Dole. *Plus ça change . . .*

12

Bobby Baker

So this gentleman said a girl with brains ought to do something else with them besides think. And he said he ought to know brains when he sees them, because he is in the senate and he spends quite a great deal of time in Washington d.c. and when he comes into contract with brains he always notices it.

—Anita Loos, *Gentlemen Prefer Blondes*

Pimp to the Senate is not a title easily worn. But somehow Larry King gives Bobby Baker just the right Lorelei Lee tone of voice. His book should be called *A Guy Like I*. Senators just kept shoving money into his hand. His stance is half swagger, half sulk—a blending of goniff with gofer. The affronted air of one who stood by his deals is perfectly sincere: he is shocked when people won't stay bought. I met Mr. Baker while I was reading his book, and remarked that he is surprisingly candid about his own actions—which included special delivery of a call girl to the home of a senator in need. "You have to give value for money," he said. It was his creed on the Hill, and it explains his success even in jail, where he was a principal ornament of Allenwood's chapter of the Jaycees.

Senators need pampering. They have the egos of movie stars. That comes out even in Bernard Asbell's rather fawning book, centered on Edmund Muskie. Asbell notes that Muskie uses his height theatrically—choosing a chair for TV that put his belly button up above the desk, putting none but short men on his staff. Bobby Baker's major insight, achieved in his teen days as a Senate page, was that it is impossible to praise a senator too much. His period of service to Lyndon Johnson is perfectly captured in the fact that he seems to have begun every sentence with a verbal salute. He did not call him Senator or Mr. Johnson or Sir, but "Leader." The senator did not stop, of course, to consider what that title is in German.

One can safely skip the childhood chapter in Baker's book, which lacks the economy of Lorelei's account while retaining its point: *"It would be strange if I turn out to be an authoress. I mean at my home near Little Rock, Arkansas, my family all wanted me to do something about my music."* But once he hits Washington, Baker is quick to ingratiate himself. In time, avoiding Bobby Baker's friends list would become as imperative as joining Richard Nixon's enemies. Among friends mentioned here, Senator Muskie is shown receiving emergency campaign funds from Baker's service. Elsewhere Bobby Baker gives Bobby Kennedy an envelope with $10,000 in Murchison money from Texas for the 1960 campaign. When Baker's troubles began, he says, a German beauty named Ellen Romesch was swiftly deported by Attorney General Kennedy, with a trusted aide to escort her and keep her mouth shut. The charge was that Mrs. Romesch had been made available to senators at one of Baker's command posts, the Quorum Club, and that she also had an affair with a Russian diplomat. What did not come out at the time was the news that one senator who sought her company by way of Baker had since then become President of the United States. It seems there was a Judith Exner before Judith Exner, playing musical beds with the Russians instead of the Mafia.

Baker has less to reveal about Lyndon Johnson than some have feared or hoped—a kickback here, an affair there. Johnson was too stingy to cut Baker in on his financial deals. "When it got down to the licklog, friendship with Lyndon Johnson was a one-way street." But what should one expect when one came in, always, to shine the boots of one's "Leader"? What Baker supplies is a glimpse of the fear under Johnson's swagger. Much of his skill at Senate maneuver came from an "overkill" need to have all-or-nothing support. He could never risk losing—which explains his withdrawal from the 1968 race. Even the remotest possibility of losing to the hated Robert Kennedy was more than he could face.

Baker has a con man's natural gift for seeing through other people's acts, and he shrewdly notes that "Ike, one of the better *political* generals the army has produced, may have sometimes out-country-boyed Lyndon Johnson when [Majority Leader] Johnson didn't know it. Ike, too, was a fine actor and not a bad manipulator of men."

Baker's rise and fall were not the result of his ties with Johnson. They came from errands run for his father-idol of power and money, Robert Kerr of Oklahoma. Kerr died with two million dollars of loose cash in his office, and there is an obvious connection between that fact and Baker's considered judgment that Kerr "had the finest mind of any I knew in public life, and I do not make exceptions for Lyndon Johnson, Jack Kennedy, or anyone else." Kerr, quick and businesslike in bribes, once told Baker, "A man who doesn't have money can't operate. Why, if I don't have at least $5,000 on me as pocket change, I'm afraid that taxi drivers won't pick me up."

Baker's book throws light on Nixon's memoirs, and takes back reflected light. Baker has now joined people like Hiss and Howard Hughes among Nixon's obsessions. Three times Nixon reflects in his book on the unfairness of Democrats in the Congress: they covered up (he says) the Baker scandal for Johnson, and then refused him the same courtesy. But in a fourth place—a diary entry from 1972—Baker suddenly emerges as the source of up-to-date information about Lyndon Johnson's last days. A number of anecdotes—and some flattery of Nixon —are reported as Bobby Baker's. How and why was Nixon communicating with Baker?

The answer to that question shows how desperate Nixon was to get something on Larry O'Brien—the reason for the Watergate break-in in the first place. O'Brien represented the Kennedys for Nixon; he was both a danger and an opportunity because of his ties with Howard Hughes. In August 1972, Nixon people threatened Baker with new trouble from the Justice Department. Then, on Labor Day, Baker got a call from Bebe Rebozo, summoning him to Key Biscayne—it was so important, Rebozo said, that Baker must come at once, even though Baker was on his way to his mother's funeral. At Key Biscayne, Rebozo got right down to business—what did Baker know about O'Brien that could help elect Nixon President? "We need to nail O'Brien. Have you heard anything about what really happened at Chappaquiddick? Did O'Brien play a big role in that?" Baker said he had nothing to give, and tried to steer Rebozo to other scandals where he might make a contribution. But Rebozo wanted O'Brien or Ted Kennedy.

After his mother's funeral, Baker got another call from Rebozo to set up a meeting with Herbert Kalmbach. "Herb Kalmbach seemed almost desperate to uncover dirt involving Larry O'Brien. . . . Kalmbach said, 'Tell me about the TFX fix . . . did O'Brien have anything to do with the TFX decision? . . . Do you have anything on him from the Johnson era?' "

It is clear that Rebozo's peremptory contact with Baker was ordered by Rebozo's friend, who acquired his anecdotes about Johnson when Rebozo reported back. The threatened prosecution of Baker was just another in the endless list of Nixon's dirty tricks—once again justified on the grounds that the target of his scheming had "got away with" prior schemes that hurt Nixon.

There is another connection between the Nixon and the Baker books. Several times in his memoirs Nixon dwells on a mysterious comment Johnson made to him while showing him around the White House—that Nixon should trust and work with J. Edgar Hoover, because Johnson could not have become President but for Hoover. Now, in Baker's book, we hear what Johnson told Baker during their last days together:

> "J. Edgar Hoover came to me shortly after I became President and said he had electronic evidence that you were mixed up with a bunch of Las Vegas gamblers. He warned me against lifting a finger to help you. I felt helpless."

The Baker scandal arose in the fall of 1963, just when Johnson succeeded to the presidency. There was urgent need for Johnson to cut ties that could embarrass his 1964 campaign. That, presumably, was the way Hoover helped Johnson become President in his own right—and another of the ways Hoover let his chiefs know what he had on them. In context, Johnson's counsel that Nixon should trust Hoover becomes an ambiguous bequest. In fact Nixon could not, finally, trust Hoover because Hoover was not devious *enough*.

The Nixon and Baker books would make good companion reading if Nixon's made good reading at all. But the ex-President's suspicious style is less convincing than the ex-convict's complacent one. Nixon said "I am not a crook" and commanded disbelief. Baker says "I was only a crook" and engages sympathy—something about whores entering heaven before Pharisees. When Baker built his senatorial bilkees a pleasure palace called "The Carousel," he gave them value—an ocean view: *"Nothing makes gentlemen get so definite as looking at nature when it is moonlight."* Baker is righteous about not "ratting" at his hearings and trial, and even that sounds better than the whining of a President who would not stand trial. *"Everyone at the trial except the District Attorney was really lovely to me and all the gentlemen in the jury all cried when my lawyer pointed at me and told them that they practically all had had either a mother or a sister."*

—*New York Review of Books,* July 20, 1978

13

Daniel Patrick Moynihan

Towson State used to be a good little teachers college in Maryland. Its wider (dubious) fame came from the fact that Governor Spiro Agnew claimed he found two Communists there in time for his vice-presidential campaign swing through Michigan. But now Towson State's president, an Eddie Fisher look-alike, has battled the term "university" out of a reluctant legislature and has thrown up a pleasure palace of sports activity. It is a school trying to dribble and high-jump before it can walk.

One way to use the athletic center's echoing spaces is a speaking program aimed more at entertainment—i.e., at *filling* the damn thing—than at education. Attractions of the current season have included Ronald Reagan and Dick Clark (who knew Eddie Fisher when *he* looked like Eddie Fisher). Also Truman Capote—who came out and melted into wet little incoherencies.

The serious part of the program was to be a debate. Something was needed to bring people in for this heavier stuff—so two senators with gossip-column followings were ordered up: Hayakawa and Moynihan. The audience was supposed to get two stars for the price of one. What the school got were two clunkers for the price of two stars. (Capote was scheduled to receive $3,500 but didn't. The senators received $3,000 each, plus limousine service from their homes.)

The senators were an odd match in many ways, including some that made it not a match at all. They are both hawks who berate the "counterculture." Both favor a cash dole to replace welfare services. Both think our principal domestic perils are inflation and big government—until either water for California or solvency for New York enters the discussion. Each is a maverick and a showman, likely to preen and prance separately before dancing into the other's arms. I called a member of the Student Government Association group that arranged the debate. Why these two? "Because one is very liberal and one is very conservative." I did not have the heart to ask which was which.

The topic of the debate was "Jimmy Carter's First Year in Office." Four panelists—two faculty members and two students—had divided up the issues and prepared questions for both senators on each issue. Members of the SGA wanted to ride over with each senator in the limousines provided by the school. Moynihan begged off, claiming earlier engagements that day, and tried to duck the dinner preceding the debate ("I do not like to eat before speaking"). He failed on the dinner, but did slip the student escort. Those who joined Hayakawa in the limousine got to talk with his aide, Gene Prat, since the senator did his famous sleeping act on the trip from Washington to Baltimore.

But so, apparently, did Moynihan. When his press secretary, Tim Russert, arrived at the Towson Club (a converted Victorian mansion perched on a lonely rise), I asked where the senator was. "Out taking a tour of the campus; he is very interested in architecture." Was he being driven or guided, I asked. "Neither; he likes to explore on his own." Then he was not touring the campus, which is several long blocks of urban highway off from where we stood; nor was he seeing much architecture, in the dark. He came in shortly, his eyes still sleepy despite the restorative. Why this double-talk, when a nap on a short ride makes as good sense as does a five-minute wake-up walk?

It is customary to call Moynihan expansive, but the word does not do him justice. He is *expanded before* he expands. Even in his sleepy stretch of attention toward the similarly reviving Hayakawa, he is florid with compliment: "God, I *love* your thesaurus, Don."

"You know that?" Hayakawa sparkles alive from the prince's magic kiss of flattery and explains to the rest of us the superiority of his synonym collection. "I wrote *essays* to explain the *degrees* of difference between the words."

Moynihan nods agreement. They seem for a moment to have changed roles—Hayakawa's face boyish with an Irish pol's self-satisfaction.

Moynihan nodding a Buddha's enlightened patronage from on high. Even when sleepy, Moynihan sparks life in those around him. The quiet rathskeller soon gets politely noisy, Moynihan rising like a moon on the gas of subtly tickled egos around him. (He asks me, "What are you doing now?" Writing about him. "I mean what you're *really* doing: Are you working on some books?" A wink of fellow writers is supposed to brush aside this minor gig at Towson State.)

It is Moynihan the student of architecture who conducts his own guides up two floors in the Auburn mansion to our dining room. (By the way, he does know his architecture, and wrote a gem of an essay on the restored Cathedral of Berlin and its contrast with a chapel at the nearby Plötzensee prison.) "What mirrors!" he says with a broad sweep of arm—pointing out things almost everyone but he and Hayakawa have seen before. "Those are *Baltimore* mirrors! I *mean, I mean . . .*" He makes a speechless large appreciation of gestures at a nude picture topping the staircase, as if sending it down again with a Duchamp stop-frame blessing. The huge cubist artifact of Auburn House turns now upon volubly Ascending Art, who is about to eat.

He declares it *Baltimore* food, to match the mirrors, and later tells his hosts he would come again just for the dinner. (A student whispers to me at table, "This is the man who doesn't like to eat before he speaks?") The subject of speech defects comes up, and Moynihan says he was sent to school to cure *his* speech defect, but now he would not part with it for anything. He commands attention by threatrical deferring; polls the table on the Panama Canal Treaty—it splits even: "We are in a den of reaction, Don," he says in feigned terror. "No wonder the candles and food are so good." He asks the nearer students' names and where they are from; gives one shy girl an instant nickname, "Schenectady," and joshes her off and on through and after the meal.

It was said, before he went to the Senate, that Moynihan was too exotic to fit into that stuffy club, where team players are preferred to colorful loners. But that judgment neglects the most fascinating aspect of Moynihan's career. He has perfected the gifts of the nonsubservient courtier. From the outset he has risen by flattering service to one patron at a time —Franklin Roosevelt, Jr., Robert Wagner, Averell Harriman, John Kennedy, Robert Kennedy, Lyndon Johnson, Richard Nixon. In all these relationships he maintained an unlikely dignity by stressing his own improbability. With Harriman, the Brahmin governor, he was the hardheaded guy from Hell's Kitchen, fighting off do-gooders. When

Governor Ribicoff made national headlines with his traffic crackdown in Connecticut, Moynihan dismissed that as cowboys-and-Indians stuff.

John F. Kennedy was notoriously distrustful of ADA liberals, who remained (he rightly suspected) weepy for Adlai. Kennedy won his own race with the help of Catholic urban machines of the shadier sort and of "bosses" like Daley in Chicago, Green in Philadelphia, Brown in California, Bailey in Connecticut, and DiSalle in Ohio. He needed a celebrant around him like Arthur Schlesinger, but he also wanted to hear a good word for the bosses—and Moynihan gave him one in his essay against reformers published three months after the inauguration. Here was a tony professor who could out-tough the Irish Mafia in his love of political deals.

Bobby Kennedy's first interest at the Justice Department—before the civil rights explosion—was juvenile delinquency, and Moynihan again shuffled statistics while telling stories about *his* days as a shoeshine boy in Times Square. Moynihan mourned both Kennedys heroically. But, like the equally mobile Dick Goodwin, he never felt other Kennedyites' reluctance to work with Lyndon Johnson. The two "New York kids" wrote the Texan's appeals to fellow Southerners' better natures.

Moynihan's slickest trick was to thread his way through Nixon's mined and barbed-wired outworks of hate for any Kennedy associate, or professor, or ADA liberal, or Ivy Leaguer, or media favorite. Moynihan was all five of those things, yet he ended up in the Oval Office jollying Nixon along with visions of his place in history (right up there with Disraeli).

Through all these essentially servile relationships, Moynihan contrived ways to bend and praise with such flourishes of fondness for himself that he seemed more the detached connoisseur than the hardworking client. He genuflected so floatingly that he seemed to be drifting by, rather than tagging along.

It is said that Evelyn Waugh could make his army salute sting like an insult. Moynihan has flattered in such a way that he could become a subtle critic, later, of his patrons—of Kennedy's activism, Johnson's Great Society and war, Nixon's policies and personality. As each patron fell away, Moynihan moved on without a break in his stride, oblivious of breaking ties, fascinated with the passing scenery: "I *mean,* I *mean . . .*" (expressive eyebrow-signaling and gestures, reinterpretable later).

The dinner party trails over to the palace of Towson State's own Kublai Khan. Dr. Richard Vatz introduces the combatants with wild quotes from Hayakawa (affirmative action is "damnable racism") and a stunning claim about Moynihan (his lecture fees one year came to $165,000—16 percent of all honoraria given to senators *or* candidates in 1976).

Hayakawa, up first, ignores the announced topic and gives us instead his standard solution for unemployment: remove the minimum wage from teenagers. He seems to think all Harlem's blacks would have jobs if we just paid them less at workplaces that don't exist. The panel is looking a bit panicky, and the moderator (he tells me later) is so angry he wants to get up and talk about breach of contract. Hayakawa, a Japanese American who praised both atom-bomb drops *and* the imprisonment of California's Nisei, ends with a plea for upbringing that will test a boy, give him challenges like those of the football field, to make him a *man*. (Harlem, it seems, offers children no challenge.)

The panel looks to Moynihan to save a foundering evening, and in a way he does. In his best barroom-brawl-winner manner, he gets up and shouts to the microphone: "I have here a forty-page book demonstrating that the first year of Jimmy Carter's administration was a *triumphant* success! My opponent's failure to challenge any of its achievements I take as a concession by *default!* Which is just as well, since *I* was not *entirely* persuaded of the case I was to make!" Then he drifts, on billows of laughter, even farther than Hayakawa from the evening's subject. Hayakawa's comments on the problem of growing up have punched one of Moynihan's continually recyclable data tapes—this one on the stress caused by earlier menstrual ages in technological societies. This was the point of his Alfred E. Stearns Lecture in 1973 (published later in *The Public Interest* and in his book of speeches, *Coping*). Out the data spin: "The age of sexual maturity has been going down and down and down. In round numbers the age of menarche was almost eighteen years in the early part of the last century. In the course of industrialization, it has dropped to below twelve years." Even the same gag: Freud would now say infantile sexuality never stops. (A whisper behind me unfairly remembers the announced topic: "So Jimmy Carter is to blame for lower menstrual ages?")

Why this divagation? Is Moynihan just showing off, proving he can take any topic thrown up and produce statistics on it? Perhaps. Earlier, Moynihan had made a point of reminding Hayakawa that he was going first, according to prior agreement. Moynihan seems no more prepared to give a special speech for this occasion than Hayakawa himself (the pay would come to $150 per minute of *prepared* text). Moynihan was content to bounce off whatever was said first—a dangerous approach considering the likelihood that Hayakawa would say nothing.

Yet the crowd takes to Moynihan, indulges his reminiscent irrelevancies. Few of them would have read the speech in either of its two published places. Beyond that, a reliance on autobiography as his political tactic serves Moynihan well. He is constantly going over his own career as a set of useful parables. He even begins the long introduction to *Coping* with a list of his eight crises and the morals to be drawn from them.

Hayakawa had cited an unemployment statistic from the forties, and Moynihan trumpets ungrammatically, "That's *my* data!" He worked on the compilation of those figures as Assistant Secretary of Labor under Kennedy. He was there. It is amazing how often he can say "I was there" in discussing modern policy. Not always at the center. But somewhere near, and visible. He has "nodded in" on more domestic-policy discussions than anyone except gray eminences who prefer to work in quiet. Moynihan has been everybody's emissary to anybody else—the voice of the streets in academe, the academic among pols, the aesthete on the hustings, the pol among the aesthetes. It has worked. He almost always has some tidbit of information useful to the little group he has just joined. He tiptoes as it were from huddle to huddle, telling what the "other guys" are going to try. And he trades the information with such an air of garrulous goodwill and bluff candor that no one thinks he might be playing a double or triple game for a single client (no matter how many patrons are also served along the way).

The "debate" ends in an orgy of senatorial courtesies. When Moynihan says Carter's welfare proposals are better than the current hodge-podge, Senator Hayakawa says he will bow to Moynihan's great expertise in this area: "I'll be with you on that." Moynihan shouts: "He's a pretty good *man*." Moynihan, in turn, agrees with his "foe" that Humphrey-Hawkins will mainly employ "lawyers, accountants, and economists." An Oriental bow: "I agree that you are a great statesman, Senator." Moynihan even promises he will reverse his stand on the minimum wage if teenage-unemployment figures do not get better in the next year or two. But even in this Gaston-Alphonse competition of co-operation, Moynihan prevails. He is sycophantic with a flair; he dominates by yielding.

There is only one sour note to the evening, and that is not struck by the combatants. A panelist asks why Carter is stronger for civil rights in South Africa than in Mozambique or Uganda. Because, says Moynihan, there is a hope for some response in South Africa. A loud boo comes from the audience. Moynihan ignores it. The bleater emits one more, a louder, prolonged boo. Moynihan leans into the microphone and mimics the cow's flat tone of his invisible Archie Bunker: "Boo *you*." There is nervous laughter, undecided—is it witty for the professor to play at Archie, or has some basic Bunkerism horned through the rhythmic stammer of surface dandyism? It is a suspicion entertained by many who watch Moynihan in action. He hits back with abrupt rhetorical coarseness, as in his name-calling with Idi Amin, and he tends to overstate in self-crippling ways, as when he said it was "no accident" that Amin was chairman of the Organization of African Unity. An accident is *exactly* what it was, and Moynihan turned a mutual embarrassment with Amin into an insult for the whole continent. It is one thing to tickle "Schenectady" with friendly teasings, but another to put one's hand, as Moynihan did, down Midge Costanza's gown and shout, "What we need is more cleavage *here* and less cleavage in the party!" The "bad boy" pose sometimes becomes a bad reality.

Despite his verbal dexterity in many situations, Moynihan seems plagued by an uncertainty of verbal tone that threatens to trip him up, time after time—as when he said it would be "dishonorable" to run for the Senate after being UN ambassador in New York. His two most famous phrases misstated by gratuitous overstatement. He was not content to claim the black family has been victimized by history, an incontestable assertion, as we later had to hear (interminably) from those defending what he actually said—that the black family was "a tangle of pathology." He could not call for a cooling-off period in overheated urban centers; he had to dandy the phrase up in deadly circumstances and call for "benign neglect." Careful but quiet watchfulness is not neglect, malign or benign, and calling it that was an affectation, not the candor Moynihan tries to make it. One does not search for a literary turn when handling a street brawl—or Moynihan's street memories have faded badly.

A basic uncertainty of gait under exquisite flourish shows up in the structure of a typical Moynihan sentence. He is so arch that his grammar founders by the third or fourth parenthesis. He learned to float before he could walk; so ten pages of crippled syntax back laboriously into the aphorism they have been aimed at from the start. Even the *Coping* speeches, which have been sieved through several editors and multiple publishings, are not combed free of basic errors like the double negative. I have a florilegium of grammatical flaws penciled on the back cover of each Moynihan book.

Moynihan's entertainment value perdures, no matter how many times he tells his Disraeli story about the man with one idea, and that one wrong. (Yes, it showed up at Towson State too—he is a man with one main joke, and that now lamed.) But his claims as a social analyst survive *despite* his writings. His métier is the conference—or, better, the patron's office, where he can jolly power along toward his favored policy of the moment.

The Towson State moderator told me he wished a brighter conservative like Bill Buckley had sparred with this evening's champion: It would not have been such a pushover. (He obviously did not know that only Jim Buckley's suicidal attempt to represent New York while disowning its claim to financial aid saved Moynihan from the kind of nondebate we had just witnessed.) Bill Buckley can charm students as skillfully as Moynihan, though he would flatter them less. But he would have come prepared and taken the announced topic seriously; he would also have given good value for his money, entirely apart from showmanship. The TV performer is more substantial, in this case, than the scholar-politician.

I thought of that as I made my way to Moynihan through a satellite cluster of admirers, and asked him later: Since the *danger* of having flair is to be considered *only* a showman, what evidence do you offer that you are a *solid* senator? He gave me a reproving and pitying stare: "That's not a very good question." I know from reports of other people's interviews that this is his way of not answering if he can—the flatterer even uses praise for condemnation ("Not really worthy of you, old boy; I had expected better, even at this little gig"). But I persisted: That is the question people ask about you constantly. "Look now, I am the first New Yorker on the finance committee in fifty years, the first *Democrat* on it in one hundred years. My predecessor put Tilden in nomination for President." That, too, is the answer he gives most often to the question he thinks is asked too often. Is that all? "I don't have to list my good qualities." Why not? Politicians cannot be shy. "*This* politician can!" he said, clothing his shyness in hauteur. It was funny on the face of it. The showman's last, least expected trick, the climax to his act: He will now proceed to disappear—into humility!

Yet he was telling an oblique truth, after all. He has succeeded by flattery and deference. He *did* get elected by the people, something the icier Buckley could not do, despite better grammar and a higher sense of ethics toward his audience. Moynihan *did* get good committee assignments, taking on new patrons in Robert Byrd and Russell Long. (Can he con Nixon, and *not* win a committee plum from Bobby Byrd?) He has done his homework on committees, serviced his constituents, been a bulldog on aid for New York City, pleased crowds at many a night like this, and kept people awake while he mouthed the soporific gospel of S. I. Hayakawa (and dozens of other pols). They said he could not possibly get along with Lyndon Johnson (who would call him a fancy pants) or with Richard Nixon (who would call him a Camelot leftover). He not only got along, he got ahead, and he is still getting. Those who bet against him in the past were wrong.

—New York, March 6, 1978

14

George Wallace

At Florida's International Speedway, cars are qualifying for the Daytona 500. Five surgeons work intently at an open hood until their patient gives a roar of healthy agony and pleases its tormentors. A drizzle, wetting the track, stretches the intervals between time-runs. Still, on occasion, one catches sight of the bright cars rolling like birds in flight up, onto, and along the track's bowl-edge. Devoted wives are in the pitside stand, over-made-up in honor of their part in this grease-and-oil branch of show business.

There are almost as many Wallace stickers as STP ones on attendant "workaday" cars' rear bumpers, and with reason: there is something obscurely patriotic about car racing, and these are George's kind of people. The program for the 500 even carries an ad for ROTC. And the driver of the pace car, had she not been struck down by a virus, would have been George Wallace's second wife, Cornelia.

Her driving is not merely *honoris causa.* She drove the pace car at the Winston 500, and got the car up to 115 miles per hour in her own practice runs. She would have gone faster, but it took her a while to build up to that speed, her first time out, and asphalt was tearing at the car's new special tires each time she went around the track; so the right front tire blew just as she hit 115. The president of the Speedway, sitting beside her, tried to grab the wheel but she coolly assured him, "It's all right. I've got it." She likes to tell people, "I'm a safe driver—I've got a governor on me."

But she really governs him, trying to couth him up, and clothe him right. In the effort she has demonstrated how intractable he is to all civilizing arts, and (at the same time) how strong a hand she can use in the softening. She will rush out of a TV control room to tilt his head up, draining shadows off the little rounded fist of his face. While he lives tearing ties off and shirttails out, she lives putting them on again, and in. It is the oddest political pairing imaginable on the Right—a male Martha Mitchell wed to a female Ronald Reagan.

He is the protégé of Alabama's ex-Governor James Folsom, "Kissin' Jim," and she is Folsom's niece; but that link makes for less resemblance than one might think. George, for one thing, never had the large air of public ease that Kissin' Jim exuded in his boozy charmin' way. Not even Cornelia gets affectionate with him in public. Teetotaler George always seems ill at ease unless on the attack.

But Cornelia is unflappably at ease in any company. George would blunder on out, after any speech, and neglect to say hello to the children of his sponsors, but Cornelia shoves them forward: "Here's the Smiths' *baby* boy, turned out to see you, and so young!" When a hostile questioner got up, in Miami, to ask how Wallace could be a law-'n'-order candidate, George, with his bad hearing, assumed it was the old charge of opposing courts, standing in doors, or attacking the buses; but just as he got launched into his set answer on those matters, his wife pulled him down toward her seat at podium-side, and said, loud enough to carry on the microphone, "He means your *wife*"—Wallace evaded the law on gubernatorial self-succession by running his first wife, poor Lurleen, soon to die of cancer—so George shifts smoothly into his standard answer on *that* one: "All you have to be to run for governor in Alabama is thirty years old and a resident. It just shows how liberal we are in our state that we don't mind voting for a woman." The new wife smiles radiant applause at the old-wife gag she "saved" for him.

Cornelia knows what George has never learned, despite many opportunities—that the South can *charm* other parts of the country; that resentment is one of its great skills, but not the only one. After all, Jim Folsom proved that Southerners were big on kissin' long before the cheek-nudge became de rigueur on Johnny Carson's show. Cornelia, in fact, modulates niftily between the two modes. Big Jim would fold 'em in a hulking down-home bear hug; but she has the practiced brush of powdered cheeks, double chins, and tall men's necks, that would look absolutely New *York* if she did not compensate with the accent, and with scatterings of that inclusive "y'all" which heads off any damning charges of the sort. Cornelia is pretty, unpretentious—her accent has been sweetened toward a beauty queen's "elocution," then mellowed with maturity (she is thirty-three years old, with two children by her former marriage). Few people can call themselves "ole-timey" with just the self-mockery that becomes self-flattery, and bring it off—she's one of them; every racing driver's dream of the girl he'd like to come home to with his Daytona trophy.

One of the tall men whose neck gets her durable kisses is Bill France, who is every year's king of the Daytona, and who feted Governor Wallace all through the 500 weekend—France is, you see, the chairman of Wallace's Florida primary committee. The whole campaign effort was run, until well into February, out of a room in the Speedway offices, and from the Holiday Inn across the street.

William H. G. France has the big flat cross-hatched face of a more intelligent Victor McLaglen; and he does not like to back losers. One of his favorite sayings is, "Quitters don't win and winners don't quit." It is a mark of Wallace's new campaign that he is attracting men of this type —men who have more than his own crude animal vigor and cunning. When I asked France who had run Wallace's Florida campaign in the last election, he dismissed the matter with a shrug, saying, "I don't know any of them [the American Independent Party people]. This is a *Democratic* election." There was a poll that showed Wallace had, in a typical area, drawn as many Republican votes, last time out, as Democratic ones—a matter that could hurt him in the Democratic primary— but France brushed off the implication: "That's why he's going to do even better this time. Most people register Democratic here, to vote for local candidates. He'll do better as a labeled Democrat." (Democratic registration still runs three to one against Republicans, though the state has gone Republican when voting for the White House.)

The importance of a France to the Wallace race in Florida shows up in the coordinator of the state campaign: Dan Warren is a lean, graying lawyer of the kind you want on your side if Hamilton Burger is prosecuting—scholarly, yet with enough ham in him to rise to courtroom melodramatics; a protégé, who became a partner, of Daytona's most celebrated criminal lawyer, William Judge, one who is taking Judge's place among those who love telling trial anecdotes. Warren was for years the state's attorney in Florida's Seventh District; as such, he was given the governor's proxy to handle civil rights disturbances through the sixties. When SCLC ministers showed up in bathing trunks, ready to swim in St. Augustine's segregated pools, Dan Warren took them swimming, though state troopers had to break cracker heads to do it. "I always figured the law could not deny a man his freedoms because of its own incompetence—you can't tell people they might cause a riot, when it is your own lack of adequate police that would make a riot possible. I marched every time with the SCLC in street clothes, with no helmet or weapon, and told them if I couldn't protect them I'd get hurt with them." When Klan leader Connie Lynch came to St. Augustine, intent on making trouble, Warren found an obscure Florida statute and arrested him. Warren was the first Southern prosecutor to get the death penalty for a white man who murdered a black. What makes a person like Warren run a state campaign for George Wallace?

"You're going to find a different kind of Wallace supporter this time than you did in the past," he answers. "Younger. And not just blue-collar types. Wallace is speaking to the businessmen now, the people worried about our trade imbalance and the shrinking dollar. When the Jaycees had their state convention here in Daytona, several candidates addressed them, but only Wallace got a standing ovation. And they polled the members on their choice for President—Wallace got 41 percent against all other candidates. This is no longer a red-neck phenomenon. Wallace has been able to take his regional populism and address it to national issues, especially those of the economy."

I ask him how, as an ex-prosecutor, he feels about Wallace's defiance of the law. "I don't think he ever broke the law—if he had, he'd have gone to jail. He has criticized the law, and tried to change it; but that is what our system is all about." I wonder, too, what he thinks, now that he is a defense attorney, of Wallace's attacks on the "permissiveness" of the courts: "I don't know what specific things he has in mind. As a prosecutor I found *Escobedo* was just unworkable; but *Miranda* struck a fair balance. I'd say our criminal courts work well now. But Wallace is mainly attacking the rulings of federal courts, and those judges *are* unlimited in their use of power."

A part of Wallace's stock speech goes this way:

> The courts can redistrict, reapportion, put a tax on, take a tax off, tell you what to do with your child, your union, your property, your business. These judges are appointed for life, and have all that power over you, and you wouldn't even recognize one of them if they came in here right now. You might be sitting next to one, but you'd never know it—yet he has all that power over your life, your child, your money, your property. They're despots, dictators.

When Wallace says "Thank God for the police," he always gets a good hand from the common folk, but it is a more serious matter when his rhetoric touches even Dan Warren, ex-president of the Florida Prosecutors' Association, friend of the whole law-enforcement community, a Southern liberal who can speak to a meeting of sheriffs and get applauded for his views on "strict but fair" administration of the law.

The Wallace campaign headquarters in Montgomery shows that the new Wallace is more than the result of Cornelia's remedial haberdashery. Charles Snider, the campaign director, is a spruce, affable Southern gentleman. He has issued strict de-kooking orders to the operation, and the campaigners' manual, which runs to 120 careful pages of instruction, warns Wallace workers against those with a single "pet issue" (polite for "race"). There are separate offices in the building for the youth vote ("Young Americans," flirting at confusion with the large national organization of right-wingers called Young Americans for Freedom), for the labor vote, for mail, for computer keying, for money raising (the Century Club), and for press relations. The brand-new building, out by the airport—its large warehouse annex filling up with campaign supplies—now houses sixty-one salaried employees and sixty-five volunteers.

Joe Azbell, who writes the candidate's material from a messy journalist's desk in this overneat setting, says, "You can look all through the campaign literature and you won't find a single mention of race." Busing counts, of course, as an educational issue, not a racial one. Back in the warehouse, when I try to pick up old campaign biographies or flyers, Azbell deftly snatches them away and says, "I'll give you things more up to date." He likes to point out that nothing in the headquarters says

"Vote for Wallace"—yet. The campaign has a careful countdown: "In a month or so, we'll unmask our slogan, and say something about voting for him then." The slogan is "Send Them a Message"—and "Vote for Wallace" is only a part of the message to be sent. "What we are doing now is conducting an educational effort, to inform people of the things being done to them by an unresponsive government." Azbell, too, affects a total ignorance of what has happened to the American Independent Party structure—all he says is, "Wallace people will vote for Wallace. Only this time, in the primaries, they'd better be registered as Democrats." Yet a huge chart in one office shows what states' ballots George is on—and all the states where there is a qualified AIP are also marked. "Well," says Azbell, who winks as often when he is straight as when he is conning you, "those are in parentheses—just for our information."

Azbell's emphasis is on all the new things about this race—like the campaign symbol, a modernistic American eagle. When any words are used with the symbol, they say that the American people need a voice, not that Wallace needs a vote. The whole pitch is to a movement larger than one man, a spontaneous surge of indignation. This is to be a movement that can use Wallace for expressing itself, not just one man using some cranky people's angers. It is an odd house that George built, where everyone speaks well of George but says he is not really all that important. "He just has the platform for voicing these grievances." If one hung around the office long enough, one might come to disbelieve that George Wallace exists. One comes back to earth listening to Joe Azbell talk on the phone to a woman from Swedish TV: "Come on down, I'll make sure you get some sun, and a little fishin' on a creek bank." Outside it is forty degrees in Montgomery, gray and sleeting.

What, I ask Azbell, who is as close to an "issue man" as Wallace seems able to stand, sets the new Wallace off from the old? "The whole country has changed. People are fed up with the government, and want to be left alone. We've been doing extensive mailings and pollings, and you know where our best response is coming from? The suburbs. We're going to surprise everybody this year. Wallace's strength, they used to say, was with the blue-collar workers. We're going to show them it's in the suburbs."

But if Wallace is promising to get government off people's backs, it is not through any right-wing talk of economic individualism. "The Guvnuh" is a tough protectionist, who wants to make it illegal for parts and products manufactured abroad to be sold here with American brand names. His formula, as everyone has always known, is not libertarian but populist—"soak the rich" in favor of the poor; and buy and build American, over against foreign aid or trade or monkey business. What is new is the growth of a middle-class politics that can feed on "proletarian" resentments. As the country seems to become one vast middle class of indeterminate size and longing, with a decent subsistence but undefined hopes, the do-gooder sense of responsibility which used to spell middle-class respectability is yielding to more naked lower-class types of demand. Vietnam may have been the end of our causes. Idealistic issues like civil rights and integration have lost their political glamor, and even Left-maverick politicians (e.g., in the short campaign of Fred Harris) now talk of "populism"—i.e., basic issues of the pocketbook—as the only viable "cause" of our time. We are not so much in an economic depression as a spiritual one, a stingy time for moral expenditure, when Nixon won office by registering 33 percent on a grievance-and-resentment scale (developed by a team of Harvard psychologists) where even Wallace—the 1968 Wallace, of naked regional hates—only ran up 41 percent. The trouble with such a politics of want is that the man who wins by exploiting grievance has a hard time holding office, since he cannot run against himself—and running *against* is the whole trick of the game. Nixon has tried to finesse this difficulty by casting himself as the underdog persecuted by students and the press, by war critics and ideologues, by an Eastern Establishment stronger than the President himself. Still, as Wallace loves to say of his tardier imitators, "If they're against it, why don't they change it? They been in power a hundred years, all together."

The last sentence does not have any meaning—not, anyway, one that I have been able to extract, either from Wallace himself or his listeners. The "they" is indeterminate, as it was in James Buckley's campaign request that "we" should at last have a senator (over against the nebulous-menacing "they"). When I asked Joe Azbell who the "them" are in the Wallace slogan "Send Them a Message," he turned eloquent interrogative hands up and out, and smiled his chessycat smile.

All Wallace campaigns have fed on ambiguity; but only now has he been able to set a national tone of such evasiveness, as all the candidates hedge their bets, on the war, on busing, on the issue of American empire. Wallace, in such a field of recanting virtue, stands forth in all his honest disregard for the decencies.

Yet that is stating it too harshly. Wallace is not so much openly indecent as indecorously honest. He knows, and does not hedge about it, that politics is basically a matter of getting. The pretense of liberalism has always been that ours is a politics of giving—but even that idealistic talk is curving down, now, toward the gut kind of vote getting. Even populism, of course, involves some trickery, mainly the managing of all necessary giving so that it looks more like getting. The worst example of this is Wallace's crusade on tax-exempt foundations, universities, and church properties—which perform services that would otherwise come out of tax money, and in any case would not cut the ordinary taxpayer's load in half, as Wallace claims, by losing their exemptions. A better example of his sleight of hand is the new Alabama plan that extends Medicare-type benefits down to the under-sixty-fives, using Blue Cross accounting, nonprofit private insurance companies, and state underwriting. It gives $250 deductible coverage to a family for $16 a year, costs the state nothing, and affords member companies an operating base from which their profit-taking work can function. Not a bad advertisement for Wallace populism.

It is no wonder Bill France was first drawn to Wallace as a businessman. The Alabama governor saw people flooding into Daytona for the races, and called France to discuss opening a track in another state. Through an intricate series of negotiations, the city of Talladega gave France an old World War II airport site on condition that he build a new airport. Wallace's part was to get state funds for access roads—so the Daytona Wallace extravaganza grew from a large web of mutual favors done, deals made, and politics mixed up with show business, patriotism, and a Southern taste for unnecessary risk, a kind of government by racetrack.

Still, many people have done favors for Bill France, yet this is the first time he has run a political campaign for one of them. France seems not only dedicated, but convinced. When I interviewed him in his trophied office at the track, an intercom crackling all the while with results of the qualifying trials, he put it flatly: "Jackson or Humphrey is not going to beat Nixon. The Democrats don't have a chance of getting into the White House without Wallace on the ticket." Why not? "He *owns* all the issues these other candidates are trying to catch up with. Look at national defense. Or crime. Or busing. Just think what they'd do for George if he went to Richmond, Virginia, right now. Christ, they wouldn't make him President, they'd make him king."

And George Wallace, after all, has already won the queen.

I first saw her, after she made a live TV appearance, being driven to a luncheon where her husband would be the speaker. In the car, I asked what she thinks of women's liberation: "I think *everyone* should do what they want to do," she said. "If I wanted to run for Congress, I would. When I want to take up flying again, I will. I just don't have the time now. As for working, I have enough trouble getting all the family together with George on the go. If we both worked, I'd never see them—unless I had a little ole television show like that one I just did. I could do that while the children were in school.

"I wanted to move down to Florida and put the children in school here for these three months, but unfortunately George has to be governor part of the time. Besides, if we lived here, he'd be out on the stump every day."

She seems to have got the impression I am a Florida journalist, since she points out the Florida charms on her loose jangly bracelet—a Miami remembrance, a city key. There is a large engraved clasp on her arm—"That's my POW bracelet. You vow not to take it off till the man whose name is on it comes home freed. I went to school with this boy—he's the son of a retired general." There is something of her power in this hard steel under the flounce of loose gossipy gold charms. She is the perfect example of that separate species, Political Wife. She is prettier than Jackie was, and far more outgoing; even more intense than Lenore Romney, without her rasping preachiness; she can out-homebody Muriel, out-movie-star Nancy, out-modest our reigning First Lady, and out-First-Lady anyone in sight. She is all the glamor queens a Mickey Rooney can ever have wanted, and her only flaw seems to be that she married the Mickey Rooney of politics, egotistic and precocious yet stunted and going nowhere.

She goes right in to the luncheon, too expertly painted to need retouching. George is still up in his room with Bill France and Dan Warren, down from Daytona, and with Bob Weller, Montgomery's man in Florida. She goes up to join them, while The Galileans warm up an audience of Miami businessmen. The Galileans are a Jesus-rock group that works out of Allapattah Baptist Church, "The TV Church," as it bills itself on cards being passed out to journalists—the place where Miami Mayor David Kennedy goes to Bible class, and where the last class of Miami policemen held their graduation. Other primary candidates have addressed this club, and more will follow Wallace, but The Galileans can get out of school for only one of these speeches, and they chose the one most likely to attract a crowd even here in Dade County, which is supposed to be the toughest place for Wallace. After a few rock hymns, Wallace comes in—and so do Lindsay picketers who have been circling the entrance. PHONY POPULIST says one sign—though it is hard to imagine who could be a more genuine populist than George. WHY NOT DEBATE? Wallace will answer that one after his talk: "Let Lindsay draw his own crowd." Alabama state troopers help hotel security men nudge the disturbers around the edge of the hall, all but a few who, instead of being skewed off to the outskirts of activity, plunge up right past the dais. Cornelia, already seated, looks up with hurt wide eyes: WALLACE HATES EVERYBODY.

Wallace, who has been told that Miami is the place in Florida where "them liberals" lurk, begins almost apologetically, telling people they do not need to worry, he's spoken to all kinds of gatherings—up North, students, you name it, "and every single club has survived my visit." He is being what counts, for him, as ingratiating—no karate-chop gestures, nor going up on his toes like a toy boxer, and easy on the rhetorical earthinesses. He hardly sounds Southern till he gets a big hand for this: "Anybody who thinks Communists will disarm just because we disarm ought to have his head bored for a holler-horn." The audience did not seem quite certain what it meant; but they enjoyed it, nonetheless. (Even George is a bit vague about this saying, which he calls "an old Florida expression"—at later speeches it tended to come out "bored *with* a holler-horn.")

The only variations in a Wallace speech are those of tone—less or more bellicose. Today is, mainly, less; but he covers the same territory as ever—it is true that he speaks from his head, and only five points are rolling around in that head today:

1. *Defense:* "We gotta get off this unilateral disarmament jag. . . . I'm not going to gamble on it. We ought to be Number One."

2. *Taxes:* "My own church owns a printing establishment that prints books with four-letter words in them and sells them for profit." Soak the rich institutions, for the little guy.

3. *Foreign Aid:* "We spent billions of dollars, and most of that went down the drain." In more raucous mood, he says down a *rat hole.* "I'm against helping people ten thousand miles away who just spit on us in the UN. The UN's a kind of Playboy Club, with a bunch of folks who don't know what to do with themselves."

4. *Law and Order:* "There were 127 policemen killed last year, more than any year in our history. The federal judges have just about destroyed the law . . . and what's so sacrosanct about a judge? He don't hang his britches on the wall and run and jump in 'em in the morning."

5. *Busing:* The big one last—"All five of the senators running for President down here voted for busing. I got the records right here."

The questions afterward are loaded, but he fields them easily (after each has been relayed into his good ear by a solicitous Billy Joe Camp, Wallace's young press man, just about as blank a human being as has ever been wound up and headed out into the world). The speaking ended, George is nervous, as always, anxious to be off. The man so alive at the podium seems lost and disanimated when crowds move up to him as individuals, wanting more of the human fire he only sheds from the platform. The body that bounced out seems to flag, waiting for new afflatus, the victim of that frenzy he inflicts, and of the frenzy's transitoriness.

But now is Cornelia's time. She has found out that the sponsoring club's president has a daughter who sings with The Galileans, so she grabs his hand and runs over to the group, just as it begins "Put your hand in the hand of the Man from Galilee." She says hello to the daughter and then, while she's on the platform anyway, joins in, scoring points for each line of the song she remembers or can pick up. When the song ends, they ask what she would like to hear. "Oh, I don't know these new songs"—by this time she has gravitated toward a microphone. "But I think the best thing that happened to this country is spiritual rock." She is attracting parts of the crowd that drifted toward the door with George. She leads the group in "Give Me That—*whump*— Ole Time Religion." On the *whump* she darts forward with a vigorous stomp of her foot and clap of her hands, setting the choreography for other Galileans (the group has only two awkward teenage boys, each too awed to shake her hand, much less lean into her light brush of New York kiss across their beardless cheeks). Even the spectators join in the chorus, and all the girls, as the *whump* comes up, do an anticipatory little dance of readiness. Longer pauses are opening in the song as people try to think up further verses: "It was *good* for the pilgrim fathers . . . It was *good* for Paul and Silas . . ." They have whumped their way down to "Make us *love* everybody . . ." and nothing else seems left until, slyness masquerading as shyness, Cornelia tentatively offers, "May we *all* vote for Wallace . . ." and everyone crows along, admiring the nerve of her prayer: "He's good enough for me."

While Cornelia spiritually rocks, her husband is spirited off for more meetings with campaign managers. Billy Joe, from behind his mask, has agreed to get two of us—a *National Observer* man and me—in to see the Guvnuh before he suits up for another delivery of the speech. We meet at an end-of-corridor "Governor's Suite" (so named, on the door) and wait in the anteroom with two Alabama troopers—one tall and dark as a movie-Western Indian chief, the other looking badly hung over, stretched out on the bed with his little pig eyes and little pig gun both peeping out from inflamed bulges. A very young advance man is scanning Miami streets through binoculars. The Wallace entourage will go up ten notches in class when the Secret Service men are assigned to him and show up looking human.

Mrs. Wallace comes out, slippered and makeup-smocked, to get something from her bag—pig-eye leans upward in deference, showing a will to rise though the flesh is weak, and advance-boy lowers his glasses. It is two hours before show time, and she is on the first, or ghost-white, layer of work around her eyelids and upper cheeks. She works more on her clothes than George will let her work on his. By the time we get into his part of the suite, a room-service meal has been finished, and we miss one of the longest-running journalistic sideshows in American politics, watching George eat while he slurps and tells everyone that he does not slurp—all we get in for is the postprandial tooth sucking he also says he does not do; and the omnipresent *sig*-ar; and the cocked ears, and miniature John L. Lewis glances out from the Mephisto eyebrows ("What's that?"—the bad hearing, long used as a gimmick, has become a trick played on himself by now). Cornelia glides by in a further stage of her self-creation (countdown time to her apparition, one hour) but George is in the same pants he wore at the luncheon and will wear this evening. Though his shirt is modish, deep-blue, and wide-collared, it is also out of his pants, and fastened by only one button, with cuffs flapping. He burrows under it to scratch his belly, pulls the sleeves up his nut-brown little muscular arms. He is a great scratcher, and the fact that Cornelia has got him to wear knee-length hose just means he has to pull his pants up very high to rub his legs around and under the socks. He is an Antaeus who needs to keep touching earth—and he is his own earth. The strength he draws is all from a sense of resisting selfhood, as if he knew he had clicked by some trick of genuineness and had to preserve it at all costs, original earth under all this attempt to pastel and tone down and civilize him. What if his strength also went off with his cruder habits? So he sucks a tooth, rubs a calf, rolls out the country-boy accent—still just George, the original item, no need to worry about that.

"You the onliest ones comin'?" Better a crowd, of course; but even three would be better than two. He intersperses his conversation with little shrugs that "this is off the record," even when he has just quoted a line from his stock speech. "When you don't have law, you get lawlessness; when you don't have order, you get disorder—off the record, of course." He almost boasts of his ignorance. "Didn't George McGovern vote against the Tonkin Gulf Resolution?" No, we assure him—only two men opposed it, both out of the Senate now. "Well, damn, I been doin' him an injustice. Ain't that sum'n?" He looks at us, almost, for approbation. It's as if he knew he would look like a bully if he did not look like a child—a vision out of old "Our Gang" movies, stogy and domed hat above, but diapers below. Then, with a switch, it is man below and child above: "They all *afraid* of me. They know I speak for the people. That's why y'all are sitting here, ain't it?" Throwing the questions back.

I ask him about the accusation, just voiced in Florida by conservative Republican John Ashbrook, that Wallace has struck a deal with Nixon not to run as an independent candidate. The talk has become very circumstantial—that the bargain was sealed when Wallace flew on Air Force One from Mobile to Birmingham. "Deal! Shee-it! The only time I was with the President was at lunch on that plane, and Louie Nunn was there, and Reubin Askew, John Bell Williams, Harry Dent, Red Blount, Tricia Nixon, and Julie, and David Eisenhower—some deal I'm going to make with all them sittin' around. Hell, Nixon didn't want me on the plane at all—thought he could come into my state and not even have to meet the Guvnuh. That ain't right. When he saw that, he invited in all those other Guvnuhs." The bitterness is not feigned. Authenticity here is the whole secret of his power. No deal was made on that flight, which atoned for Wallace's presence by being turned into an airborne Southern governors' conference.

"What kinda deal can I make with the President of the United States? What am *I* supposed to get out of it?" The talk goes that a federal investigation into alleged financial irregularities of Gerald Wallace, George's brother, was dropped, so that Wallace would surrender to Nixon the five states he won in 1968. "You think they'd let me off if they could find anything on me? Don't you know you gotta be clean as they come to go up against the big fellas? They spent millions of dollars investigatin' me. If they had anything on me or my family, they'd use that to get *rid* of me, not make no *deal*. Besides, if they had something on Gerald, I wouldn't let that stop me from speaking up for the people. You think it's all that much fun campaignin' at my age—the airplanes, the threats on your life?" He is honestly aggrieved now, and at his best. He fears planes, especially the smaller ones he has been jumping around in. On a typical night in Lakeland, police received three phone calls saying he would be killed, or the old hangar where he was speaking would be blown up. "It ain't all that much fun, y'know. You gotta *b'lieve* in somethin' to do this."

And what is it that he believes? That George Wallace has had a raw deal. This feeling of being a loser is what makes him so convincing to other losers. He only "works" as a tribune of the persecuted. But that sets up contrary tugs in him: he keeps winning by losing—which means that if he ever really won, he'd be lost. A nagging sense of this seems to keep him from using the fine machinery at his disposal in the Montgomery campaign office. If he used Madison Avenue techniques, what good would he be? *That's* what he's *against*. Dress him ever so fine, he's still going to deliver his stock speech and attack "people who dress good, and part their hair real good, and wear good clothes, and that makes them an *authority* on everything" (pitter-patter of amused applause from elaborately gowned and coiffed Cornelia). He is the Guvnuh, but he won't let that get in the way of his grievances. He is eloquent about his persecutions, immediate or by proxy: "Our National Guard General, he's bein' indicted by some pimply-faced federal attorney, looks maybe seventeen, because he collected some campaign funds —some for me, some for my opponents. He was just hedgin' his bets. But the same man resigned his commission to train men for the Bay of Pigs. He was down there in South America in his plane, ready to go, but the order never came. Hell, the man's a hero—I wuddna gone off like that, with nobody knowin'. Would you? Then some pimply-faced teenager comes along and tries to nail him for a measly few thousand dollars."

His need to be wronged made him seize with delight on Larry O'Brien's threat not to assign his party hotel space in Miami, unless he pledged to support the candidate chosen by the convention. "These the people who are against your signing a loyalty oath to say you're against communism. They say they's some question about my bein' a Democrat. But Humphrey's run on different party tickets. And look at Lindsay. He's run on every party in New York 'ceptin' the Communist Party." He still uses these lines in his speech, though O'Brien has backed down. We ask if that affair hasn't been settled. "Oh, I think one of the boys got a telegram or somethin'." He doesn't want to let go of a grievance.

"I still might be kicked out of the convention down here," he says almost wistfully. Wallace called the Alabama Legislature back into special session to pass a unit-vote rule on the Alabama primary—it was defeated because it would probably have disqualified the Alabama delegation before the credentials committee; and that seems to have been Wallace's aim all along. He is constantly asked whether he will run as an independent if he cannot get the Democratic nomination in July. He says he does not know—"I got no master plan"—and he probably doesn't. But the one way to guarantee his running, his fighting back, is for the Democrats to question his credentials or expel his delegates. He is getting older, and does find it harder to run these days—he has canceled most of his smaller engagements in Florida, speaks only to large gatherings. The fuel he needs to keep him going is mistreatment by "the big fellas." He is at a loss when treated well.

It is often said that Wallace is just a charming scoundrel in the classic Southern mold, a "good ole boy." But the essence of a good ole boy is complaisance. He'll hang you a nigger in no time, but he don't *mean* nothin' by it—he was just bein' sociable. But Wallace is never very sociable; his life has a sour and joyless air to it, reflected in the glum company he keeps, his teetotaling pointless austerity, the aimless air he has when he gets away from politics—and the very limited definition he has given to his "politickin'." Public malevolence is his lifeblood. Lacking it, he would not know what to do with himself. He is a good hater, which good ole boys never were. They were good forgetters; and George remembers—a slight on Air Force One, just as much as an attack on his table manners. And, with pugnacity, he remembers the attack by persisting in the bad manners. He won't "climb," at least not thet-a-way. He's cold-sober serious on this point. (Southern politicians often can't handle their booze, but without it they're unbearable.)

Oddly enough, this saves him from the calculated graces of those who would use him. A great deal of political energy will be expended, in the seventies, to create a populism of the suburbs, to fit a set of outsiders' resentments to the "insiders" of Middle America, to unite the old success ethic with losers' anger. The tone will be that of self-pity maintained in conditions of sufficiency, or even of relative affluence. That is why Wallace is not attacked by the other candidates as a populist, but as "a phony populist." Populism is a property they might want to lay claim to themselves.

But they are the phony populists, taking the old anti-city rhetoric of the country and using it in the suburbs, trying to convince the winners in our rather empty race that they are actually losers, and that it is somebody else's fault. Wallace should be ideal for all these purposes.

As he points out, most of the Democrats are belatedly saying what he said first—about neighborhoods and local rights and busing and crime and defense. Winner Cornelia and loser George symbolize the merger that a hundred political managers want to bring off this year—a union of stylishness and sullenness, of respectability and resentment, of suburbs and blue collars. But Wallace is the prisoner of his own rather narrow honesty—this Antaeus needs a certain kind of earth, and it is only found on a tiny plot of ground.

That was made clear the very night of his Miami appearances. An obsequious fellow arrived, toward the end of our interview, to dance attendance on the Guvnuh, even answering some questions for him. "This here's one of my legislators." One, as it turns out, of three pro-Wallace Jews in the Alabama Legislature. From the other two, Wallace only got telegrams today, but this man came all the way to Miami for the Guvnuh's speech before the Young Men's and Young Women's Hebrew Association. It was not a friendly audience, and there was speaker trouble at the outset—the hostiles applauded the arrival of a wooden Coke carton for George to stand on and reach the microphone. Still, Wallace did surprisingly well, with people who had been tough on Mayor Lindsay earlier in the month. Grudging YMHA officials conceded as much—but he could not leave well enough alone. After the talk, Cornelia casually asked one official if he knew a prominent Miami jeweler—and he did. "Oh, say hello to him for me. We used to go skeet-shooting together." Just the right degree of nudge—she didn't have to come right out and *say* some of her best friends are Jewish. But blundering George *did* have to. He broke into the conversation: "I hope you don't mind my bringin' a Jewish legislator down here with me. I just wanted to show you I'm supported by *all* races and creeds. Was that all right? Did I do right?" The embarrassed man had to cough and comply, saying, yes, it was perfectly all right for him to bring his Jew down here with him. The parting was an awkward one.

Wallace has such great talent as a demagogue, it is lucky for us that he lacks the human graces. That makes him incapable of our politicians' smoother deceits. He has all the hostility of Southern defeat with none of its chivalry or romance. No matter how bad he gets, he stays open and blatant, ill-fitted to the designs of those who would use him, and finally above such manipulation. The slick ones are trying to take him over, but he is too resolutely dumb to be took—they'll never clean up the old crotch-scratcher. That is why, though we may be in for some discreeter Northern crackerism, whether it is called "ethnic politics" or "the new populism," at least it won't be led by George Wallace.

—*New York,* March 6, 1972

AFTERWORD

This piece had not been long in print when Wallace was shot. All bets were off—though one security man would remember the article vividly when next we met. And Joe Azbell swore at me with great virtuosity for five minutes nonstop when I returned to his office—so heatedly that a secretary ran in, afraid a fight was going on. "What's the matter?" she said. "This man wrote mean things about our governor." She relaxed, while admitting it was a shocking thing to do. "He can't have met the governor." "Oh, he's met him, all right." She stood there awkwardly, till I asked Azbell, "Shall we each do our work?" "Of course," he said affably, his duty discharged. The next time we met would be over drinks in Kansas City. It is good to work with professionals.

Cornelia Wallace was cordial when we met again. She clearly did not mind articles critical of her husband, so long as they recognized her formidable qualities. The marriage was bound to end when each ceased to be a vehicle for the other's ambitions.

15

Bert Lance

Atlanta, September 7–9

British theologian Austin Farrar, discussing the chronology of the three Synoptic Gospels, compared them to staggering drunks who push and

shove each other around, each sometimes leading and sometimes leaning, in a tug-of-war mutual dependence. In retrospect, Bert Lance,
Jimmy Carter, and the Atlanta Establishment look like such interdependent beneficiaries of rambling near-collapse.

Governor Carter was never popular with the Atlanta potentates,
whose candidate (Carl Sanders) he defeated. But he knew how to use
the prosperous showcase of the South as his own emblem and setting.
Visiting journalists, members of the Trilateral Commission, were
housed in the Atlanta governor's mansion, and shown around the ra-
cially "moderate" city that (narrowly) bought off prejudice with dividends.

And Bert Lance, Carter's director of transportation, was popular
with Atlanta's businessmen. As a small-town banker (where he married
into the bank's ownership), he had dealt with the four big money
houses of Atlanta; and as Governor Carter's dispenser of county-road
patronage, he combined good business and good politics in a way the
capital admired. When Carter ran Lance as his successor, Lance got the
newspaper and Chamber of Commerce support that had been denied
Carter.

Yet Lance was defeated, as Sanders had been, by South Georgia
voters who do not like lavish-spending bankers. In 1970 Carter cast
Sanders as "Cufflinks Carl." In 1974 the winning candidate, George
Busbee, attacked "Loophole Lance," the man who was running his own
campaign out of legal chinks in his own bank (an argument the United
States Senate would learn to take seriously, at last, after ignoring it in
Lance's confirmation hearings). Busbee goaded Lance into disclosure of
his assets—over three million dollars (if Lance was not kiting the figure
then as he did later on). It says something about a political figure when
he claims more wealth than everyone knows is good for his campaign.
Georgia populism is more a matter of resentment than of program, but
not the less powerful for that. Lance's boss knew enough to call himself
a peanut "farmer," not a peanut processor.

Lance lost, after sinking half a million dollars of his own money in
his own campaign; but he thought the loss a gain. He had invested in
Atlanta, and in having his name recognized there. He moved to town as
president of the fifth-largest bank (trailing the big four by a long way),
and bought a sixty-room *Gone With the Wind* pillared mansion (with
fourteen bathrooms). His wife named it "Butterfly Manna"—Butterfly
because that is her private pious-flirty symbol, and Manna because the

Lord had dropped it on them overnight (but can *butterfly* manna fly off again after landing?).

Lance blitzed Atlanta, doing bank commercials on TV as if he were still running for office. He seemed a reviving breath of hot air just when the wind had gone again in Atlanta. This was 1975, and the boom was over. The buildings still rose, but slowly; and their offices stood empty. Real estate investment trusts were folding, and banks looked twice at the kinds of loan they used to push out at investors. It was time to rethink the creed of that man most responsible for Atlanta's boom in the first place.

Mayor Ivan Allen is given the popular credit for Atlanta's surge in the sixties. But the one who persuaded Allen to run in the first place—using cash, that greatest persuader—was banker Mills B. Lane, who had made his Citizens and Southern Bank not only the biggest bank in the South, but the hottest bank in the Sunbelt. It had by his retirement eighty offices in Georgia, and stock in forty out-of-state banks; and it moved to the rapid beat of Lane's confident expansiveness: "We're a department store for money."

Lane, like Lance, came from a small-bank family background. And Lane's life was a series of leaps forward on the big risk. In 1947, as a bank vice-president, he proposed an easy loan policy for car purchasers (his father had grown soybeans for the Ford cars' synthetic steering wheels). When bank superiors turned Lane down, he walked off the job. Six days later he could only be lured back as president. The car loans remained dear to him. Years later, he pedaled a kiddy car into a board meeting to dramatize his love of quick turnover in the car-loan business.

Lane always linked bank and civic activities. If you lend a man money, he said, the only way to make sure he will pay you back is to make sure he is able to pay. When it looked as if South Georgia tobacco farmers could not meet their notes because of a dying crop, Lane assembled a train of tank cars filled with water, and carried liquid manna down through the dying fields. Then he financed, out of what could be saved from the first loans, an irrigation system that paid returns to the bank over many years. Lane used to say a banker who could boast he never lost a dime just didn't know how to use dimes. You *ought* to lose a dime to make a dollar.

And no one could doubt that he made dollars. The drawling South has its fast-talkers, and they tend toward the cash piety of "Reverend

Ike." Mills Lane was out in his own bank lobby greeting people as often as he could be. He always answered his own phone, and often answered it with his battle call: "It's a wonderful world. Can I sell you some money?" He was one of the first theatrical bankers. When he offered "Instant Money," he said, "Why should a person be able to walk into a department store and charge hundreds of dollars worth of goods if he cannot come into a bank and take out the same cash equivalent on his signature?"

He joined shrewd dealing with folksy drama—sheep in the bank lobby to encourage investment in Georgia textiles; a reenactment of the St. Valentine's day massacre to talk about "killing" costs. He was the con man as cute ingratiator—"Let's do bidness" he would baby-talk high rollers. He sent marked C and S helicopters flitting about the city, rushing checks and cash at a breakneck pace to suggest the pace of the town's urgencies. *"Resurgens"* says one Atlanta bank's emblem, and the South's promise to rise again seemed redeemed in the Mills Lane Atlanta of the sixties.

But this Second Coming was in for a Second Sinking. In 1971 Lane retired early, at age fifty-nine, after a heart attack. His young successor, Richard Kattell, seemed ready to carry Lane's own standard. A rumpled and friendly hard worker, Kattell answered his own phone, prowled the lobby, leaped on and off the helicopters. Lane had given out tics with the "It's a Wonderful World" slogan. Kattell wore a special gold tiepin with the tongue-twisting slogan YCDBSOYA. You were *supposed* to ask, so he could tell you it meant: "You can't do business sitting on your rump."

But the 1974 recession hit Atlanta's overheated economy with particular force. Land bidding had gone up so fast that no return on it was valuable except a shopping center—and there were no new small businesses to fill the centers up. Banks were cutting back just as Lance lost his race for governor and came to town as the new Mills Lane, big and jolly and ready to take risks in the worst times as well as the best. He became president of the NBG (National Bank of Georgia). It made sense for Lance's bank to pick up some of the big money now that C and S was giving harder terms. Lance wanted to set up his own holding company as Lane had; and the market was good if he had the cash for it. He decided to take advantage of a slow market as if he had the resources for quick trading. Maybe the rules he played by would *become* the rules of the game, at least for him.

Lance might have carried it off, if he had stayed and played a concentrated game; if he had kept his own shaky financial ambitions separate from this fast play; if he had not mixed his own political ambitions, and those of Jimmy Carter, with his bank expansion plans. But those were too many ifs in a situation where the addition of one more if was foolish. After all, the bank plan was iffy in itself. Lance had bought subsidiary banks over the slow price others were offering, expanded his bank's out-of-state activity, and stocked NBG with political officers in twice the number expected of a bank his size.

Lance looked at first like another Lane. Lance had tethered bulls at his bank, to equal Lane's sheep in the lobby. But Lane lived carefully. When he did, at last, buy a sensational yacht, he sacrificed his beloved antique car collection to the purchase. Lance, by contrast, wanted everything at once. He came to Atlanta still jumping ahead of and around the huge personal debts he ran up in his 1974 campaign for governor. He instantly added to these the big loans from New York and Chicago that let him buy (with his partners) controlling stock in NBG. His bank having loaned $4.5 million to Jimmy Carter's peanut business, he was also helping Carter's presidential campaign. And all the while Lance was expanding his own scale of living—a private plane, four grand homes in Georgia alone, gaudy parties. Any one of his risky lines of effort could have been salvaged, if he had worked at it. But each put the other in jeopardy. Lance did not heed one of the maxims Lane quoted from his father: "Tend thy own house and thy house will tend thee." Lance was too busy buying new houses to tend any one of them.

It all happened overnight—'74, his expensive campaign for governor; '75, the Atlanta bank presidency and purchase, on borrowed funds; '76, bank expansion and the Carter campaign; '77, the jump to Washington. He was not ready for any of those four additions to his own precarious financial state. Yet each seemed to promise unlimited opportunity, while draining his day-to-day resources. Each was a mistake, in context; and none greater than the move to Washington. Carter asked a man in mid-trapeze-swing if he could walk away on air, and the man answered "Sure." Carter obviously did not know how close Lance had been shaving his multiple commitments when he offered him a job in the Administration. But Lance either knew or should have known; and he made his crowning mistake when he went to Washington.

His own bank's board resented this decision. Lance had been president for less than two years, had spread the bank's resources, and had not paid off his own loans for buying in. One member of the board, Alvin Cates, said: "When we interviewed him for the job of president, he made it clear to us that he was prepared to devote his future to making NBG successful." Charlie Kirbo, staying in Atlanta, is powerful by his ties to Washington. Lance, with his assets sold or in a blind trust, could not help the bank bluff through some of the riskier things he had dreamed up.

So the board called back a former president, Robert Guyton, who had resigned before Lance's time. With assurances that he was boss this time, Guyton began a severe retrenching, dismissing Lance's busy expansion team of political advisers. He wrote off $2.3 million in bad real estate debts and canceled the quarterly dividend. That is what sent Lance's bank stock plummeting, so that he had to ask for an extension on the divestiture requirements of Carter's regime. And that in turn is what made the Senate begin to reconsider the easy ride it gave Lance through his confirmation hearings.

Lance probably cannot oust Guyton, now, even if he retains the stock he is currently trying to sell. I asked Guyton if he was appointed on terms that precluded Lance's return. Without answering directly what he called an "inappropriate" question, Guyton said he was "dedicated to stay with the bank *whatever the circumstances.*" The White House tried to limit the Lance controversy to matters of illegal or unethical conduct, applying rules of evidence as if this were a criminal proceeding. It is a neat maneuver—on the one hand, you *call for* rules of criminal evidence, then you denounce any discussion of Lance's shortcomings as aimed at *criminal* conduct. One by one, Lance might defend most of his actions, the shifts and near-squeaks and precipice-dancing. Faced with the most damaging matters, where his "good word" is in question, he pleads inadvertence (easy to do *because* he had so many balls in the air)—e.g., he says he did not notice that he put up the same collateral to a second bank, though the first one was sending him dunning letters for the dividends on that stock collateral.

But Lance's acts must be judged by the main political rule, as worse than wrong—they were dumb. He thought no tomorrow could catch up with his hairbreadth yesterdays. He risked various properties, in a rising scale, until he risked a whole Administration on the permissiveness of his creditors and the public. His bank purchase was balanced precariously on his gubernatorial campaign; and then he balanced Carter's campaign and his own Washington job on those overtoppling bases. He left his own bank in trouble, and damaged the reputation of Atlanta banking in general—all to make arguments, in Washington, for sound budgeting and careful bookkeeping! It was the classic case of a drunk preaching abstinence. And he counted on his staggering partners to keep him standing—Atlanta could not afford to call him on his debts, and Carter would stave off threats by his mere position. The tipplers might reel, but never fall.

It is understandable that Carter did not see through Lance's bluff. He was the prime bluffee. But he had the right (and maybe the duty) to feel betrayed, precisely as McGovern was betrayed by Eagleton. The issue was not whether the subordinate did something wrong to begin with. Eagleton very properly sought medical care. Lance, before the Senate and House hearings began, was not clearly indictable on any of his breakneck deals. But both men misrepresented themselves as "safe" for appointment to very sensitive spots, and gave as little information as possible to the leaders who went out on a limb for them.

It is far less easy to defend Carter's long protectiveness of Lance. Faced with good reason to suspect hidden truths (whether criminal or not does not matter), he did not—like Eisenhower, faced with the stories of Nixon's slush fund—pursue the facts himself and remain judiciously above the fray. Instead, Carter overreacted to the damaging comptroller's report, saying it made him "proud" of Lance—proud of over fifty irregularities, questionable practices, "unsound" banking use of overdrafts and correspondent accounts. Attacks on Lance were taken as attacks on Carter. Jody Powell Ron-Zieglered serious questions into a mockery of "third-rate-burglary" responses when he discounted the story of half-million-dollar overdrafts, juggled accounts, and reusable collateral.

In this way Carter's first crisis seemed to confirm all prior guesses about his private weak points. He has been, in the process, simultaneously provincial and amateur and pious and stubborn. Lance conned him; so he doggedly maintained, for his own reputation's sake, that Lance couldn't be a con man. The danger with an "outsider" campaign like Carter's is amateurishness. Bright aides like Jordan and Powell could learn the mechanics of a campaign: youth is an advantage where the main peril is fighting last year's war. But it takes larger and longer experience to know the difference between substance and bluff—between, say, Mills Lane and Bert Lance—in the mingled areas of politics and financial speculation. Carter thought he had, in Lance, his one adult and equal who could talk to businessmen in their language.

Regional pride was involved. Carter came to feel that Lance had to be as sound, in his own way, as Carter was—Lance challenging Wall Street from small-town Georgia origins, just as Jimmy challenged Washington. Lance seemed to combine the best of Carter's own two worlds—a hick who could outplay the city boys. But Carter would not face up to the obverse of that trick's impact: its brilliance, if it works, is offset by equal weights of risked indignity if it doesn't. *What if the good old boy got took—*what then?

That contingency remained, perilously, unimaginable for Jimmy Carter—as critics said his own loss of the presidency would have been unmanageable for him because unimaginable. One of the pressures working at the back of Carter's mind was that regional sensitivity voiced by the Atlanta *Constitution*'s chief political columnist, Bill Shipp, when the attack on Lance began: "There is a feeling in much of Washington that somehow a fellow from the mountains of North Georgia simply is not able—and should not be allowed—to have a decisive role in guiding the government of the United States." That is the feeling Carter had run up against in his own race, or imagined he was running up against. Shipp later backed off, asked Lance to come back home for his own good; but Carter hung on to Lance, harming his friend as well as himself every minute he did so. The interlinked stumble was no longer a matter of standing together, but of standing alone or falling together. The Atlanta Establishment sobered to that fact with surprising speed, and lost its earlier pride in Lance. It no longer backs him. Carter was far too slow in sobering to the same realization.

The proud man from a soiled background develops a special prickliness, a quirky honor amid compromise. That could be seen in Truman (Carter's favorite President). Retaining his personal honesty as one of Pendergast's functionaries, Truman made his own integrity exist as a thing apart; and then, illogically, thought it should reflect back on his associates *insofar* as they were *his* associates. He was better than his surroundings, and that fact should stand warrant *for* his surroundings! I. F. Stone indelibly described all the Wimpies who trooped into the White House with Harry Vaughn. Yet Truman took each attack on Vaughn as a personal affront.

Carter, from soil stained with racism and provincial Southern ways, has similarly taken each criticism of Lance as an attack on him. Add that Lance (unlike the shrewder Charlie Kirbo, who "tends his house") has shared foxholes with Carter, as a member of his Georgia administration, as a candidate defending that administration's record, and as a member of his Washington family. Add that Carter designated Lance his Georgia successor—with some of that displaced parental pride bestowed on such designees (e.g., Eisenhower's for Bob Anderson, or Nixon's for John Connally).

Add small-town-Georgia shared background, and similarly close family ties. Add, most of all, piety. Lance blends overt evangelism with business as Carter blends it with politics. These men have prayed and read the Bible together. Anyone who has seen Harold Hughes rush to defend a "brother" like Chuck Colson knows what strong ties these can be. Visitors to the Lance farm at Calhoun are forced to tour and admire his wife's chapel building there. One such visitor told me: "It's uncomfortable as hell. What do you do? Are you supposed to start praying or meditating? Or just stand around saying it's real nice?" (Mrs. Lance takes down poetic dictation from God, bad spelling and all. A Georgia journalist, considering the results, told me, "God*damn* but God's style has been deterioratin' of late—and just when we could have used some of the good stuff He used to write: *And the sea returned to his strength when the morning appeared.*")

Lance hit Carter on his blind side, where all his blind spots are accumulated—his region, religion, judgment of friends, efficiency, pride in management, and reluctance to admit mistakes. Lance was half hero to him, and half crony. He was the part of Atlanta he had taken captive, and the part of little-town Georgia he knew from the outset—part of his roots he could take to Washington and never be ashamed of in the most sophisticated company. Lance's was a success that made Carter's success less an anomaly. Lance was a bit of a clown, a valet to whom his leader was a hero—but something of a mentor, too; a Kirbo who would travel, a Georgia teammate who was not a kid.

The White House's defense of Lance first emphasized ethics rather than intellect. That emphasis had special meaning for Jimmy Carter. Protective as he is of a *virtuous* reputation, he is edgier yet at any hint that he might be dumb. His relations with the press have always been sourest on both sides when Carter felt his intellectual inferiors were treating *him* as the inferior. The charge that hurts most to a man who has been helped by a pose of country-boy naiveté is the charge of country-boy stupidity. That is why the thing that will rankle with Carter, always, is not that Lance got took, that he overreached himself. Carter, so shrewd and deft behind the ambush of peanut farmer hedges, will forever resent the knowledge that the ambush became in this instance a trap—the knowledge that he was the trickster's victim, that the "took" good old boy took him.

Bert Lance is not important in himself. He overshot himself in Washington—as he would have done, somewhere down the road, in Atlanta. Lance matters politically only as he shows the regional flaws still left in Carter, the failure to graduate up from a local figure to a national one. The political significance of Lance is easily put: Does he end Carter's political education, or further it? A big question, considering the tasks Carter must graduate up to, and the serious misjudgments he made (and defended too long) with regard to Lance. After all, he tried to keep Lance on when bank practices were themselves a sore point, the focus of understandable resentments—internationally, because of a rickety loan structure arising from "recycled" Arab oil money to weak third-world nations; regionally, because of New York banks, which tried to dump city securities while pretending to support them; and nationally, as cozy arrangements for overdrafts and favored interbank deals throughout the United States were revealed by the Lance affair. Lance became a walking parable of many things Carter must oppose and try to remedy—and he kept thinking he could mount his opposition not only despite Lance but, incredibly, *through* Lance!

He tried to keep him in a post as important as it was exposed. The Office of Management and the Budget, a Nixon creation for the reorganizing of government, is resented by the departments it oversees, and is subject to bureaucratic infighting when its director becomes a political cripple. In this case, it is feckless to say Lance would draw on *presidential* vigor when that is just what he was draining away through his own lesion-spot in the Administration.

Those were the issues still waiting resolution when I left Atlanta recently—driving out to the airport past Braves Stadium, built with Mills Lane money in the sixties. Lane put up the cash before there was a team to play there, or city authorization for the project. He took a bet on the future, and there the result sits like a space saucer tethered on white milliped struts. It touched down lightly, like butterfly manna, and the paint is peeling from the struts now. The trouble with manna is simply put. It melts. Bert Lance didn't know that. More important for the rest of us, he conned Jimmy Carter into thinking it would last through the night. It never does.

—*New York Review of Books,* September 29, 1977

16

Jerry Brown

It was his ambition to be an oriental philosopher; but he was always a very Yankee sort of oriental. Even in the peculiar attitude in which he stood to money, his system of personal economics, as we may call it, he displayed a vast amount of truly down-East calculation, and he adopted poverty like a piece of business.

—R. L. Stevenson, of Thoreau

We had been talking about E. F. Schumacher, the author of *Small Is Beautiful*. "You'll be interested to know he is a convert." Pause. I didn't get it. "Convert from what to what?" I asked. "He's a Catholic." Then I got it. He was talking in code, and did not know that particular code is no longer in use.

Perhaps I should have caught on. We had, after all, been talking about our Jesuit seminary days, back in the fifties. We had remembered prayers "for the conversion of Russia," said after every mass. Conversion meant one thing then—acceptance of the one Church. But I had just come from a series of meetings where conversion was a hot topic, and the conversions involved were those of Jimmy Carter and Charles Colson. Besides, the people I go to mass with do not use "convert" in the way Jerry Brown still does. It takes a dropout from the Church to talk in outmoded ways without knowing it.

He knows, to be sure, that much of what he liked in the old Church is gone. "Not eating meat on Friday was like Jews wearing the yarmulke—it gave one a sense of tradition. All the binding symbols are disappearing. Even the Trappists—I used to go to their monastery, but now they can talk, and eat any time of the day, and even watch television." So Brown prefers to "hide out" at Richard Baker's Zen center, where the life is more self-consciously disciplined, structured, drilled. Some think Brown is interested in the East because of a taste for mysticism. They forget that the political leaders who most intrigue him are Mao Tse-tung and Ho Chi Minh. It is the ability to organize a people to puritanical effort and standards that he admires. He reduces Eastern mystery to seminary slogans: *Age quod agis*—the least mystical of Western spiritualities.

Brown rarely goes to mass any more—though some friends have urged him to "hit the communion rail" as a way of retaining ethnic support. He had the Sufi Choir perform at his gubernatorial prayer breakfast. Last Easter, he tried to find a Latin mass he could attend, but the only group offering one was schismatic—right-wing diehards separated from Church authority. It was too risky politically, even for a dabbler in Sufi liturgies.

Brown seems to remember with fondness only the hard things of religion, the external discipline. His preferred rite would be all Ash Wednesday, with no Easter. There are men with a positive taste for self-denial. Thoreau was one. Emerson said he found Thoreau always readier to say no than yes. Stevenson, in one of his acute psychological studies, described the man's "negative superiorities," his ascetic self-indulgence: "When we go on to find the same man, on the same or similar grounds, abstain from nearly everything that his neighbors innocently and pleasurably use, and from the rubs and trials of human society itself into the bargain, we recognize that valetudinarian healthfulness which is more delicate than sickness itself."

Much has been made, and rightly, of Brown's spare style in office—no mansion, no limousine, no banquets. He is accused of theatrically displaying his lack of display or credited with the sacrifice of something he might want. But he is no doubt telling the simple truth when he says he prefers to live thus—always has, and always will. Others love the perks of office and hate the duties. He loves the duties, hates the perks. He has cenobitic appetites.

It goes beyond personal preference. His taste is made a norm for others. Donald Burns, a close friend from Brown's Yale Law School days, told me: "I have a nice home, but quite modest. Yet when Jerry visits, I'm sure he thinks I have too many creature comforts." And he added: "He is incorruptible, because there's nothing he wants." It surprised people that Brown launched himself so well and enthusiastically into the campaign rites of Maryland, moving from Muhammad Ali to

"Polock Johnny's." He found ways to boast of the effort's very rigors, tireless in his claim that he does not tire. He kept telling reporters how long it had been since he ate, how long since he slept. He even told us how long it had been since he bought a suit ("before I was governor").

I asked him about the campaign silliness, and he spoke again in code: *Mea maxima penitentia vita communis*. I had forgotten, until he reminded me, that the phrase came from a patron of Jesuit novices, the seventeenth-century seminarian who died at the age of twenty-two, John Berchmans. I do not know if young Berchmans was really the prig that came through to us in the seminary; I do know I would not treasure up or try to live by any of the slogans peddled to us in his name. In fact, as I talked with Brown, it dawned on me that all the things I hated in the seminary he seemed to prize—right down to the compulsory readings in a crazed thesaurus of miracles (leaning heavily on worms) assembled by one Alphonsus Rodriguez (*not* the A.R. of the Hopkins sonnet). The Brown religion was not only all a thing of rigors, but of the particular rigors of the 1950s.

Yet he claims to be a voice of the sixties, the only presidential candidate formed in that decade's creative fires. As he put it to students at Towson State University: "I started in politics in the 1960s, in the civil rights movement, in the antiwar movement. I marched with Cesar Chavez." To those at the May 1976 Black Delegates' Congress in Charlotte, he said: "I represent the generation that came of age in the civil rights movement, in the anti-Vietnam war movement. . . . I'm a new generation."

But when one looks to the bases for Brown's claim to be the voice of the young in the sixties, they seem fairly slim. He refers to his "Berkeley days," and during his Maryland campaign the New York *Times* mentioned his "three years" there. It was three *terms*—spring of 1960 through spring of 1961. That was three years before the Berkeley rebellion began with the Free Speech Movement. I asked Brown about this, and he admitted, yes, the only excitement during his stay was the holding of HUAC hearings in Oakland. "But that was right across from I. House [International House], where I lived." Did he take any part in the demonstrations or reaction to them? No. Brown was able to graduate in three terms because Berkeley accepted his credits—mainly in Latin and Greek—from the seminary. Though he took his undergraduate degree with a classics major, he told me he had only one Greek course at Berkeley, in which he read only one play. He could remember the name of the play, but not of the professor. He says he is still working his way through Mark Schorer's reading list from his English course—a slow journey.

When Brown moved to Yale Law School in 1961, he was still intellectually provincial enough to seek out the other two ex-Jesuit seminarians in his class as his closest friends (and one of them as his roommate). Both would later work for him in California. It was during this period, many articles on Brown have asserted, that the young law student went South to work in the civil rights movement. I asked him about that. In the spring of 1962, during the Easter break, Brown did travel to Mississippi to look at the unrest there. As the son of California's governor, he visited with Governor Barnett, and he met some of the demonstrators. He did no civil rights "work" at the time, and he never went back. Bill Coffin, Yale's chaplain, was organizing Yale efforts in the South at that time. Brown told me he never even met Coffin in his three years at Yale—which was impossible if he was interested in doing any civil rights work.

We come to his next claim, his work in the antiwar movement. That seems confined to backing Eugene McCarthy's campaign in California in 1968. So the sixties experience he continually refers to shrinks to some knowledge of Cesar Chavez before his own election as governor.

Brown's real experience in the sixties was that of any Ivy League lawyer laying the foundations of a career—he clerked for the California Supreme Court from his Yale graduation until 1966, when he went into the corporate firm of Tuttle and Taylor in Los Angeles. He ran for Community College Board in 1968, and for secretary of state in 1970. It was a fast rise through normal channels—not hard for the son of the state's first sixties governor (Pat Brown's terms of office ran from 1958 to 1966). Jerry Brown's younger sister put her feet on the rungs of the same *cursus honorum,* with election to the Los Angeles Board of Education in 1975.

Here we come, of course, to the one debt Brown is most anxious to deny—that he is really just a politician following in his father's footsteps. He was not "radicalized" by any sixties experience, or even profoundly affected by one (with a possible exception in 1960, to which I must return). That is the mildly surprising thing about someone his age, in his place, with his background. The Zen mumbo-jumbo, churned out on demand by his freaky chauffeur Jacques Barzaghi (aka Lorenzo), was simply protective California coloration for a bachelor lawyer making his career. Now that Tom Hayden and Sam Brown are trying to look "straight" and electable, both find good words for a man who was always electable but did not move in their earlier world.

Brown himself, despite the vague litany of sixties experiences, makes no directly false claims about his distance from the things he "observed." He was nearly invisible at Yale and in the McCarthy campaign. When he first attracted notice, it was *only* as his father's son—and there's the point. The youth issue was a way of playing down the obvious debt he owed Pat Brown. He exaggerates every other aspect of his life, to minimize that one. He talks easily, often, and at length about his three and a half years in the seminary, his little more than a year at Berkeley, his three years at Yale, his few months in Chile (which took him away from his father's 1966 campaign). But he never talks, if he can avoid it, of the eighteen formative years in his father's house.

When he does refer to his father, he misleads, no doubt unintentionally. Asked why he will not live in the governor's mansion Reagan built, he says he lived in the old governor's mansion and never thought much of the experience. But he never did *live* there. He visited, just as little as he could. Pat Brown was elected governor after Jerry entered the seminary. And when the son decided to leave the seminary, he did not have his parents pick him up—he called a friend who had left before. Then he went to Berkeley and Yale, living on campus, before taking up his legal duties as an adult. He briefly visited his father's mansion as a man in his middle twenties—hardly, one would think, a traumatic experience. In the same way, when explaining why he drives a Plymouth rather than the official limousine, Brown remembers going to a baseball game, in his father's limousine, through a crowd which pounded on the windows. The impression he leaves is that it was the governor's limousine. But if this happened in his childhood, it had to be the district attorney's car or the attorney general's; and the reaction to the experience seems deliberately or otherwise exaggerated.

Robert Scheer, in the commentary he wrote on his interview with Brown for *Playboy,* said: "The closest Brown has come to rebellion in his personal life was joining a Jesuit seminary for three and a half years of virtual silence and daily penance for sins he had not yet had time or opportunity to commit. Given his prior Catholic school education, this was about as rebellious as an eagle scout's enlisting in the Army." That was no doubt true of most Catholics brought up in the period of fifties religiosity—especially those of us who went (like Brown) to Jesuit high schools and came to admire the "scholastics" (college-age seminarians on teacher-service to the schools) as bright and idealistic young men. Catholic families were proud to have a boy go off to the seminary in those days—and we went in unprecedented numbers.

But Brown's was not the typical Catholic family Scheer seems to imagine. Jerry Brown's mother, like her mother before her, was a "convert" all right—but *from* Catholicism to the Episcopal religion. Bernice Layne came from a non-Catholic family; and she dutifully bundled him off to church without ever joining him there. More than that, his grandmother on his father's side was noted for her acerbic anti-Catholicism, and she had harsh things to say about her grandson's decision to become a priest. Pat Brown's own ethnic-politics religion seems to have been played off in Jerry's mind against the fervor of Jesuit scholastics he met at school. Then, when he graduated from St. Ignatius High, his father proved just as opposed to his entering the seminary as did the non-Catholic members of the family.

Jerry was, after all, the only son of the four children. So great was opposition to his choice that he yielded to his parents' entreaty that he take a year in college to think it over—but he took that year at a Jesuit college, Santa Clara, and entered the seminary a year later than most of his novitiate classmates. So it *was* a kind of rebellion—one that became more pronounced with time. Brown took his first ("simple") vows the spring of Pat Brown's inauguration as governor, and a classmate of his told me Jerry *did* feel uneasy about the limousine when it showed up for periodic visits at the "juniorate" (second stage of Jesuit training).

Then, in 1960, Jerry Brown made his first and only remembered attempt to influence Pat Brown's decisions. The governor had inherited the sticky problem of death-row author Caryl Chessman, who had delayed execution with appeals through the reigns of Earl Warren and Goodwin Knight. Time was running out for him, and sympathy was running high. But so was the call for his death. Pat Brown is a sentimentally good-natured man, but one who resisted political pressure poorly. With Chessman, he got the worst of both worlds, first issuing a stay of execution, then turning down final clemency. Pat Brown dates to this sequence his reputation as "a tower of jelly," and implicitly blames his son for urging him—at first from the seminary—to grant the stay of execution.

The night before Chessman died Jerry called his father from I. House, urging him to spare the man—unsuccessfully. Donald Burns, Jerry's roommate at Yale, said Pat would call New Haven to discuss later clemency issue requests (which Brown became rather free in granting). But even in the long bull sessions Jerry loved at Yale, he did not bring up what was said in those phone calls. "Now that I come to think of it, that sounds odd," Burns added.

The younger Brown, still opposed to capital punishment, is very skeptical of rehabilitation schemes, and has said punishment should be just that—some people are not amenable to reason. Asked if he would respond to an Attica as Rockefeller had, he said he would use "no more and no less force than the situation required." He seems to resent his father's wavering incompetence in handling the Chessman case as much as his yielding to political pressure at the end. Jerry is critical of sentimentality and likes to boast of his "clear mind" in crises. If Pat Brown was going to let Chessman die anyway, he should not have played with the hope of clemency. Asked point-blank what he would do in such a position, Jerry answered: "I would make the decision and not agonize over it." The sentence is a ringing condemnation of his father.

So *one* sixties event *did* shape Brown's later actions—the Chessman execution. But only because it crystallized a long struggle to repeat his father's career without being like his father. Over and over, that is his real guiding principle. The other norms that he suggests—his Jesuit days, sixties activism, Ivy League law study—are part of an elaborate smokescreen laid around political drives and emotional revulsions. Brown shows an open contempt for political hacks, and gets furious at their attempts to influence him.

Just before his first campaign trip to Maryland, Brown gave a press conference to announce his new program against banks' redlining of loans to minorities. He spoke with a passion and mordancy that surprised Sacramento reporters, used to his fey and teasing manner. Don Burns, who had come up with Brown on the plane from Los Angeles that morning, said the transformation had nothing to do with redlining. In some time snatched between briefings on the redlining proposal and arrangements for his Maryland trip, Brown had taken several calls from California pols and money men, who offered him backing for the presidency—then set their terms. Brown got madder at each call, and was in high oratorical dudgeon with Burns before they went out to face reporters. He was angry, but not at redlining bankers. This kind of displaced fury against political ways serves his political ends. As a politician, he is a very good hater.

The other Brown, who so intrigues and eludes people, is insubstantial. Brown the intellectual can dazzle the easily dazzled—Marquis Childs fairly swoons at his dialectical skills. On "Firing Line," Brown

and William Buckley traded Latin tags, each pretending to know something about Thomas Aquinas. To have read a book is to pass for an intellectual in politics. Brown favors Catholic mentors—like Marshall
McLuhan, Ivan Illich, and E. F. Schumacher; but there is a faddism to
his interests, fed by personal seminars with these authors. When I asked
his very bright legal counsel, Anthony Kline, for an example of Brown's
intellectual discernment, he referred me to a toast given at the reception
for Japan's emperor. "It was just a few sentences Brown wrote out at
the table, but it said so much." I got a copy of the toast from the governor's press office, and found that he said, "Centuries of separation are
gradually dissolving into a global village." Citing the second-best-
known tag from Marshall McLuhan in 1975 is hardly a sign of intellectual prowess.

I asked Brown himself if he had written anything he felt proud of, either before or after becoming governor. "No, I only gave one lecture in
my life—it was on this as an Alexandrian age." Those who care for excerpts can look in the collection of his *Thoughts* (City Lights, 1976) on
page 50—he seems to think that Alaric lived in the Alexandrian age.
"One thing I'm proud of is the Arts Council Bill. Gary Snyder and I
wrote it up in Nevada City one night. We wanted to make it a model of
good English." I later asked Barzaghi, who drove Brown to Snyder's
telephone-less cabin, about that night. "Everyone else went to sleep, but
I stayed up to watch them. It was an act of creation." After that advance billing, I was amazed to read the measure—it intermits the bureaucratic with the gaseous.

When I asked Brown about his Catholicism, he said, "I am interested
in Catholic theology." I asked what he had read recently, and he said
Thomas Merton—which suggests either arrested development or his living off fifties capital for too long. He sometimes quotes Teilhard de
Chardin on "convergence." Since I devoted a chapter of my last book
to Teilhard, I tried to talk to him on the subject, but he said: "Oh, I
don't feel I really understand him."

Some assume that Brown's presence in a Jesuit seminary must have
familiarized him with St. Thomas or St. Augustine. But the early stages
of Jesuit formation were, in the 1950s, intellectually stultifying. For two
years nothing was read but ascetic books like *The Imitation of Christ*
and semiliterate biographies of obscure Jesuits. The Bible was not studied. St. Thomas was first encountered in one's fourth year—and by that
time Brown had left the Society. The aim of the early years was to
"mortify" the intellect, to empty it before refilling it with knowledge on

a Renaissance scheme that tried to "reenact" the order of natural-to-supernatural learning (arts-philosophy-theology). Brown got midway through the first step (arts). One classic that *was* studied in the novitiate was the *Spiritual Exercises* of Saint Ignatius, and some have tried to explain Brown in terms of that book. The California historian Kevin Starr even tried to predict the successive years of Brown's reign according to the "weeks" in the Ignatian *Exercises*.

But the last place to understand the *Exercises,* in the 1950s, was a Jesuit seminary. The *Exercises* were used by Ignatius, one on one, while counseling individuals in the choice of a vocation. They were meant to develop initiative in an order that broke all the rules of monastic permanence, discipline, silence, and contemplation. Ignatius, though something of a natural mystic himself, fashioned a team of individuals to counter northern reformers on their own terms. From the outset, his fellow Spaniards tried to fix times of prayer and monastic practices on his order. He savagely fought this off while he lived, but a series of ascetic Spanish generals totally changed the nature of his order shortly after his death. What came through in the Jesuit houses by the 1950s was almost the exact opposite of what Ignatius intended. What we got was a mixture of American religiosity with Spanish ascetic weirdness. As I pointed out, Brown seemed to like the rigors—but from a natural bent of his own, one that preceded and survived novitiate "taking of the discipline."

None of what I say reflects on Jerry Brown as a politician. Indeed, I am arguing that he should be judged precisely as a politician, not as an intellectual. The only reason he encourages the displacement of these priorities is to prevent people from thinking he is a politician in the same sense that his father was. As a politician he has major claims to make. His own orderliness, the talents he draws on, the high goals he aims at, come across not in bull sessions about Teilhard but in his marathon budget draftings, which he conducted in a cold fury of proud competence.

In putting together the Farm Labor Bill, which provided for collective bargaining in the fields, Brown showed all the political arts of wheedling and head-banging. He got growers and workers together, and brought the legislature along with him. The bill was written by Rose Bird, Brown's appointee as secretary of agriculture—he uses women's talents very well in government. (The administration of the farm program has fallen apart because Brown appointed a board that too openly favored Chavez—but it will in time be pieced together again, and the law is there, now, to be implemented.) One reason old-style pols resent him is the way he beats them at their own game.

Brown's distaste for hacks has led him to bring in an extraordinarily talented group of administrators, many of them from California's legal-aid programs or the universities. They are fashioning imaginative programs—for tax reform, court reform, highway measures, and child care. If Brown is no intellectual himself, he is at least not afraid of them. Besides, his moves make political sense. Most of those appointed are his age or younger, with no previous political ties. They are totally dependent on him, with no prior debts to others and no claims they can make against him. They have been willing to put up with the unorthodox style of Brown's governing. This style derives less from the publicized reasons—Zen meditation or monkish reclusiveness—than from two of Brown's most stubborn traits: his single-mindedness, and the fact that he is a night person.

Brown, with his passion for order and a "clear mind," likes to tackle problems one at a time, working on each exclusively, while other decisions are put off, other areas ignored. "I *focus*," he told me. "That's the way I do things. I *zero in*." He boasts about the long hours he serves at the office, but some associates say the early hours are put in for appearance' sake and not much is going on. Brown warms up to a problem slowly and begins to get enthusiastic only toward nightfall. "At six I'm just getting started," he told me—and aides have found that out in midnight sessions and early morning phone calls. Those who are not willing to put up with this schedule are considered somehow "soft" or not quite loyal. He has criticized an aide who put his family life above his job.

Brown makes great demands on others, but resents any demand made on him—not only by his father's friends and other hacks, but by his own assembled team. He acts as if close emotional ties would cloud the thing he prizes most, his clear mind. In Stevenson's words: "A man who must separate himself from his neighbors' habits in order to be happy, is in much the same case with one who requires to take opium for the same purpose." Brown must not only *be* independent, but be *seen* to be independent. When rumors spread during his race for governor that aides Tom Quinn and Richard Maullin were directing Brown, the result was a rapid downgrading of their importance. When a journalist started calling his pretentious "Kato" (Barzaghi) Brown's Svengali, the driver was seen much less often with his boss. Brown, who loves order and stability, is solicitous after the marriages of friends—but one such friend says he asks how things are going with an air of expectation that things must not be going well.

Brown told me, "I fly down from San Francisco, where almost no one I know is divorced, to Los Angeles, where almost everyone is divorced, and I wonder about the future of this society." Old friends of the saltier sort make it clear Brown is no homosexual; but some doubt he could ever give out the hostages to another that love or marriage requires. He speaks often of his own family as a stable one, but he prefers to eat Christmas dinner with a high school chum, not with his father or mother. He has a genuine compassion for the poor and mistreated, but he seems to feel it best from some height of ordered action. He intends to help the disadvantaged, but without the sentimental ties that might make him waver like his father; look a fool; let his heart cloud his head.

Stevenson admired, but with a shudder, the Thoreau who wrote: "It is not that we love to be alone, but that we love to soar, and when we do soar the company grows thinner and thinner till there is none at all." Thoreau goes on, immediately after, to say: "Use all the society that will abet you," and Brown proves he can do that. Despite his bristling at hacks, he can deal with distant unthreatening "bosses" like the crippled band that helped him out in Maryland. Still, the man who can reflect on a campaign crowd and murmur *Mea maxima penitentia vita communis* must always expect little shivers in among the earned praise of his excellences.

He is no saint, no mystic, no scholar; he is a politician. But he is a different kind of politician. A better kind? Perhaps. He has his compulsions, but so do old-time politicians—compulsive womanizers, or boozers, or boasters. Brown has certainly done well, so far, by California—and been rewarded with immense popularity. The enthusiasm spilled over into his Maryland victory. Brown not only means to soar but has begun to.

—*New York Review of Books,* June 10, 1976

AFTERWORD

When this article appeared, some people close to Brown said they thought the clash with his father was exaggerated; but I got the best confirmation of my claims when Brown's father told me he liked the article very much. But I do think I exaggerated the son's political skills.

Tommy d'Alesandro III, Baltimore's retired mayor, Brown's host during the Maryland primary, told me about a day he spent campaigning with him. "After one stop, Brown got back in the car, beside me in the front seat. His aides piled into the back, telling him how great he was doing, what good crowds, how much they loved him. After this had been going on awhile, he turned and asked me, 'How do you think I did, Mr. Mayor?' I answered, 'Well, Mr. Governor, I think you're a prick.' He looked surprised. 'Why?' 'I'll have to spend the night on the phone apologizing to all the people who turned that crowd out for you. You didn't shake hands with one of them, or thank them. You stood off on the side trading cracks with some smartass journalists.'" Tommy said he meant to shock Brown into shaking some of the pols' hands at the next turnout. But he had no success. Brown went on as before. That night Tommy took him to a banquet. "I said, 'Do me a favor, Jerry—don't shake a single hand up on the dais tonight.' 'Why?' 'Then I can say you had nothing personal against all those people you passed by. You just never shake anyone's hand.'" By forswearing his father's tutelage, Brown deprived himself of the d'Alesandro wisdom.

V

Conventions

||||||||||||||||||||||||||||||||||||||||||||||||||||||||||||||||||||||||||||||||||||||||||||||||||||||||

At eight national political conventions I have heard journalists go through the unvarying ritual of complaint at how boring conventions are. I think the blame for this self-hypnosis rests with Mencken, who taught journalists to think of conventions in the category of revivals and hangings. But the excitement is of a quieter sort, the slow crowding of outsize egos toward a compromise, like trying to cram too many blown-up balloons into too small a box. The convention's platform, for instance, is called irrelevant—and so it is, as a product. It is the process of accommodating this or that faction, in the drafting of the platform, that matters to the party's inner health.

Some make the mistake of thinking that the only function of a national convention is the choosing of national candidates. But delegates to the affair have each their private agenda—people they want to see, projects they want to push, careers to be launched or rescued. It is a multitiered swarm of political egos doing what they do best, pushing as hard as possible within the constraints of a final cooperation. The authority on this matter is not Henry Mencken but Samuel Johnson: Anyone tired of conventions is tired of life.

||||||||||||||||||||||||||||||||||||||||||||||||||||||||||||||||||||||||||||||||||||||||||||||||||||||||

17

The Best Reporter

As national conventions come around, journalists read again the funny outburst H. L. Mencken wrote after the 103-ballot ordeal of the Democrats in 1924, when the exhausted delegates picked the only candidate (John W. Davis) who had dodged every single issue. Mencken has an undeserved reputation as the classic reporter of our conventions. That title really belongs to Murat Halstead of the Cincinnati *Commercial*, who arrived at the 1856 Democratic convention to admire the skills and deplore the principles of Stephen A. Douglas:

> An exposed political empyric; a dishonest truckler for unsound popularity; a false pretender to notions of honor, and a foul-mouthed bully self-convicted of cowardice; though a coat of whitewash a foot in thickness would not cause him to pass for a gentleman, it cannot be denied that he will make a most admirable candidate.

Only twenty-six years old in 1856, Halstead was already part owner of the paper he worked for as reporter, editorial writer, and occasional managing editor. He saw his first conventions in 1856. There were four that year, and he covered all of them, along with a Republican preconvention. In 1860 he would report the *seven* conventions of a nation splitting up over slavery even as it tried to put together "unity" tickets. His reports on these "caucuses" (as they were still called) made him famous. Brought out in book form, they remain the delight of all Lincoln scholars (though Halstead claimed, with good reason, that Honest Abe stole the nomination in Chicago).

At the 1908 conventions the absence of white-haired Halstead was mourned by journalists—he had died that July at the age of seventy-eight, having covered every convention since February of 1856, when he traveled to the Know-Nothing meeting in New York. He even helped rig the 1872 Liberal Republican convention, held in Cincinnati, his own backyard. But, as often happens with journalists, his first reports were sharpest, those of 1856. The 1860 book gets most attention because of the stakes in that critical year. But in 1856 Halstead's native shrewdness was keyed up by excitement and indignation at his first contact with political bargaining on a national scale. He exulted in his power to use that still-fresh instrument, the telegraph, to let the nation in on the way it is bartered for where politicians gather. He looked up from the clotted madness at the Democratic meeting to observe with satisfaction that "across the upper northwestern corner of the hall were two bright copper wires, through which the continent listened."

Halstead did not watch conventions as mere circuses, like Mencken. He knew that slavery was the issue politicians were trying to avoid, and their success in 1856 gave his wit a bitter sting: "The politics of the country have, in a great degree, fallen into the hands of individuals who have proved themselves worthless for any other purpose. . . . In tender consideration of his interests, the politician is ever a trimmer—ever ready, by a sale of himself, to realize upon the capital he has acquired."

He watched the "buzzards and camp-followers, pimps and prostitutes" roll into Cincinnati for the Democrats' convention; but what fascinated and appalled him were the schemings of the pols: "Party leaders are putting their heads together in out-of-the-way places, laying plans to forestall the action of the assembly, and to gather into their own hands the means to control the nominations, or to sell out at the best possible price to those whose prospects seem better than their own." He saw that James Buchanan was bound to get the nomination— and he saw why:

> To occupy doctrinally a place certain, is totally foreign to his character. . . . He stands nowhere. He never did stand anywhere. . . . The chance to give him, pending the canvass, either a Northern or Southern face, or both at once if it shall be deemed expedient, is too good to be thrown away. Having been about half a century in public life—one half of which he spent in the bosom of the Federal, the other in that of the Democratic Party—this experienced and veteran camp-follower has learned, if nothing else, the benefits of non-committalism, and the art of holding his dish right side up, whatever may be the direction of the shower.

When Buchanan prevailed by virtue of these nonvirtues, his rivals were trotted out for the immemorial rite of party unity: "With a refinement of cruelty worthy of a darker age, poor old Cass, stupid Pierce, and bitter Douglas were hauled forth to amuse the crowd by a hollow pretence of happiness at the auspicious result."

When the Democrats were finished in 1856, Halstead turned back to New York for a second convention of the anti-immigrant Know-Nothings—the so-called Whig Americans, who were pretending they still existed as a threat to other parties:

> Once upon a time, when the brains were knocked out of a thing, death would follow in a reasonable space. Now, we have something both brained and disemboweled, yet it mouths at us, and insists that it is intensely alive. We have heard of persons running a few steps after their heads were chopped off, but this thing of the tail and hind legs merely of an animal entering regularly into a race has heretofore been unheard of.

Part of the artificial life of the party poured from the thick pocketbook of George Law, a construction tycoon who was hoping that donations would lead to his own nomination:

> If George lived in our town, we would take him for a heavy pork merchant, who had both slaughtered and eaten, shipped and digested, huge hosts of ye hog. He had a large amount of manliness in him before the New York *Herald* seduced him into becoming a politician, and there may be enough of common sense yet remaining to him to direct the use of his immense fortune to nobler purposes than in carrying on a vain squabble for office, and the petty glories of a one-horse politician.

But the Know-Nothings passed over Law to drag ex-President Millard Fillmore out of his deserved retirement.

> Mr. Fillmore was acting President of the United States. How came he so? By accident. Not a man of the body by which he was selected for the Vice Presidency, thought of him as a President—not a man who voted for him did so with a view to his fitness for the successorship. Who or what he was, few knew, and fewer cared. He was a mere consequent. He stopped a hole in the ticket; and, elevated by a casualty, constituted a government as secondhand as his character and official state.

By 1860 the thirty-year-old Halstead knew more about politics and had a tenser scene to survey. The divided Democrats needed two conventions that year before agreeing on Stephen A. Douglas; the new Republican party would shortly discover Abraham Lincoln and take charge of the disintegrating Union. But some of the edge was going out of Halstead's descriptions. Only at the opening 1860 convention, that of the Democrats in Charleston, did he pause for a vivid sketch or glimpsed vignette of the sort that animated his 1856 coverage. Here is a page one might read in Twain's *Gilded Age,* with the added attraction that it records an actual encounter where men put up slaves (and their own souls) for votes:

Passing along, we find a tall portly man in glossy black, with a bad stoop in the shoulders, a new stovepipe hat retaining in places the original shine, a bright red face out of which look brilliant eyes, carrying in his right hand, as if it were a mace, a huge gold-headed cane—it is Colonel James L. Orr of South Carolina, late Speaker of the House, now suspected of Douglas inclinations and of a willingness to be either President or Vice President of the United States. He is in the midst of a confidential talk with a burly, piratical-looking person in a gray business suit, the sack coat making him look even more squatty than he really is. The features of this individual are a little on the bulldog order. He does not look like a man of much intellect, but is evidently a marked man—a man of energy and perseverance, of strength and strategy. Ponderous as he is, he moves lightly. Fat as he is, he is restless, and as he smokes his cigar, he consumes it with furious, incessant whiffs. It is Young America, otherwise George N. Sanders [a Douglas manager]. And so, Mr. Orr, we see how the cat is jumping with you. You would have no objections to be second choice of the Douglas men—not a bit. You would be willing to take the Vice Presidency at the hands of the Douglas Democracy, wouldn't you?

We see political aides scuttling about claiming that their man really *did* mean what he said, or, as strenuously, that he *didn't:*

A Douglas man or two have strayed down here, and are trying to explain that Douglas doesn't really mean anything by popular sovereignty. "He had to talk that pretty strong to get back to the Senate." The people must be talked to violently about something—might as well say popular sovereignty to them as anything else. "Douglas would leave it all to the Courts at last. The Courts will fix it all right. Let us drop this immaterial issue and go in for the strongest man—Stephen A. Douglas."

The Chicago "Wigwam," specially constructed to hold the Republican convention, introduces the note that has made of our national conventions (in Murray Kempton's words) "a feast of reason":

The curiosity of the town—next to the Wigwam—is a bowie-knife seven feet long, weighing over forty pounds. It bears on one side the inscription, *Presented to John F. Potter by the Republicans of Missouri.* On the other side is this motto, *Will always keep a "Pryor" engagement.*

Halstead saw Lincoln's men pack the house, delay proceedings, bribe governors with office, and otherwise sabotage the chances of New York senator William H. Seward, the convention's first choice: "The fact of the Convention was the defeat of Seward rather than the nomination of Lincoln." Lincoln's home-state organizers had learned how to turn demonstration into intimidation—which goaded the desperate Seward men to frenzied effort:

> Hundreds of persons stopped their ears in pain. The shouting was absolutely frantic, shrill and wild. No Comanches, no panthers ever gave screams with more infernal intensity. Looking from the stage over the vast amphitheater, nothing was to be seen below but thousands of hats—a black, mighty swarm of hats—flying with the velocity of hornets over a mass of human heads, most of the mouths of which were open.

Halstead saw some of the deals made during the crucial nights of bribes, and sensed others being struck behind closed doors:

> I saw [Indiana Governor] Henry S. Lane at one o'clock, pale and haggard, walking as if for a wager from one caucus room to another. He had been toiling with desperation to bring the Indiana delegation to go as a unit for Lincoln.

Halstead should have been following Judge David Davis as he handed out public office to any hack who could help Lincoln's cause. But Halstead was perhaps too disgusted with the Chicago spectacle to pursue every detail with his usual gusto.

Another thing he did not see was the candidate himself. Though Halstead traveled the entire country in that intense political year of split-away conventions and unraveling parties, he never laid eyes on Lincoln —certainly not at the convention held in the Rail Splitter's own state. Lincoln did not want to be around when the deals went down—he would pretend not to know about them, though he honored every one. Halstead, himself an antislavery Republican, knew Seward's record was more forthright than Lincoln's; and no newspaperman could approve of Lincoln's conduct after the nomination—he refused to make a single campaign speech or answer a single editor's written questions, on the ground that this would be "divisive." We reserve an exalted role for Lincoln in our national history, one Halstead had no reason to expect from the absentee briber of the Chicago convention. But Halstead himself deserves a bit of praise for intruding on the privacy of men who were settling our national fate in the weird intimacies of a mob.

Some—and I am one of them—think we may have gone too far in "opening" our politics to amateurs, so that we lose the professionals' skills as well as their greed. National conventions seem doomed to become rubber stamps now, with everything settled long before they convene. But a young Murat Halstead may be out there this minute, preparing to kick up a fuss about that state of affairs. The TV cameras have replaced the two bright wires, but nothing can replace the kind of brain that Halstead carried from city to city, convention to convention. And even he could not improve on the twelve classic convention reports he wrote in 1856 and 1860.

—Quest, September 1980

18

Miniconvention, '74

During the Democrats' first midterm convention, held at Kansas City in early December, pilgrims sought out two shrines—the Harry Truman Library, full of relics from Democratic days of glory; and the Holiday Inn suite where George Wallace was staying, himself a relic, tended and dressed, wheeled around and displayed, like a political Infant of Prague.

Indeed, the dressing of him caused some problems. It is hard to prod him out of his Montgomery cave these days; he agreed at last, and reluctantly, to attend the convention on Tuesday, two days before the opening—and only then because his man Charles Snider telephoned him from Kansas City that all the delegates would think he was dying if he did not show up. He arrived Thursday afternoon, determined to go back Friday morning—he brought only one suit and one shirt. But all along the way he had been brightening back toward life. His reception at the airport and motel cheered him further—by the time he reached the Truman Library for a bash closed to the press (all but three pool reporters), he rolled into the building bubbly with refound confidence: "Make way for the leading contender!"

It was not an empty boast. That very day Gallup had released a poll that showed Wallace ahead of all others among Democrats (19 percent, to 11 for Humphrey and 10 for the putative "front-runner," Scoop

Jackson). Even more surprising, Wallace led all other Democrats among independents, the very people that "reforms" of the sort promised at Kansas City were supposed to be attracting: among them he won 24 percent of the people's loyalty, to 12 percent for Jackson. He came in among the first three with 32 percent of the Democrats and 42 percent of the independents. He was unacceptable, however, to 29 percent of the Democrats and 27 percent of independents.

There, in symbol, is the Democrats' dilemma. The party desperately needs to regain the South; yet the cost is also desperately high. One way or other, the leaders seem doomed to write off a necessary element of victory—19 to 24 percent if Wallace is repudiated, 27 to 29 percent if he is embraced. The convention would live uneasily with those facts, and with Wallace himself, all through the weekend—for he did not fly back Friday morning, or (as later announced) Friday afternoon. His staff had, with his permission, bought him a new shirt in Kansas City— and quietly sent for another suit to be flown up from Montgomery. He held court all through Saturday, and almost had to be crammed into his airplane on Sunday afternoon, when the affair was almost over. By that time he was back in the air with his own buoyancy, a balloon that had to be hauled down to earth again long enough to strap him into the plane's seat.

Wallace is afraid to campaign; but he only lives for crowds, and for his power to sway them—not to run would be, for him, a subtler form of suicide. All his staff is devoted to making him President as a way of compelling him to live. It makes for extraordinary personal dedication in his entourage. One of his bodyguards was shot with him, but is back at his post. Another who was near him in Laurel was asked in Kansas City: "You'd suck the governor's dick if he asked you, wouldn't you?" This brute of dedication was taken, momentarily, off guard. At last he said, fervently: "I just pray God he never asks!" The Wallace crew has changed far less than any other major candidate's over the years. They are still with him:

—Bill France, the millionaire racetrack owner; he is to good old boys what Ziegfeld was to Broadway columnists, or Colonel Tom Parker to aspiring Elvises—King of the Real World of risk and overnight fame.

—Charles Snider, handsome in a Ted Baxter way, gliding everywhere without seeming to move his legs, lest they disturb his perfect coiffure.

—Billie Joe Camp, the perfect press secretary who says nothing and says it inoffensively. When they make a life-size Ron Ziegler doll, and try to pretty it up, it will look and act just like "Bi'y Joe."

—Joe Azbell, the genially educated bullshitter and wordsmith of the organization, who admires Murray Kempton's style and apes Rod McKuen's. He brews the black magic of phrases like "Send Them a Message." In Kansas City, it took him no time at all to find a pocket of Chicanos prepared to hear Wallace out, and then he gave them a pitch in his own soft Spanish. Azbell has been a one-man J. Walter Thompson agency (only better) for his perennial candidate; and his skills are underpaid, no matter what he gets.

—And, always, the hovering hard beauty of Cornelia. She, too, came without a three-day supply of clothes, and went to Macy's for some shopping. When she got back, she had acquired new buoyancy herself. She picked up a Macy's catalog, flipped through it, and unerringly focused in on her own image—it was one of the display photos in the section advertising picture frames.

After the governor had been wheeled into the Truman Library, Cornelia boasted that her uncle, Jim Folsom, had been the only major Alabama politician to back Truman in the 1948 election. All the others had gone with Strom Thurmond's "Dixiecrats." She had, without realizing it, just let the skeleton out of the Democrats' closet. The reason these delegates were here in 1974 was to be found down in the Truman Library's basement, where a continuous film celebrates the weird election of 1948. That is when the Roosevelt coalition began to come apart, tugged in two directions by the Henry Wallace Northern breakaway (presaging a later McGovernism) and the South's walkout from the national convention.

Part of the unreality in Kansas City came from the Democrats' tendency to explain their presence in town as a result of the divided 1972 convention, of the unfinished business of McGovernite "reforms"— really, the question whether those reforms should now be re-formed out of existence. Another (even more misleading) opinion traced this mini-convention to the angers and violence of the 1968 convention. One fond myth of liberals is that Daley's head-busting in Chicago hurt Humphrey's election chances. It helped them—just as Agnew's violent rhetoric helped Nixon. It is true that the formal motion to revise rules was adopted in Chicago—as it had been adopted in Atlantic City four years earlier. But all these efforts were attempts to deal with a strain that first appeared in 1948.

The Democrats' national trouble is very simply put. They lost the presidency when they lost the South. That largest homogeneous bloc of voters gives any party an enormous advantage. With the South's electoral votes assured, it is comparatively easy to shop around among the other states for the necessary complement that spells victory. The importance of the Solid South to the Democratic party was for years celebrated in the mandatory choice of a Southerner for the vice-presidential nominee. Roosevelt ran with Jack Garner, Truman with Alben Barkley, Stevenson with John Sparkman (in 1952) and Estes Kefauver (in 1956), Kennedy with Lyndon Johnson. It is true that Roosevelt ran in 1940 with Henry Wallace; but he had previously offered the position to two Southerners—Cordell Hull and Jimmy Byrnes—who each turned him down. Besides, with war threatening in Europe, the belligerent Southerners could be relied on; the "isolationist" Midwest had to be won over, and Roosevelt assured the critics of his choice that Wallace would help get the farm vote. In the wartime election of 1944, Roosevelt took Harry Truman as a symbol of the Senate's efforts to keep the military honest.

Roosevelt was able to keep the industrial North and agrarian South together on economic issues—protection, welfare, industrial aid—and, later, on war issues. But there was never an ideological consanguinity between the Northern liberals and those presentable mild segregationists they had to run with. In 1948, that truth began to make itself felt. Through the fifties Democrats fooled themselves that Eisenhower was a passing fad, a freak of popularity not permanently upsetting party alignments. But Eisenhower—and even Nixon, in 1960—was making serious inroads into the South. In 1964, the crushing of Barry Goldwater obscured the most important fact of that campaign—that the South had bolted from the Democratic party, even though it was running its first Southerner for President since Woodrow Wilson. The challenge of the Mississippi Freedom Democratic party represented the plight of Democrats; it was compromising that issue, not kids in the street, that truly haunted the 1968 convention.

Democrats flattered themselves that Nixon "almost blew" the 1968 campaign. They neglect the real factor that made the difference between Nixon's early lead and the close election: while Wallace's third party was denying Nixon the South, labor swung Northern Wallacites back to Humphrey—only barely, and with effort (and because Humphrey was identified with Daley and a hard line on kids and war protestors). And the real lesson to be learned, down in the library basement, about Truman's survival of the party's first unraveling is that a gritty and somewhat demagogic appeal had only worked because of Dewey's pallid effort (he really did blow it, by thinking it unblowable).

Party reform only arises as an issue when the party is losing. Nobody seriously questions a winning combination. The comedy of recent Democratic reform is that it misconceived its task. If the party was losing because the South had defected, reform's job was to reenlist that sector. Instead, reformers listened to their own ethical-culture rhetoric, and read the South even more firmly out—along with the very elements that had given the Democrats a photo-finish chance in 1968 (i.e., labor and Daley). These "reforms" were so suicidal that a complex apparatus had to be assembled for gently extruding them from the party without ever quite disowning them.

The machinery had three main parts: (a) a Mikulski commission to set new rules for the 1976 convention, (b) a Sanford commission to prepare for the miniconvention, and (c) the Kansas miniconvention to adopt a party charter. The program, at each step, would be to ease off the 1972 guidelines with as little damage as possible—to leave the reformers with their rhetoric and little else. When stage (a) staggered to a relatively harmless conclusion, the Democratic governors tried to use its formulas to obviate stages (b) and (c). On the symbolically critical point of minority "quotas" among the delegates, they offered "the Mikulski language" for the charter, even though Sanford was supposed to be preparing the charter. (Sanford's own commission had broken down in its summer meeting.)

By and large the strategy worked, even though the black caucus, with women's support, gained a further technical point beyond the Mikulski language (on burden of proof about unbalanced delegations). The two major sides claimed victory, Jackson's man Ben Wattenberg acting as pleased as Willie Brown did for the blacks. Only some labor diehards, like Al Barkan, complained that they had been betrayed by Robert Strauss, that shrewd Jewish Texan and nonideological operator. Strauss knew the charter was something the party had rather to survive than to achieve; and it survived, to fight (itself) again another day. Asked why the party had to go through the ordeal of public self-criticism in Kansas City, Strauss said: "Being chairman of the Democratic party is like making love to a gorilla. You stop when *it* is ready." After all the picky attention to language, the main job of the convention was to get Daley and Jesse Jackson back under the same roof, and on waving terms, with McGovern and Wallace.

And Wallace was the real test. He did not go to the convention floor—

arranging that would have meant an earlier decision to come in the first place. It was a mistake not to put in a brief appearance there; everybody got a warm embrace in that hall. But if Wallace did not go to the floor, the floor did everything it could to go to him. Daley called up. Delegates clogged the approaches to his suite. And Senator Robert Byrd of West Virginia, the keynote speaker, came to consult very visibly with Wallace on the party's future.

Actually, Byrd meant the consulting to be more visible than it was. As word spread of his meeting, a flock of journalists gathered in his wake, went up in the elevator with him to the twenty-seventh floor, down the hall, and to the door. There Wallace's loving bodyguard interposed, bulkily, his body—and singled me out for a healthier shove than the occasion seemed to demand: "You got in once. You won't again!" Another journalist asked him, that night, over drinks, why he shoved me: "Because he called me pig-eyed." Oh, yeah; I had forgot.

Byrd's aide was still trying to get us in as witnesses to the summit meeting; but Wallace's people said no; only a few photographers were allowed to enter, after the two had been carefully posed, Byrd on a lower seat than Wallace—two small Southern boys from dirt-poor towns putting on an act for the city boys. With their slithery combination of dry eyes and wet lips, they almost winked at each other between furtive glances toward the camera. Appearances are almost everything to most politicians—to Wallace they are everything. He must never be caught on camera in pain or weariness. Montgomery journalists have learned the one sure way to talk with Wallace whenever they want to: just tell Billy Joe they hear the Guvnuh is feelin' pohly: he is soon on the phone, "I hear you got me *dyin'* any day now." In Kansas City, taking a breath mint to counter the ever-present bite of cigar, Wallace flourished the package so all could see it: "Now don't you go saying that Wallace has to take pain pills. It's just a little old *Certs*." The photographers filed obediently out after Wallace had done his vigorous look from the wheelchair.

Byrd's aide had promised us an impromptu press conference after the talk inside. But what had first been billed as a ten-minute call dragged on, more than half an hour, approaching an hour. Flowers were delivered, and messengers came and went; then two men came sheepishly down the hall, old men not looking comfortable in their ties and shined shoes, one man rather sickly (he was, in fact, on his way to the hospital, but had to put that off until he had seen George). After a whisper to the burly guard, they went in. Wallace broke off his talk with Byrd: "I have to greet some old friends." One of the men drew out of his crinkly suit five brand-new twenty-dollar bills: "It's old Bad Boy here to see you again." Wallace put his arm around Bad Boy, who was crying.

Later, I found out that the place to meet Bad Boy Brown, retired wrestler and ex-cabdriver, is at Allen's Bar. Allen, the owner, was the other man in the hall, now in a hospital. Until recently, the West Side political club, a dim remnant of the old Pendergast organization, had met next door to Allen's, and political talk is still a staple in the bar. It is all Wallace talk.

Bad Boy is a miracle of survival. Soft-voiced and courteous, his face is a palimpsest of every kind of violence. The ear on one side looks half-swallowed, and it is totally innocent of cartilage or stiffness. The other ear, though, is fresh-looking as an artificial flower: "Plastic. Feel it!" Bad Boy's nose goes in and around where most stick out; it is a fleshy little whirlpool in the middle of his face. After World War II, Bad Boy had been banned in most states for his wrestling tactics; but in Alabama he met a young lawyer named George Wallace, who took up his case and got him reinstated. He has kept in vague touch since, and even now he squeezes more tears out of that clenched face when he talks of the assassination attempt in Laurel, Maryland. When Wallace came to Kansas City to speak at a television station two years back, Bad Boy made his way through the crowd and dropped five twenties on his lap, laboriously saved from social security payments—the ritual he had repeated that morning in the Holiday Inn.

Joe Azbell is a great hero at Allen's Bar, one of those places he has a genius for sniffing out in each town. "Who's for Wallace in here?" Everybody—or everybody with the nerve to answer at all. Several men, including one of the bartenders, are planning a trip to Montgomery. Can Joe get them in to see the governor? Sure. "Yesterday, when I was in here, I think everyone I talked to compared Wallace to just one man. Who was that?" he asks a drinker standing by. "Mr. Pendergast." That isn't what Joe wanted: "Yesterday it was Truman they were all comparing him to." "Same thing."

In a place like Allen's Bar, outside the Deep South, there is an empathy with Wallace that is oddly touching. Rough men are not ashamed to weep over the attack on Wallace. "He is on our side. He knows what it's like. He takes on the big guys." At Dixon's Chile Parlor, there are autographed pictures of Harry Truman eating the eponymous chile. Joe asks if the proprietors would like a signed picture of Wallace to put there alongside Truman's. "Sure would!"

Wallace, like the South he has come to symbolize, originally stood apart from the rest of the nation on the issue of race. But, also like the South, he appealed to the rest of that nation at some level of gut respect for blunt defiance. He has all the paradoxical charm of crudity, that drag on the heart that bold ruffians possess. Truman had just a crucial touch of it, and it carried him through. But Southerners come by it with a special ease. There is an earthiness, a warmth of human contact, a realism about their own bullshit, that looks good in a time of "cool" TV packaging for candidates. A hug for Bad Boy Brown is good public relations precisely because it did not start as public relations.

American politics is the South's revenge for the Civil War. Congress was wrapped up and delivered to its ancient Southern chairmen. National elections turn quietly on the "givens" of the South while suspenseful attention is directed to swing states that make a difference at the last minute. The national press is disproportionately stocked with Southern boys, bred up in the fascination with the lurid game. The South is another country, not only for race relations but for religion, patriotism, family, and fighting. It has preserved what Democrats are trying to foster elsewhere—"ethnicity," neighborhood, populism, a blend of tradition and rebelliousness. One measure of this was the way that people, dithering back toward some kind of populism in Kansas City, hoped to run against Hoover one more time. BATH buttons were popular (Back Again To Hoover), and a great deal of talk returned to Old Dealer Hubert Humphrey. But the New Deal was launched with the South's support, and Hubert prompted the walkout back in 1948. Humphrey is not the one to defuse the Wallace threat.

Nor are most of the Southern candidates who put themselves on display against the pale green-and-orange art deco interior of the K.C. Convention Hall. The candidates rolled around those curvilinear ramps and bumped the chrome fixtures like too many balls clicking down the slope of a pinball machine, sharpies and smoothies who made unconvincing good old boys: Lloyd Bentsen, Dale Bumpers, Terry Sanford, Reubin Askew, Jimmy Carter, Bobby Byrd. Most of them reflect the Northern liberal's hope that a Southerner will come along who talks just like the New York *Times,* a man to lead his benighted neighbors into civilization. College president, nuclear physicist, superlib—they will even accept a tailor's dummy like Lloyd Bentsen if they have to. They are missing the point—a point made most easily by saying this: the ideal Democratic candidate in 1976 would be Sam Ervin twelve years younger. If you want patriotism and religion and connectedness, you have to take them with all their dangers and ambivalence, and find a man good enough to control them from within. And if you want all that, the South is the place to find it. Southerners are fighting fools, but not fanatics; they know what it is to lose, and even do it gracefully at times. Their politics is rough and realistic, and roguish. Lyndon Johnson could have taught us this, if he had not been obsessed with Kennedys and stuck to the Vietnam tarbaby. Stylish black operators like Andy Young know it, and their time is coming. (Young is backing Jimmy Carter this year; someday it may be the other way around.)

Democrats have spent all the years since 1948 trying to make the South join the rest of the country. Maybe their only hope is to give in and join the South. Not the Wallace South, surely. But not Lloyd Bentsen's, either. Isn't there some young Sam Ervin coming up?

—*New York Review of Books,* January 23, 1975

AFTERWORD

Someone was coming up, of course—Carter. I interviewed him in his motel room at Kansas City, and had dinner with his aides Betty Rainwater and Jerry Rafshoon. But I was one of the few journalists who paid any attention to Carter, and I did not trust my own instincts enough to think *this* was the Southerner the party so clearly needed. As you may have noticed, I was still buying his "nuclear physicist" line. In retrospect, it is too bad Carter did not have a touch more of Sam Ervin to him. Just a touch.

19

Democrats, '76

New York: I badly needed to rinse my mind of political rhetoric. Of Barbara Jordan: "Now we must look to the future." Of Jimmy Carter: "It is a *dis*-grace to the *hu*-man race." Of Fritz Mondale: well, just read his speech.

It was raining when the delegates left town on Friday (munching "big apples" distributed by the city as they checked out of their hotels); but the rain slacked off in time for Joseph Papp's free production of *Henry V* in Central Park that night. I would drive out bad rhetoric with good. The tall buildings, faintly visible over the park's trees, were cottony at the top—pleasant contrast to the lights (dimming by their very brilliance) at Madison Square Garden, that oval stretched tympanum of hopes and fears on which Carter tested midriff reactions all week.

The play got off with its own kind of gavel-bang, a cannon shot; off which the Prologue played for comedy. I settled in on the wet bench, expecting to forget Democratic delegates for two hours or so. But a kingly dictum caught my mind and nagged at it, one of the quiet prose sayings that reveal a virtuous Machiavellian at work behind King Henry's more bellicose arias. He is telling his officers how to treat the conquered: "When lenity and cruelty play for a kingdom, the gentler gamester is the soonest winner."

It was a captive party that met in New York. A fairly willing captive, by the end; but captive still—in thrall to a gentler gamester. Udall and Brown delegates chafed at the thought of John Glenn for Vice-President. They showed this resistance in a studied coolness toward Glenn, even before he began his keynote address—and in the way they went crazy over the slow and exaggerated pronunciations, the long chewed words and Everett Dirksen pauses, of Barbara Jordan. Carter could afford to read this signal drummed off the collective diaphragm—especially since it confirmed what he already suspected of Glenn. Carter yields where it costs nothing. He is a master of calibrated lenities.

But he is tough where it counts. The last holdout against his campaign was played with and dismissed. Jerry Brown made this easy for Carter by acting the sullen fool—but even the more adroit Edward Kennedy was gamestered aside by Carter. His hot-and-cold mastery of control came out best in a postnomination meeting with the California delegation. On the one hand, Carter desired a show of unity with the last semiholdouts (he had pulled together the riven New Jersey delegation to get its unanimous vote). But he also meant to show who was in charge. He stole Brown's troops out from under him, mainly by gentleness.

Brown was dour and minimal in his introduction; but Carter, who had been trading coal-hearth fires of bright smile with John Tunney on the dais, used lenity to win. He opened with a graciousness that shamed Brown's affronted prima donna act: "I want to thank the people of California for keeping Governor Brown so long out of the presidential race. I hate to think what would have happened to my campaign if he had come in before New Hampshire instead of before Maryland." But Carter's lenity is an accommodation of his power drive, not a denial of it. He went on to use his slogan of the week: "I did not intend to lose the nomination, and I do not intend to lose the election." In other words, though he hates to think of an earlier entry by Brown into the primaries, that would have entailed new strategies for winning, not any *thought* of losing.

The convention was so uneventful that its critics—mainly journalists—were tempted to call it "rigged": Carter's people would not allow the airing of dissent. But the truth is there was precious little dissent to be had for airing in New York. Feminists tried to raise the quota issue for future conventions. But that echo of 1972 was not a welcome one for most delegates. Besides, the women did not have their allies from the past—black leaders could do little, since their troops had gone for Carter in the primaries; quotas for the young lost their urgency with the disappearance of the Vietnam issue on campuses. The women's first demand (fifty-fifty representation) was not serious; it was raised to give them bargaining leverage for talk of female appointments in a Carter administration. Bella Abzug upstaged Daniel Patrick Moynihan at the convention, and praised Carter's compromise measure in the hope that he will be helpful in her fall Senate campaign. Barbara Mikulski spoke for many women when she said: "Women are like Israel. We are not weak—just vulnerable. We do not aim at nuclear war, but at dazzling raids. We accomplished our raid here."

The other issues—a larger "miniconvention" in 1978, uniform primaries, amnesty rather than pardon—were matters of degree, not clashes of principle. The abortion challenge of Missouri was not supported by any sizable number of delegates, as the McCormack vote on Wednesday proved. Carter's floor machinery, assembled for tougher purposes, had to contain such debates—if for no other reason, to stave off boredom. But the real reason trouble was avoided was plain lack of fire on the troublemakers' side. The party preferred its pleasant captive state to feckless revolt. The captor's lenity was all it asked for. The grumbling of some Chicanos and other ethnic groups was aimed at getting Carter's attention, not challenging his dominance.

The thoroughness of Carter's victory left his critics with little to say that was not silly. Some of those who accused him of "overkill" management in New York were also predicting he would prove lackadaisical, like Dewey, in the fall campaign. He was charged simultaneously with trying too hard and trying too little. His tight circle of devotees was compared to Nixon's band of Prussians. But Nixon used Haldeman and Ehrlichman to seal himself off from the electorate; and he was watching even these watchers—he trusted no one. Carter the campaigner demands total loyalty because he delegates so *much* of his task. He does this, not to raise barriers between himself and the people, but to free him for maximum contact with them.

The one thing he will not delegate is campaigning, shaking hands. His whole approach to politics—the "full court press" of personal contact, the use of his family, the staying in others' homes—derives from a supreme confidence that the more people see and know of him, the more they will admire. He talks easily of his own "character" as the test of this campaign. He is convinced that he is not only a better politician than the next man but a better person. That can infuriate, especially if he is right—and he may be.

Other campaigners want to retire from the public at night; to be alone with themselves or their friends—or with booze, or with a woman. Carter stayed in supporters' homes, disciplined enough to sleep, read, and write in such circumstances. He is a submariner always on duty. It is eerie, that control. He wrote his own campaign biography, in snatches, on commercial flights before he had a campaign plane. Its clear prose reflects the orderly mind at work:

> In later years my father bought a steam-driven mill for grinding the [sugar] cane stalks and for heating an inclined pan. The juice was piped continuously to the top of the pan and would run slowly back and forth between baffles from one side to the other, being boiled by heat from the steam plant. The inclination of the pan would determine how fast the juice ran down and, therefore, how long it was boiled, and the juice was changed into syrup before it reached the bottom of the twenty-foot-long slanted pan.

I know a Milton scholar who claims he will vote for Carter because he is the only politician he knows who uses the subjunctive properly.

One of the ways Carter subtly put Governor Brown in his place before the California delegation was to remark: "I have never gone to a governor's office—or a congressman's or a mayor's—to ask for the support of their constituents. I have gone to the constituents." There is little Carter does by accident. It was surely no accident that he began his list, in this place, with a governor's office. He will defer, but only from strength. One of the most revealing comments in the primary season came when Senator Kennedy first aired misgivings about Carter's fuzziness on issues. Carter snapped: "I don't have to kiss his ass to be President." He asks for votes, but does not beg. On the day after his nomination, he said, "I needed a factory-shift line this morning to get the campaign out of the hotel suites and back to where the people are waiting to be recognized."

Recognized by him—he thinks of handshaking as granting the people an audience. He talks of playing with the dozens of black children who islanded his boyhood home in Archery, outside Plains. His sister, in her book, reveals something of the reality behind that early exposure: she says she did not realize, till she had grown up, that she won games because blacks understood they *had* to lose. Carter, all smiles and hugs and instant friendship with voters, never looks silly as he does the things that demean most politicians. He carries with him a calm knowledge of his worth—just as he did in intimate contact with black girls and boys. It is a family trait: his mother, asked how she gets along with Plains blacks, said she gets along beautifully: "They know how much I have done for them." (It was fascinating, in New York, to watch liberals apologizing for their embrace of Carter by saying they *really* liked his mother.)

If Carter was to represent the real South, and not just be a front man for Northern liberals (like Terry Sanford or Reubin Askew), he needed most or all of the following traits—and Carter possesses them all in supreme degree.

1. He should be a military man. A military record helps almost any American politician running for national office—look how McGovern's supporters emphasized his bombing days from World War II. But the South is the most bellicose region of the country (it would call itself the most patriotic). Nixon won Strom Thurmond's support in 1968 less from hints at racial toughness than by commitment to ABMs. Southerners fought and died in disproportionate numbers in Vietnam. Those who grant Carter a little leeway in his first campaigns on the issue of black rights—the maneuvering room that was given William Fulbright—do not see why he had to express a hedged concern for Lieutenant Calley, a fear for national security when the Pentagon Papers were leaked, a doubt about McGovern's toughness.

I do not fear bellicose acts would follow such political noises. Carter is not only a military man, with nothing to prove; he will, if elected, be the first Annapolis graduate to hold the presidency. (Grant and Eisenhower, from West Point, are the only earlier Presidents from a national military academy.) Eisenhower stood up to the Pentagon better than any other postwar President. He saw the military establishment from the inside, as a professional making it his career. That experience can be more disillusioning than inspiring; and Carter has shown a healthy skepticism about the top-heavy military staffs. He thus gets the best of two worlds—a Southern respect for the military without the awe that naval amateurs like the two Roosevelts displayed.

2. A Southern voice should, ideally, be a rural voice. Even as it undergoes rapid urbanization, the South still thinks of itself as rural, and likes a sense of its farm roots. Many jokes were made, when Carter started his run, about a peanut farmer as President. But Southern politicians must affect a country style even when they do not come by it honestly. Carter knows this. He did not, with success, move from South Georgia to Atlanta, as his friend Charles Kirbo did. (Kirbo compensates by driving a pickup truck around the big city.) People argue whether Carter is a "real" populist; but populism is more a matter of symbols than of ideology in America. It is a style that either stimulates or allays the rural envies. Carter, because of his self-confidence, does not mobilize resentment as Nixon did. When he says, "We are as good as any others," he really means he is better; and the better voters think of themselves, the less demanding they become.

3. The Southerner must be a religious man. The South is the last refuge of sin in America, of devils and the fight against devils. Its politicians still quote the Bible. Carter's late partner would not process peanuts on the Sabbath. The church was the center of social life when Carter was growing up. He is not a fanatic, but a sincere member of the mainstream of Southern belief. This not only gives him a compelling identification with the South's third of our voting population; it makes him an appealing figure to the substantial evangelical movement all across the country. And thus another historical anomaly is resolved. The evangelical tradition in America, seen recently as opposed to progress and change, has been a reforming influence through most of our history.

Paul Kleppner in *The Cross of Culture,* his brilliant analysis of the populist elections in the 1890s, separates American religion into two political types, the ritualist and the pietist. The ritualist believes in hierarchy, social structure, compartmentalization, and an intellectual theology. Episcopalians, Catholics, Lutherans, and Jews are all ritualists in Kleppner's terms. In varying degrees these faiths have accepted secularization, departed from biblical fundamentalism, and been thought of as "enlightened" and progressive. Pietists, by contrast, stress a personal experience of the Spirit, one not mediated by priests, ceremony, or theological argument. They descend from the Puritans, Congregationalists, Baptists, Quakers, and they call for constant reform, revival, awakening, "new lights." Like the Puritans and dissenters in England, they were allied with many forces of political reform and liberal change—with abolitionism, temperance, wider suffrage, easier money. The combination of populism and fundamentalism in William Jennings Bryan was not a personal quirk but the natural result of social configurations in his time. Pietists have always tended to think of themselves as the *true* Americans, formed entirely by this country's experience—as opposed to "foreigners" like the German Lutherans, Irish Catholics, or Russian Jews.

In recent times the increase of secularism seemed to favor the cool and intellectual style of ritualists. Pietists were still there. Billy Graham showed up in polls as the most admired man apart from the President—but otherwise the pietists made little noise. But ritualists seem to have grown so secular and pale today that they do not feed the spiritual hungers of our time. Pietism was making an astounding impact in the sixties even as ritualists proclaimed "the death of God." And the impact occurred on familiar turf, amid forces of change and reform. The sixties style of protest emphasized personal experiences of conversion ("radicalization"), "witnessing," and emotional assembly. Martin Luther King and his SCLC set the tone. Preachers and hymns turned anti-war marches into modern revivals. Communes were the new monasticism. This atmosphere helps to explain the appeal of the Jesus People, or Pentecostals, or Moon cultists, or Eastern contemplatives, to so many young people.

Evangelical religion of the more traditional sort has been growing in numbers and intensity. The National Opinion Research Center recently found that over a third of Americans claim to have had a mystical experience (the born-again feeling), with the proportion approaching 50 percent among middle-aged middle-class Protestants. Educated guesses now put evangelical Christians at between sixty and eighty million people in this country, with a large fringe of sympathizers or the well-disposed. This makes it far the largest religious group in the country. The person mobilizing that group has a broader political base than Kennedy had in his fifty million Catholic supporters. Carter often says that Martin Luther King made his campaign possible. He means that King broke down the racist isolation of the South. But King as a symbol made Carter's rise easier for reasons that go beyond the civil rights movement. King was partly a cause, and partly the result, of the evangelicals' reentry into reform politics. Carter could not have risen so far so fast without the energies of this large social shift behind his effort. Pietists tend to vote for persons of distinct moral character, unlike ritualists, who concentrate on intellectual "issues."

Only yesterday, it seems, too overt religiosity would have been considered risky in a presidential candidate. Kennedy's lukewarm Catholicism was acceptable because it was irrelevant. But Catholicism is ritualist and "foreign." The *native* religion looks almost like a necessity this campaign year. President Ford, an Episcopalian by formal denomination, has an evangelical mentor, the well-known preacher Billy Zeoli; and, as a congressman, Ford was a devout frequenter of evangelical prayer meetings on the Hill. He has spoken to almost every large gathering of evangelicals this year, where he reminds audiences that his son is in divinity school. Reagan has an ex-all-American evangelical pastor, Donn Moomaw, who has conveyed the message that Reagan is "one of us." Indeed, in a recent interview Reagan said: "In my own experience there came a time when there developed a new relationship with God, and it grew out of need. So, yes, I have had an experience that could be described as 'born again.'" (Off to the right even of Ford and Reagan, evangelicals like Carl McIntire and Billy James Hargis have lost influence, though John Conlan of Arizona is taking up some of their "crusade.")

The energies of evangelical revival help to explain not only Carter's appeal to others but his understanding of himself—the unembarrassed use of his own character as the prime issue, the need for personal contact, the demand for personal trust; the instant familiarity and first names and diminutives (it is *Billy* Graham, not William); the discipline and puritanical attitude toward wasted time; the framing of political matters in moral terms. When Carter traveled as a preacher of the Gospel, his style of witnessing, of living with the people, was essentially the mode of his campaign. When he ran for governor, he brought friends and neighbors from his own county (Sumter) to vouch for him around the state. In the primaries, he brought Georgians to other states for the same purpose.

Pietism has often gone hand in hand with political shrewdness in America—and it certainly does in Carter. I thought of that, too, as I watched Papp's production of *Henry V,* a production whose flow and jumble of scene changes create a deliberate chaos through which the king can move in a circle of ordering calm. This king jokes a bit, as he did when prince. But he also prays; and he executes men at each step of his way—the conspirators at home, the prisoners abroad, and even old Bardolph (who was hanged in grisly effigy off the backdrop of Papp's stage). Those who complain of Carter's cold eyes would be just as upset if he were a wild-eyed hot-gospeler. As it is, he likes order and creates it, sometimes ruthlessly. As Assemblyman Willie Brown put it, complaining of the way Jerry Brown left his California delegation dangling and uninstructed, "You cash no loser's tickets at Jimmy Carter's window."

Even while Carter graciously praised Governor Brown, he mocked him. Referring to Brown's ascetic style, Carter said, "As we were saying upstairs in his *suite*—I mean, his *room* . . ." He could not finish—the Californians laughed and applauded. Brown had not stayed in the hotel with their delegation but in a fleabag with Cesar Chavez. Carter talks a lot about love, and presumably means it; but he is a businessman who talks about profit just as earnestly. King Henry woos fetchingly when he meets the Princess Katherine; but he keeps counting the realms he wants in the very midst of his courting: "I am content, so the maiden cities you talk of may wait on her; so the maid that stood in the way for my wish shall show me the way to my will."

—*New York Review of Books,* August 5, 1976

AFTERWORD

Carter's faith-healing sister, Ruth Stapleton, held a small press conference in her suite at the Democratic convention, and I asked her if she ever spoke in tongues. She hesitated, grimaced "Oh-oh, here I go," then admitted that she did. Truth weighed more than her fear of hurting Jimmy's chances with a "kooky" religious confession. There was a great deal of scare talk about Carter's religion before he entered office, and practically nothing afterward. Not, I think, because Carter was not sincere in his hope, for example, to continue teaching Bible classes as President. Time and logistics quickly shove that to the bottom of a modern President's schedule. In some ways Carter ran a less religious presidency than did his predecessors—there were, for instance, no prayer services in the White House. But when he did try to marshal religion for political effect—calling on *theists* of the Middle East to unite against the Ayatollah Khomeini, whom he called a heretic—people ignored him. Lack of a religion may slightly damage a presidential candidate in America; but presence of one is not supposed to matter much. In 1980 his fellow evangelicals deserted Carter for Reagan over mainly secular discontents like the Panama Canal "giveaway."

20

Republicans, '76

An unconsummated hate can be almost as unsettling as an unrequited love. After all its decisions were made, the Republican convention was nonetheless inconclusive. Reagan's forces won every battle, and lost the war. They humiliated Ford without defeating him. The President crawled through to the nomination on his knees; and, while he was down there anyway, embraced Dole as his Vice-President. The ticket needed balance, and Mr. Ford is obviously nice.

The scattered impact of it all was augmented by the convention's physical setting. Kansas City tends to wander, like the attention of its visitors. Near the mushy junction of the Kaw River with the Missouri, Kansas City survived when other landings had been washed away; it straddles rock outcroppings. This meant gullies had to be bridged or ignored as the city reached out to neighbor bastions. The result is episodic, like the convention that chased around the city trying to find itself.

Whenever Kansas City picks a rocky height and decides to build on it, it builds big. Tom Pendergast encouraged that, since the biggest buildings used most of his Red-D-Mix cement—including Harry Truman's Jackson County Courthouse. Truman liked to boast he brought in outside (read: honest) architects to construct that pile. Pendergast didn't care: he would pour cement to anybody's plan. He even poured a broad deep bed for a creek in the fake "Seville" called Country Club Plaza. His style continues in sedater monuments. When Joyce Hall (of Hallmark valentines and dollar signs) wanted to put something on Signboard Hill, he poured more concrete than most state highways contain. The resulting Crown Center is waste and scary as a Chirico above, but a labyrinth of boutique-warrens down below. The hotel lodged in the "complex" is plashy with fake waterfalls—three fountains in a coin. The President stayed there, where the week's craziest party was held—and also the best party.

Begin with the best. Under the twisted sail-canvas shelters of Crown Center's plaza, Carl Privaterra held a *Festa Italiana* on the opening weekend of the convention. The Italian community—which furnished hit men to Pendergast, like Johnny Lazia and Charley Carolla—sang and danced more engagingly than anyone else would do all week. Unfortunately, Mr. Privaterra went on to the other party, inside the President's hotel, where he was asked to lead the Pledge of Allegiance, and forgot "under God" before "indivisible." The patriots were reduced to a gabble and disjunct patter toward the end.

This party was held by the "U.S. Citizens' Congress," with Secretaries William Simon and Earl Butz as guests of honor. Rabbi Baruch Korff, a leader of the Citizens' Congress, embraced both men as they came in, and lobbied delegates to voice a demand that he, the good rabbi, address the convention on President Nixon's behalf. Republicans are wonderfully polite—they let the rabbi babble, and tried to think of ways to help him. Mr. Butz spent much of his speech praising Mr. Simon. Then, when Simon praised him in return, the Secretary of Agriculture threw pennies up over the podium to the Secretary of the Treasury—who obligingly stooped to pick them up. Both secretaries praised the free enterprise system, which is the mystical doctrine binding Republicans together. What they describe has never been seen on land or sea, but they solemnly promise each other to "preserve" it. This makes them, they think, "conservatives," though what they want to conserve is the most revolutionary force on the face of the globe. Capitalism is expansive and risk-taking; it accumulates resources to make large leaps into the unknown. When the resources cannot be gathered any other way (e.g., from peasants), a government takes charge of the project—this form of capitalism is called communism, and is the mortal enemy of the "capitalists" who sit and listen to Bill Simon tell them how private enterprise made them what they are today.

Earl Butz told the assembled wealthy that farmers are the hope of America because they have learned that you cannot make "Bossy" stop giving milk on weekends, and because no evil union has arisen to put two drivers on every tractor, the way they put three men in a railroad engine. He concluded this hymn to private initiative by adding that the government would not embargo wheat sales to Russia any more, making sure the competitive capitalist revolutions fight each other with both hands in each other's pockets.

Republicans are terribly engaging; they are the only people who still believe the slogans we were all brought up on. Democrats talk just as much nonsense; but they know it is nonsense, and that ages them. Republicans are young forever. They go on thinking the capital-accumulation revolution is "conservative." When they are not cursing "the State," they are asking It to censor books and movies, enforce the Sabbath, and keep booze in brown paper bags. The revolutionary nature of this enterprise never strikes them at all; not even when they have drafted a party platform calling for wholesale rewriting of the Constitution—with five brand-new amendments to begin with. They are permanent revolutionaries, the warriors against reality; and the forlorn quality of their battle makes them more likable than the sordid traffickers in fact.

The only problem with Republicans is that their naiveté puts them at the mercy of crude self-promoters like Baruch Korff. Korff rose, toward the end of the evening, to offer Bill Simon to the convention for Vice-President. That explained why Simon was there—as he would be in delegation after delegation, chasing off Pat Boone, the Reagan crew's perennial teenager, with praise for Gerald Ford. But the price was a high one to pay. Conservatives put fine minds like that of Simon to the most demanding tasks—like finding reasons to praise Ford.

Praising Reagan was comparatively easy. After all, he had the good taste not to know his own vice-presidential candidate, whose name he kept mispronouncing. I was in the crowd that welcomed Mr. Reagan with such good grace that it did not even gag when Richard Schweiker got up to say he had opened the door to liberal Republicans (that forlorn crew represented by dyspeptic types like my own senator "Mack" Mathias). Murray Kempton thought this sounded like Nixon's overtures to Mao. Schweiker was decently hidden much of the week, but not quite forgotten. He took the bloom off conservatives' real claim to honor—that they have always had the decency to despise Nelson Rockefeller. What good is accomplished by endlessly insulting Rockefeller (who takes the insults with a grateful smile) if you are just going to bribe Richard Schweiker to give up his enthusiasm for COPE? The good old hates were oddly confused, and remained unconsummated.

The convention had no real focus. It was almost as individualist as Republican myths would have us believe the country is. The true story had to be chased all over Missouri and half of Kansas, wherever the Republicans' national committee had found room to stash an uncommitted delegate. The outcome was preordained: the individual, unsupported, will yield in time to the greatest power. Ford, despite all his efforts, simply could not throw away his party's nomination. There is too much power in residence at that address on Pennsylvania Avenue. Delegate after delegate gave up his idealism to the comic promise of power from a man who does not know the first thing about it. The closest thing to a focus for this convention was the Ramada Inn where Mississippi played the shameless flirt. There Clarke Reed kept bidding up the price for his own sale, until Harry Dent was forced to seek more reliable brokers in that delegation. John Sears covered his own failure with the Mississippi delegates by describing it as an act of mercy —if any more bids were made to the state, delegate by delegate, he felt the whole motel full of them would drop dead of fatigue. If you can't buy it, you might as well call it damaged merchandise.

Since no single journalist could equal the President's resources in holding the hand of all the scattered delegates—much less wangle an invitation to go with each one to the White House over recent weeks— most of us settled for the John Sears sideshow. Sears kept the attention of "the media" by ignoring facts, in the best Republican tradition. He made the improbable Ron look good by juxtaposing him with the impossible Schweiker. "This is so crazy he must have something in mind," most journalists said, for want of anything else to say or think as the dreary mopping up of the last dozen delegates or so went on. Sears amused us by proving he could commit every indignity upon the President of the United States short of unseating him. He used the nomination of Schweiker to argue: "If you think what we did is crazy, just wait till you see what Jerry does." Sears had already let Jesse Helms rewrite the platform at will; then—to add a final note of contempt—he cooked up a last-minute addition insulting the President's foreign policy. The President simply held his nose, and swallowed. Even when Ford won, he booted away victory: after defeating the Sears ploy of announcing vice-presidential choices ahead of time, Ford paused, at the very moment of leaving Kansas City, to call this a pretty good idea—which proves that it isn't.

The myth this convention gave birth to is that John Sears almost tricked the Reagan people, who love to lose, into winning. But the only people Sears really tricked were journalists, who had to justify all those daily sessions in the Municipal Auditorium. Reagan's people do not want to win; they want to believe. If Sears had found a way to win, they would have found a way to stop him. As it is, there was no way to be found. Anyone who loses to Ford obviously *had* to lose.

The night trip to the other side of the tracks, where Kemper Arena flexed its exoskeletal spine and elbows of pipe, was a formality. Almost all the buying and selling was done by the time the purchases arrived, properly ticketed for admission. Ford's people tested the tolerance of the convention on Tony Orlando before daring to slip Henry Kissinger into a seat. Mr. Orlando was the Sammy Davis, Jr., of this convention. Davis, you remember, smothered Richard Nixon with kisses in the small of his back, while the stunned President wondered what could be attached to the black hands that had grabbed him. Mr. Orlando used the President's wife to sell another million or so records of his song about a convict returning from prison. Connoisseurs of Republican conventions might remember that Mrs. Magruder tied a yellow ribbon around a tree at her front door when Jeb was sprung.

Without much serious business to conduct, the convention settled down to serious screaming. The ritual drone of Reagan's bull-roarers was punctuated with little bleeps and high squirts of sound from the

President's horns. Hilaire Belloc talked of French armies returning defeated, "having accomplished nothing but an epic." The Reaganites accomplished little at their convention but the liveliest farting contest since Aristophanes. The length and intensity of the losers' cheers proved that, though the party's small slow brain stayed with the President, its big dumb heart was Ron's. Republicans are always giving up their *real* candidate in order to win—and then they lose. They surrender their ideals, but not enough.

This dumb-out at the O-K Corral crawled to its tortoise versus tortoise dead-heat conclusion inconclusively. Ford contrived to find, for his vice-presidential candidate, another Ford, plus meanness. It was a slow learner's late and bad imitation of the John Sears strategy: the choice looked so bad it must be brilliant. It was bad enough, in fact, to silence Reaganites, without satisfying them. They were not even allowed the comfort of real contempt at the end. They felt obliged to keep kicking Ford, but affectionately. He had defeated them by refusing to fight. Even when these delegates tried to blow their brains out down long thin kazoos, the only thing they could freight the air with was sound. They left Kansas City disconsolate, knowing that Carter had not only stolen the South from them (Clarke Reed could not really play Strom Thurmond to this convention—all he had were delegates, not electoral votes); Carter had also pocketed God and taken Him away for November use. Republicans have always thought that at least they were losing in good company—God's.

Some people talk not only of the Republicans losing in November, but of their disappearing. I can't believe it. The Republicans are as impossible, and indispensable, as the national anthem. One of them, at that party for aged adolescents called together as a "Citizens' Congress," told me, "Jimmy Carter will defeat himself. He doesn't *look* like a President. Imagine him conducting a formal state funeral." Republicans have to be kept around for occasions like state funerals. They preserve our memory of an America that never was. We need them. We all need our illusions. Who else can give us lines like this, from Robert ("Bob") Dole's acceptance speech?—"Wherever tyranny reigns in the world, it reigns through the instruments of government." That is on a par with: "Wherever men tell lies, it is through the instrument of language." Still, there is a good deal to be said for not having a President. It is typical of the Republicans' lack of acquaintance with reality that, having long called for that state of affairs, they refuse to appreciate it when Ford gives it to them.

—*New York Review of Books,* September 16, 1976

AFTERWORD

My son served as a courier at this convention, sitting for long stretches of it next to John Dean, who was covering the event for *Rolling Stone.* After watching Earl Butz pitch pennies at the Secretary of the Treasury, my son especially enjoyed Dean's story of the flight back to Washington from Kansas City, on which Butz told the racist joke that cost him his job. On nomination night, I was commenting for the CBC radio network, when the kazoo demonstration made it impossible to be heard through the microphones in our booth. So the CBC correspondent and I strung mikes into a small bathroom in the booth, and my son John pushed messages under the door to keep us informed of events on the floor. There was something appropriate to that claustrophobic setting.

21

Miniconvention, '78

Scott Fitzgerald once mocked a man who "tried to get up a ship's party on a ferryboat." That was the story of the Democrats' 1978 midterm conference in Memphis last December. Some signed on for a pleasure cruise and found themselves working a White House shift in the presidential boiler room—a shakedown cruise for 1980. The President, who was accused last summer of managerial incompetence, overmanaged a nonevent—which made it an event of sorts. Dissidents in Memphis did not go away doubting Carter's manipulative skills.

Of course, a midterm conference held by his own party is a no-win proposition for any President. If he doesn't control it, he loses. If he *has* to control it, he loses. If he barely controls, or overcontrols—and Carter did both—his very victory becomes an embarrassment. Given the rules, Carter did as well as he could, which was not very.

Once a candidate is chosen at a nominating convention, the whole gathering celebrates that candidate in a ritual of party healing. Defeated rivals are reintegrated into the campaign effort. But at a midterm conference (or miniconvention)—an anomaly introduced four years ago to cure even worse anomalies in the Democratic party—there is no overt competition with the official party head, and therefore no admitted healing process. Wounds fester, neglected; friends stab, with smiles. Pols pretend they are not plying their trade. The White House army of two hundred or so had to claim it had turned out in force to "listen" rather than stage-manage.

Consider the President. Security measures—perhaps overdone, with a memory that Memphis is one of those assassination sites starred on our political map—made the meager physical resources of the town revolve around Carter's appearances on Friday and Saturday (press credentials only mentioned those days; suggesting, with good cause, that the official voting proceedings of Sunday did not matter in the manager's eyes). But delegates did not share the cops' priorities. When Carter was trundled from workshop to workshop, he had to sit there, partly ignored, partly insulted, partly lectured to. The rules of the game made him summon up a simper of attentiveness throughout, St. Sebastian pretending he likes nothing more dearly than arrows.

Carter's long address on Friday night was earnest: he tried to be forceful—which is not his mode. His power tends to come in parentheses. The heavy-handed promotional movie produced by Rafshoon did all the wrong things—pounding away with a song, speeding up the soundtrack so that Carter was seen in 33⅓ while talking at a 45 RPM gabble. But the real problem in Memphis was not so much personal— Carter's inability to deal convincingly with party backslapping sessions. It had more or do with the Democrats' need to talk underdog ideology from an overdog position. There was endless discussion, from the podium and on the floor, of a struggle for the party's soul. It was less a problem of soul than of turf—the need for a majority party to face the limits that go with that power. Lords of the manor make unconvincing Robin Hoods.

If there is one thing that unites conservatives and liberals at this moment in our political babble, it is the claim that Democrats are riding a new mood by capturing Republican issues. That, we are told, was the meaning of 1978's elections. Budgetary stinginess, the birthright of the Right, was voiced by the party of the New Deal, making it win power by a loss of principle. Even big business came bearing gifts to the *Democratic* incumbents or favorites, on the assumption that the power to be stingy with the poor (and generous toward defense industries and capital gains) is more important than the love of being stingy.

Some commentators think Democrats show a set of party skills denied Republicans. That is unlikely. There is simply no skill like an assured seat. Democrats respond to constituents best because they have a majority of the constituents and a lock on the constituent-servicing mechanisms of incumbency. In Memphis, many praised "the party of Franklin Roosevelt" as a way of criticizing Carter for planned cutbacks. (They will probably end up serving him in the long run, when he finds he cannot cut much anyway, and will say his own party would not let him.) The only ones who emphasized "the party of Lyndon Johnson" were Kennedy and Mondale. Kennedy said budget cuts would tear the party to pieces, as the Vietnam war had. Mondale said failure to cut the budget would tear it apart, as Vietnam had. Each was trying to save the party from being Lyndon Johnson's.

But ideology matters less for professional Democrats than does the role of the majority party. (We normally have an institutional majority party. The minority party endures, not by competition from a roughly equal status, but by being the majority party in certain enclaves.) Being the majority party is not easy. The broader the constituency, the harder it is to be both inclusive and consistent. Indeed a minority party has certain kinds of romantic advantage—freedom to sound principled, mobility to make raids. It has so little to lose. The majority party can grumble and squabble, but few want to break up a winning (if anomalous) combination. They have too much to lose—which is why, in the long run, the majority party loses it all.

Carter was criticized in Memphis for listening to the polls, to returns favoring Proposition 13 in its several guises. Claiming to speak for "the people," the labor Left and its allies went against what the people are saying in today's polls. Carter had to listen, without any intention of heeding. President Eisenhower talked of the rhetorical advantage given third-world countries as "the tyranny of the weak"—which did not bother him, since he thought the threat mainly rhetorical. Memphis was, in party terms, a time for verbal tyrannizing by the weak.

The weak, of course, did not see it that way. Meeting at a chilly exit from the Memphis convention hall, Michael Harrington (with the sad, gray boyishness of revolution-gone-incremental) and Tom Hayden (in his prep-school tweeds) plotted for two minutes the overthrow of Carter from within: a future recession will lead to one of two things—wage and price controls, angering the conservatives Carter is pandering to; or no controls, and economic disaster. Jerry Brown will then dance feyly into the contest, and Teddy Kennedy will be given license to intervene without being called a spoiler. Done, neatly, and done.

It was a vision to console, after Hayden's press conference, which was sparsely attended, and after Harrington had to call his off. Progressives at the conference had to settle, mainly, for dreams. Their large caucus Thursday night (the eve of the conference proper) showed why. The two Frasers—Don, the ousted representative, and Doug, the United Auto Workers chief—were on the dais explaining what favors they had wrested from the National Committee (mainly the right to get new resolutions to the floor on Sunday). The crowd was restive. Ron Dellums, tall, black, and looking like Cesar Romero, asked for a spot on the dais, said "People are going to sleep in this room," suggested that something be done, did not say what that might be, and left the dais. A resentful delegate asked why changes in the rules of procedure—to allow protesting voices and resolutions to be heard—could not be discussed until the last day of the conference. Another moved that the Frasers arrange with the DNC for a move to suspend the rules, on Friday night, in order to discuss the rules.

Don Fraser spoke against the idea mildly, and Doug Fraser spoke against it emphatically: "We are on the threshold of making a big mistake, I warn you." Mike Harrington had the grace to look embarrassed as his fellows carried Administration water to put out their own fires. Despite resistance from the dais, the floor voted overwhelmingly to arrange for a Friday night suspension of the rules. Don Fraser promised, weakly, to work for it (weakly). Afterward, Doug Fraser, with his concave, shovel-pan, red face, said:

> We have a written agreement with [party chairman] John White to let our petitions go to the top of the agenda—so long as we limit their number to what we had already mentioned. He can call it off now—say we didn't keep our side of the bargain. That's why I spoke against the new move, to show I wasn't part of it. I have to make sure I'll get a hearing—I'm going to run some tough language on health care past Hamilton Jordan tonight.

Turf protecting is contagious.

The progressives' "hard shot" was a resolution that would pledge the party to "an adequate budget to meet human needs, in no case less than the current budget for human and social services." It was fashioned as a trap—if Carter's people attacked it, they would be Scrooges; if they accepted, they would have to swallow their budget-cutting talk. As on most matters, Carter's people—after harassing moves preliminary to the conference—tried on-spot compromise to deflate protests. By Saturday night, the negotiations boiled down to one word—the White House would support the resolution if "critical" was substituted for "current" services. The Frasers could not give in; they had compromised too many other things to get a hard budget resolution to the floor. What, anyway, are "critical" services? It would depend on the crisis—or anyone's definition of a crisis. No deal.

The White House, claiming the progressives were acting in bad faith, turned loose its awesome whip forces to say the resolution was just a cheap shot to embarrass the President. A reference to the Humphrey-Hawkins bill was inflated to a charge that Carter was breaking the law. Even the party chairman, John White, professionally friendly, let his smile fade when he spoke against this resolution from the floor. He argued that "no cuts in current services" meant Carter could not cancel inefficient programs. I observed to him that the resolution's sponsors said any *redistribution* of money among programs would fit its demand that the current *level* of expenditure be maintained. "The resolution does not say that," White snapped. It does not preclude it, I answered. "You're getting tricky." He was getting tired.

By Sunday, White House forces were making promises to delegates, asking friends to speak against the budget resolution, calling its authors conspirators against their own President. It worked. Jim Wall, liberal editor of *Christian Century,* had earlier expressed misgivings over the planned cutbacks. But Sunday he told me he would vote against the resolution. Why? "It is improperly motivated." Aimed at Carter? "Yes." So this is a loyalty oath? "Call it a vote of confidence." Wall's delegation, from Illinois, was extravagantly confident—it voted against, 72 to 0.

At the outset the White House folk seemed all to be waiting for Teddy. They had packed the health panel with eleven people (including Joseph Califano and Stuart Eizenstat from the Administration) to contain the danger of Kennedy's appearance at that workshop. But Kennedy dominated the entire proceeding. He came with a carefully prepared speech (copies available to the press), program charts, and the distilled arguments of his long effort to create a national health plan.

After Califano read some figures, Kennedy rose with anticipatory little "Mr. Chairman" throat-clearings; then, head down over his text, tugged hard at the speech as if it were a tent peg. Toward the end, he took off his glasses, stepped forward from the podium, his shout carrying without the mike, and sailed off with the uptorn tent as his balloon, happy with rhetorical angers. The crowd was shouting too. Lester Kinsolving, designated jerk of the press corps, tried to scream and sneer at the same time: "And when do you announce your candidacy?"

Subsequent panel exchanges were jovial on Kennedy's part (he hoped Eizenstat would speak to the President as he had to this audience about the needs of the poor), and snippy on Eizenstat's part ("I don't need to make speeches to the President of the United States"). Kennedy won big personally, but he created a trapped excitement going nowhere. He could promise health care without getting into the full debate on inflation. In the same way, the progressives' petition to maintain social services did not add what most of them believe is a necessary corollary —wage and price controls. That would give too many hostages to Administration spokesmen, stirring domestic myths that are the counterpart of our anti-Communist fear in foreign affairs.

So Fred Kahn and Charles Schultze went without challenge when they dismissed controls at the inflation workshop (one of the two attended by Carter). Carter's people want to cure inflation with turns of language, a kind of bluff and countermagic. They attacked controls as if they were the *Nixon* controls—crassly imposed for electoral advantage and crudely removed when they had served that purpose. But Kahn (and Carter) promise controls without controls—controls by subtle threat, indirect reward, tax compensation, contract juggling, verbal play (Kahn's specialty). The American myth of individualism has bemused people with the view that our economy is not controlled. We have all kinds of uncoordinated controls—of the money supply, of tax revenue, of government contracts; of subsidized education and retirement and farming and drilling and arming and arts and scholarship and what *not?*

What we *don't* have is a rationale for control, which would involve an admission of control—on the way to openness and efficiency. The present indirect, unadmitted, arbitrary controls are not only inefficient but self-crippling—as opposed to legislated controls, democratically aired and equitably arrived at. Kahn and Schultze won't say that of course; if they did, they wouldn't be where they are. But what mattered most at the conference was the inability or unwillingness of either Fraser to say that. Even Harrington and Hayden skirted the topic; the left had to be "responsible," Don Fraser told the Thursday night caucus. Why? To get defeated over one word in one resolution that made no sense standing alone? Meanwhile, the Administration will ineptly control in order to fight controls, just as it arms to bring about disarmament.

Actually, arming-in-order-to-disarm got some vigorous criticism at the conference—thanks largely to Senator John Culver of Iowa. Carter tried to signal his domestic and foreign priorities by attending just two workshops—the one on inflation and the one on disarmament. At the inflation workshop, he was lectured to by eight people: one question was directed elsewhere and, when the chairman asked for questions to the President, none were forthcoming. Before the arms control panel, Carter was asked three questions; but six questions went to others, mainly Culver, who made a very good case for the SALT treaty *without* the increased defense spending Carter's people think necessary to placate the right. Carter at last had to quibble with Culver's long and well-applauded statements. They were especially galling since the senator was Ted Kennedy's roommate at Harvard and served for a while as his legislative assistant. Culver stole Carter's own issue from him, forced Carter almost to attack his own disarmament goals.

The ranking of potential candidates, which had no formal place on the conference agenda, cannot be avoided wherever politicians gather. One thing Memphis did was reverse the perception of Carter held as recently as six months ago. Then it was said that he might have been a good candidate, but he was not much of a President. At Memphis, there was a grudging admission that he was holding his own as President, coupled with a wish (expressed even by some of his supporters) that he made a better candidate. Jerry Brown's people, in and around the California delegation, were clearly pleased by much of what they saw. (The Administration's indirect response was to tell anti-California jokes: Why does it take fifteen Californians to unscrew a lightbulb?—One to unscrew, and fourteen to *"share the experience."*)

John White kept representing the very existence of the Memphis conference as an act of great courage on Carter's part. (Actually, he could not avoid it.) But then, when White spoke against the budget resolution, he said the future of such conferences was at stake. In other words, it is courageous for the President to risk criticism; but if any criticism actually occurs, the midterm conference will be discontinued. Some courage.

White would like to discontinue such an expensive and risky undertaking in any event. Only the progressives were presumed to want the thing to go on; and one wonders why. They went away boasting of a moral victory because their big resolution lost by only 60 percent to 40. But what would the resolution have done if it were passed? And to get it on the floor at all, the progressives had to mute their voices, control their own hotheads, beg and borrow from the Administration. As I said at the outset, the most tangible gain from this whole exercise was for the White House team that got a trial run for the 1980 convention. Turf protecting is a fruitless exercise on the other guy's turf.

Then why gather at all? The answers I got from delegates who were glad to be in Memphis had little to do with national policy or presidential politics. Delegate A got appointed to the National Committee's urban panel. Delegate B met a useful party official. Delegate C pushed his local project with state politicians conveniently gathered in one caucus room or a few hotel suites. Delegate D was a city official anxious to keep in touch with party activists. These concrete motives for being in Memphis should make us pause at the analysts' lament over party decline and electoral apathy. America has always been a nation with low voter turnout, a function in part of size and heterogeneity. Our compromising, nonideological parties do not meet the crisis standard that musters an entire nation to the polls. The broadening of the electorate in recent years—admitting eighteen-year-olds, registering new groups of blacks, Chicanos, and Puerto Ricans—drives down the turnout statistics even when greater numbers vote. Easier registration is bound to include those less concerned or less conditioned to vote. America's semipermanent minority party puzzles some by its ability to survive continual defeat. The delegates' desire to be in Memphis may suggest that *both* parties are almost unkillable near their roots, where the party worker has special needs and ambitions to pursue. Even the minority party has tiny fiefdoms and enclaves with their own imperatives.

These minor forms of turf protection tend to strengthen turf protection at the top; so Carter weathered this conference as he has managed most of his crises, large and small, so far—the winner without the air of being a winner. Carter at midterm is at a stage that might be called congratulated peril or triumphant collapse. Camp David brought peace near, but not fully within grasp. China is almost recognized, at the price of having Taiwan almost abandoned (each claim is tempered with the other). SALT is barely concludable as a draft and almost unpassable as a treaty. Inflation is almost being fought, the budget almost cut. Through the veering winds of recent polls, Carter has set a tacking course; so his goals are sighted off somewhere to port or starboard, never straight ahead.

Still, the performance of the party's Left in Memphis makes one wonder what better course could be set. Put Kennedy in office, a liberal President with a conservative mood in the country—could he muster support for his health plan, for a SALT treaty? Carter does not *want* to fight inflation with controls on wages and prices. Kennedy probably *could* not even if he wanted to.

Carter remains a cipher in his own exciting times, at a pause of risk and opportunity, less mysterious than dim. In Memphis, as part of Sunday's full-court press over the budget resolution, Hamilton Jordan circulated in his respectable new blue suit, affable, chewing ice from a paper cup. When he threw his arm impulsively around a delegate's shoulder, his hand landed right before my eyes. The fingernails were bitten deep into worried indentations. A parable?

—*New York Review of Books,* February 8, 1979

VI

Presidents

The presidency needs a good deal of ignoring, which we deny it. We increase the things we lament by lavishing so much attention on these officials, as if everything in American life depended on the man in the Oval Office. But we cannot stop talking the presidency back into the center of things—I no less than you.

22

Truman

Trumania has taken over. Harry's icon was moved into Ford's Oval Office along with the Wolverine football. In December of 1974, Democrats went on regular pilgrimages to the Truman Library from their Kansas City "miniconvention." But Republicans have chosen Kansas City as the site of *their* convention next summer, and the pilgrimages will be even more fervent as Gerald Ford tries to prove, with Truman's help, that the dumb shall inherit the earth. Truman, the most partisan of men, has become a saint for all parties. He transcends not only party but the generation gap—old-timers yearn back with longing toward his brimstone speeches, and teenagers sing along with the rock group Chicago, "Harry, could you please come home?" He even bridges the ideology gap. William Shannon, in the New York *Times,* tries to rescue Harry from misuse by Republicans; and William Buckley, one of Truman's most scathing critics during his lifetime, discovered the uses of vulgarity in the Nixon years and tried to fashion a cult of Spiro Agnew as the resurrected Truman. The Truman religion has its bible as well as its shrine—Merle Miller's *Plain Speaking* is the sacred word; Ford left a copy in plain view when reporters came to see him. And the cult has a liturgy—the James Whitmore show, repeated in person or on film. Our revenant Harry would look very strange to those who knew him most of his years; but ours is exactly the image he labored to create by the end of his life.

Truman convinced himself, finally, that Jimmy Cagney had moved into the White House on the day FDR broke his promise to himself and died. To go by his memory of events, one would think that all he ever did as President was "give 'em hell"—not only Republicans and Congresses, but all kinds of "striped-pants boys." At the end of 1946, he called in his Secretary of State, James Byrnes, and blistered him for being too nice to the Russians: "I'm tired of babying the Soviets." Again, in 1946, Truman called Bernard Baruch into his office, when he thought Baruch was getting too big for his britches, and let him know "I had no intention of having him tell me what his job should be." Baruch should speak up only for "policy approved by me."

There is only one thing wrong with these accounts of how the little guy hitched up his pants with his wrists and told off the stuffed shirts: neither of the confrontations happened. Byrnes and Baruch both claim that they would have resigned on the spot if any such meetings had taken place, and ancillary evidence confirms their version. But these inventions are comparatively mild—they were advanced in Truman's notes and memoirs, where some documentary evidence controlled his imagination. To enter fully into the Truman fantasies, one needs to jolly the old man along, page after page, with Merle Miller. (Edmund Wilson said that Carl Sandburg's ministrations to President Lincoln picked up where those of John Wilkes Booth left off. In *Plain Speaking,* Mr. Miller takes up where Paul Hume left off.) Thanks to his sycophancy, we are taken into the mental workshop where Truman constructed all those mythical encounters with Goliath.

A good instance is Truman's first meeting with Douglas MacArthur. All the details are evinced—how the President's plane arrived over Wake Island at the same time as the general's; how the general moseyed about in midair, trying to make the President land first and greet *him;* how Truman personally ordered him out of the air; how MacArthur slipped off, on the ground, and let Truman's plane land out of his presence; how Truman stayed in the plane until he forced MacArthur out of his hole—"I wasn't going to have one of my generals embarrass the President of the United States." (One of *my* generals—they're *all* my helicopters, son.) How the general showed up forty-five minutes late for their first formal meeting and was tongue-lashed to his knees:

> When he walked in, I took one look at him, and I said, "Now you look here. I've come halfway across the world to meet you, but don't worry about that. I just want you to know I don't give a good goddamn what you do or think about Harry Truman, but don't you ever again keep your Commander in Chief waiting. Is that clear?" His face got as red as a beet, but he said—he indicated that he understood what I was talking about, and we went on from there. . . . He was just like a little puppy at that meeting. I don't know which was worse, the way he acted in public or the way he kissed my ass at that meeting.

It is a very satisfying morality play, but it bears no relationship to other accounts of the meeting (e.g., General Omar Bradley's). It is even at odds with Truman's own account in his memoirs. Since Truman and his ghost (William Hillman) had documentary memory nudgers in front of them while working on the memoirs, that account should be considered more authoritative. Mr. Miller likes to believe that Truman was being diplomatic in his earlier and less colorful account of most things, but if so, what happens to the famous candor and the refusal to indulge in tactful little deceptions? At any rate, Miller consciously decided not to confront the President with the disparity: "I wanted to ask Mr. Truman about that discrepancy and many others, but I never did. The reason I never did was that I wanted our conversations to continue." So much for the notion that Mr. Truman was just a reg'lar fella who did not put on airs. On the contrary, you could not question him even to the extent of quoting his own previous words without risking a haughty dismissal (and, in Mr. Miller's case, risking the prospect of losing a lucrative TV series). Miller likes to suggest that the conversations were just chatty confidings between friends; but here and elsewhere the real truth peeks out—that Truman rambled along in self-gratulation because the unwritten ground rules were clear to both participants in these interviews. To wit, that Miller would never ask a single challenging question, and, if he did, that Truman would walk out on him and never come back.

What really happened at Wake Island? Truman had been advised not to meet with MacArthur in Washington, since the general, right after the successful Inchon landing, was at the height of his fame and power. Turned loose in Washington, MacArthur would have been a journalists' delight, overshadowing a President low in the polls at the time. So Truman went all the way to Wake to consult with, and to pin another medal on, the hero. The trip looked like a pilgrimage—the very reason Dean Acheson refused to go along—but the President could control conditions on the island. He ordered MacArthur to show up without press aides but took along a planeload of writers and photographers himself (thirty-five in all)—plus a collection of "stars" to keep the journalists busy going and coming: General Bradley, Army Secretary Frank Pace, Philip Jessup, Averell Harriman, Dean Rusk. The party filled three Constellations, which were the only arriving planes in the air at 6 A.M. on the day of the meeting. General MacArthur had flown in the afternoon before and could not have been ordered out of the air. He did not slink off to a nearby building but came out of the one where he had slept, after being entertained the evening before by Admiral Radford. MacArthur did not greet the first two planes that landed; but neither did he keep the *Independence* waiting on the runway. When its door opened, he was at the ramp. He was not late to any of the three meetings (which took place in a span of five hours).

These details were supplied to the New York *Times,* after Miller's book came out, by Robert Sherrod, *Time*'s man on that trip; and they have been confirmed by Ambassador Jessup. MacArthur was not late to any meeting and was dealt with by Truman in the most deferential way. Vivid detail after vivid detail has been created to display Truman as he wanted to be, not as he was. ("Harry, all we get is *lies,*" wails the rock group Chicago.)

Well, say defenders of the Miller book, Truman was getting old by that time, and maybe his memory failed him. True, but the problem seems to be vigor of imagination rather than weakness of memory. His accounts of the Byrnes and Baruch meetings, more careful than the MacArthur story but heightened in exactly the same way, were confected long before senility set in. Besides, if his memory had gone so spectacularly bad, why pay any attention to his version of other events even more distant?

One might argue that Truman's story has a "symbolic" truth—he was trying to cut MacArthur down to size (hence the meeting at Wake rather than in Washington); and he may have sensed, this early, that he would have to break the general someday. In other words, he can forget all the large facts to achieve delicacy in small impressions. It must, indeed, be granted that Truman's basic desires were speaking here, rather than devotion to historical detail. But the Truman thus revealed stands a long way off from the cult figure that Harry has recently become. We are told that Truman was not a stickler for privilege or protocol—but he invents details to insist on his office and the deference owed to it. He felt that deference was denied him, and he had to go fight for it. We are told that Truman was a good "team player"—yet he gives no credit to his crew of stars and journalists for whatever deference MacArthur finally did show. No, all by himself, in person, by talking tough, Truman extracted an abject (ass-kissing) humility from the man who had defied him before the assembled luminaries. Truman would, in time, break MacArthur, for good reason, at a distance; but that is not enough for Truman's ego. He wants always to tell the big guys off in person. It is a fantasy constantly indulged, as when he imagines his meeting with Paul Hume, the music critic so disrespectful of a President that he told the truth about his daughter's voice: "I never met you, but if I do you'll need a new nose and plenty of beefsteak and perhaps a supporter below."

Eisenhower, too, affronted Truman's sense of his own dignity—by not getting out of his car to greet him on inaugural day when Eisenhower picked Truman up at the White House. But Truman says this is just because Eisenhower had been humiliated in one of those private meetings Truman loves to "remember." Truman reproached him for appearing with Senator Jenner, after Jenner had called General Marshall a front man for traitors: "I'd let him know in no uncertain terms how I felt, and I think he was ashamed." It is another of those encounters that Truman has created. Eisenhower met with Truman only once between the Jenner incident and the inauguration. This was on November 18, and all accounts show that Eisenhower came reluctantly and was uncommunicative, showing a "frozen grimness," according to Truman's *Memoirs,* "taciturn to the point of surliness," according to Dean Acheson. If Truman had dared upbraid him for the Jenner affair, or even refer to it, Eisenhower—far more than a Byrnes or a Baruch—would have stalked out. Truman, in his pre-Miller accounts, stresses how polite he was to Eisenhower, giving no excuse for surliness.

When Truman runs out of imagined conquests in his past, he makes up the kind he *might* have undertaken had he stayed in the White House longer. Eisenhower, whom he calls "no better than a coward," is attacked for losing Cuba (as Truman had been attacked for losing China): "When Castro came to power down in Cuba, Ike just sat on his ass and acted like, if he didn't notice what was going on down there, why, maybe Castro would go away or something." Poor Ike was not as gutsy as Truman, as good at facing down all the phonies: "Now when Castro came into power, if I'd been President, I'd have picked up the phone and called him direct in Havana. I wouldn't have gone through protocol or anything like that." Protocol only matters when it must be observed toward the President of the United States. Truman summons Castro to Washington, warns him against Russia, and offers American patronage. Castro, of course, agrees on the spot. But that is not enough. He is not yet kissing ass. "Well, he'd have thanked me, and we'd have talked awhile, and then as he got up to go, I'd have said to him, 'Now, Fidel, I've told you what we'll do for you. There's one thing you can do for me. Would you get a shave and a haircut and take a bath?'" Old Harry knew that old Fidel was just a hippie.

Ass kissing played a very large role in the diplomatic analyses of Mr.

Truman. It is the dream of the bullied man to reveal himself, at last, as capable of outbullying anyone. Truman had been often bullied—by Tom Pendergast, who got him elected when he was a thirty-eight-year-old failure; Pendergast, who practically made him his chauffeur at the 1936 convention and sometimes called him "my office boy." He was bullied by Roosevelt, who found his own Pendergast connection embarrassing: FDR received Truman late and coldly when he came to Washington and actively tried to defeat him when he came up for reelection (Truman retaliated by organizing delegates against Roosevelt's choice for Vice-President in 1940). When, in 1944, Roosevelt let Truman know through intermediaries that he could be the Vice-President, Truman stubbornly refused to give an answer until Roosevelt asked him in person. Roosevelt came through Chicago during the convention without seeing Truman—and still no answer. At last, Truman got the great man to call him on the telephone from his train.

When Truman did lash out, at last, it was in the unmistakable accents of the closet bully—e.g., to Paul Hume: "You sound like a frustrated man who never made a success." How dare you talk up to a man who made it, like me? It is the boasting of a man whose first real success, election as a county administrator by grace of a machine's backing, did not come until the age of thirty-eight. Most of Truman's early reading was of the "great man" variety. His favorite book was Plutarch's *Lives*, which he read as a do-it-yourself manual. Other boys of his era and region grew up with Horatio Alger dreams of becoming business magnates. He dreamed of the captains and the kings.

When Miller broached the idea of having Truman visit Hiroshima for a TV show on the atom bomb's development, Truman's response was instant and typical: "I'll go to Japan if that's what you want. But I won't kiss their ass." His world is neatly divided into kissers and kissees, and he means henceforth—in *fact*, at last, as well as fantasy—to be a kissee. Because he was for so long a kisser—the honest front man and patronage channeler—for Pendergast.

When Miller asked Truman whether he had any second thoughts about dropping the two nuclear bombs, Truman snapped back: "Never." Only the weak are allowed to doubt their own wisdom. The heroes in Plutarch were decisive. Self-doubt is for failures, for those who make no big success. When Robert Oppenheimer came to the White House to plead for international arms control, he said that those responsible for the bomb had blood on their hands. When he left, Truman said: "Don't ever bring that damn fool in here again. He didn't set that bomb off. I did. This kind of sniveling makes me sick." He counted it his proudest boast, at the end, that he had taught himself never to entertain the least doubt about himself. Miller, with a straight face: "Mr. President, it constantly amazes me that you seem always to know what is the right thing to do."

Where did he acquire this infallible wisdom? By reading books. Miller celebrates that fact as if reading were Truman's private discovery. He is even awed when Truman tells him he read three thousand books as a boy. (Come to think of it, that is a strange thing—not *reading* the books, if he did, but *counting* them.) There is no better argument for the admitted constraints of formal education than the stance of the self-educated. Franklin Roosevelt was a terrible student at Harvard College and even worse at Harvard Law School. He flunked courses; he did not even try to finish his law degree after getting safely by the bar exam. But the mere passage through Harvard made it unnecessary for him to prove, every day, that he could read without moving his lips. Truman was still proving himself to Miller late in his seventies. "Old Cicero—his full name was Marcus Tullius Cicero, you know." The revelation staggered Miller.

But Miller finds his great proof of Truman's scholarship in another book, not from direct interviews. Cabell Phillips claims that Truman told William Hillman, while he was preparing his papers: "The people around him [Alexander the Great] made him think he was immortal, and he found that thirty-three quarts of wine was too much for any man, and it killed him at Babylon." Hillman tried to check this story with the Library of Congress and at first found no link between wine and Alexander's death. But then a librarian found the detail in a book only checked out once in the preceding twenty years—by Harry Truman in 1939. This is used to prove Truman's "prodigious memory for historical detail."

Hillman, Phillips, and Miller are all bowled over by the exact reference to *thirty-three* quarts, which suggests that Truman's real genius was for finding the right kind of listener. The Alexander story comes from an ancient gossip, Ephippus, who said that Alexander was polished off by a three-quart cup of wine. The tale has no particular historic weight, even in its first form. In some later graphic (or mechanical) heightening, it has even less authority. All the story proves is that Truman had the self-educated man's inordinate awe for anything written in a book. One should remember that point when cultists praise Truman for his profound knowledge of history. Miller: "I've learned more American history during the time we've been talking than all the time I was in school, including college."

Including *college:* Miller really knows how to please a guy. Truman is the only President of this century not to have been graduated from college. He had to watch his boyhood sweetheart, Bess, go off to college, while he went to work. He later claimed that his wisdom came from plowing behind a mule from the age of twenty-two to thirty-two; but he dreamed for decades of repairing the omission forced on him by poverty. He enrolled at a commercial college; he lasted through two years of night school studying law; he hoped to attend Georgetown at night while acting as a senator by day. Only in later years did he turn his deprivation into an achievement. He claimed he got much more out of private reading than he would have under expert guidance—and I guess that is true. With guidance he would never have learned that Alexander died from thirty-three quarts of wine. "The thing I found out from reading was that there is damn little information in most schoolbooks that is worth a damn."

There was an inner demand in Harry Truman to prove that he was smarter than all his opponents. They were not merely wrong on this or that matter; they were ignoramuses who knew less than shrewd old Harry, the secret reader. Adlai Stevenson thought he knew a lot, but he was not so bright in the things that matter: "He was a very smart fella, but there were some things he just never got through his head and one of them was how to talk to people. . . . That fellow didn't know the first thing about campaigning, and he didn't learn anything either." In this respect, there was little to choose between Stevenson and Eisenhower: "Ike didn't know anything, and all the time he was in office he didn't learn a thing." MacArthur, of course, was "a dumb son of a bitch." Truman thought of President Kennedy as "the boy" and told Miller, when Attorney General Robert Kennedy visited his library, "I just don't like that boy, and I never will. . . . Now they say young Bobby has changed for the better. They never say anybody's changed for the worse. They say he's changed for the better, and maybe he has, but what I can never understand and never will if I live to be a hundred is why it takes so long these days for somebody to learn the difference between right and wrong."

Miller gives Truman high marks for despising Richard Nixon, but that seems an achievement within anybody's reach. What is surprising is Truman's ability to sustain contempt for people who were no longer any threat to him (if they were in the first place). It is not enough to say that Paul Hume was insufficiently deferential to a President's daughter. Truman must add, years later: "He didn't know a thing about music." Truman can think of nothing good to say about anyone who crossed him in any way. Or anything bad about those who helped him: "Tom Pendergast was a good and loyal friend, and he never asked me to do a dishonest deed, and so I went to his funeral." If Pendergast never asked anything dishonest of Truman, it did not matter what he instructed others (like Jim Farley) to do (like saving his murderer friend Johnny Lazia). One of the more interesting twists in the Truman legend is the way his attendance as Vice-President at a hood's funeral has become an act of virtue. No wonder Agnew's acolytes tried to claim he was a Truman redivivus.

When Truman took his toy "general" Harry Vaughn to Potsdam for the President's only meeting with Stalin (who reminded him of Pendergast), General Vaughn discovered the rate of exchange on the Berlin black market and traded the clothes out of his suitcase for "a couple thousand bucks." It was the best deal made at that conference. Vaughn was one of the many "Wimpys" (in I. F. Stone's phrase) who set the real tone around Truman's White House. It was always a wonder that Truman could live at home only in Pendergast atmospheres while staying uncorrupted—just as he loved the boozy Irish Catholic bores of his Battery D without sharing any of his underlings' flaws. Sinclair Lewis found in his George Babbitt a *nostalgie* for a *boue* he would not actually touch—so much that he felt more loyalty to his transient "fast" friends than to his business associates. Truman's friends called this loyalty. Dean Acheson probably saw the truth of the matter: he played incessantly on Truman's vanity. What mattered was how nice you were to Harry, and Acheson was invariably, courtierlike, nice. Truman was uncrossable, it's true. But he was minuetable to death.

Acheson had seen Byrnes and Baruch fall from the President's favor and watched General Marshall retain that favor by mere military bearing toward his Commander in Chief. Thenceforth, Acheson, who bristled at the haughty way Roosevelt had treated him, calibrated to a quarter inch the degree of butlerlike bow that Truman could accept, in fantasy, as equivalent to ass kissing. The rewards were great—they made Acheson, in Diane Clemens's phrase, the "de facto President for foreign affairs."

But how could this county-clerk type, with his swarm of cronies, adopt the overall view of Acheson, the silky Groton-Yale product with his British mustache and cultivated disdain? There is no mystery. Who was the next President to defer almost slavishly to expertise from "the best and the brightest"? It was Pedernales Lyndon, with his own brand of crony. Lyndon hated "you liberals" and stood in the shadow of his dead predecessor just as Truman had. But both were awed by their predecessors' "brain trusts." Truman could talk in the same envious populist rhetoric that Joseph McCarthy did, mocking "the striped-pants set." It was Truman, after all, who said of Adlai Stevenson: "The real trouble with Stevenson is that he's no better than a regular sissy." When Robert Alan Aurthur asked William Hillman, in 1960, why Truman hated John Kennedy so, the answer was Lyndon's own, caught in some time warp: "The boss doesn't like wealthy Northeast elitists." This made Truman, among other things, a perfect front man for the wealthy elitists who established the cold war, pursuing "enlightened" internationalism by stirring up the xenophobes. His hick air was as useful to them as his personal honesty had been to Pendergast.

The classic story about the commencement of the cold war is of Truman's meeting with isolationist Senator Vandenberg. Vandenberg was given the price tag for "rescuing" Greece, and he answered: "If that's what you want, there's only one way to get it. That is to make a personal appearance before Congress and scare hell out of the country." It is often forgotten that Vandenberg had been very truculent at that meeting until Acheson intervened to paint in vivid colors the danger of Communist "takeovers" around the world. Vandenberg knew how scaring hell out of the country would work because Acheson had just scared the hell out of him.

So they came, in fast sequence: the atomic-secrecy program extended into peace, the federal-employee loyalty program, the Attorney General's list, the establishment of the CIA, the State Department disloyalty firings, the alien-deportation and loyalty-passport programs, the revoking of Pentagon press credentials, J. Edgar Hoover's propaganda Freedom Train, the Smith Act prosecutions. Most of these things were deployed in the crucial year when they helped coax Congress along in implementation of the Truman Doctrine and formulation of the Marshall Plan. They all preceded the Hiss case and Russia's explosion of the bomb. Most of them preceded by three years the onslaught of Joseph McCarthy, and they prepared the way for it.

Truman's acquiescence in the schemes of the best and the brightest is far more understandable and forgivable than Lyndon Johnson's. Few doubted the words of the experts who had run our war industry, our secret nuclear program, our OSS and FBI at war. But the consequences of

Truman's actions were far more grave. He put the nation in a state of permanent war mobilization, with all the paraphernalia of secrecy, censorship, the draft, rush mobilization of new weapons systems (beginning with the hydrogen bomb), and crisis indoctrination. What is surprising is not so much that the excesses of McCarthyism occurred as that Truman is credited with opposing those excesses, to the exclusion of noticing that he supported the norm from which they arose. As Murray Kempton noticed long ago, "McCarthyism was only Trumanism carried to its logical conclusion." Attorney General Tom Clark, with all the power of the executive branch behind him, did as much harm, over the years, as Joseph McCarthy—and Truman rewarded *him* with a seat on the Supreme Court. Clark deported, prosecuted, censored, and agitated; he published the Attorney General's list, the most coercive use of government power against freedom of organization in that whole dark period. But he was loyal to Truman; he went out and plugged for the Marshall Plan on orders. And loyalty to his own person was Truman's highest test of public service.

It was always easy to buy off his ego, for those willing to do it. He quickly believed in myths handed him ready-made—like that of his own spontaneity and naturalness. Throughout his later years he loved to mock the Madison Avenue types who took over the presidency after he left. Stevenson and Eisenhower and Kennedy did not know how to talk straight out with their own words—they used experts to give them a cosmetic front. But Truman had undergone extensive coaching in the White House. Leonard Reinsch was called in from radio to slow down his rapid-fire speech patterns. When Truman read speeches, his mechanical lilt made them hard to understand. He could not correct this no matter how many hours he worked at it. Then Clark Clifford, Truman's Acheson for the 1948 campaign, gave up on that approach. He noticed that natural pitch patterns made Truman's extempore speech easier to understand, even though he spoke as fast as he did when reading. So Clifford prepared him to speak extempore, from notes, to the American Society of Newspaper Editors, and he rehearsed him to make it sound *more* spontaneous. That well-prepared "spontaneity" became the basis of Truman's new campaign style in 1948, elaborated by his coaches when he was already sixty-four years old. What was extraordinary was not Truman's naturalness, but his ability to learn, that late in life, a new public style.

He loved the act so well that he spent the next decade and a half perfecting it. We have heard a hundred times the tale of modest Harry going home to Independence, carrying his own suitcase up into the attic. No fleshpots of the Waldorf Towers for him, no great farm in Gettysburg. "And that was all there was to it," Truman liked to say as he ended the story. Well, not quite all. Truman was the first President to build his own shrine while he lived. The Truman Library was set up from the outset to be a show as well as a research center, and for a very good reason: Harry was often there to put on the show. There was a multimedia re-creation of his 1948 campaign, a cannon he could fondle from his Battery D days, his cars, his piano (he sat down and banged it). He devoted his declining years to a vulgar liturgy of self-adulation—mocking all later Presidents, praising his own humility, telling us by rote how spontaneous he was. Miller himself is our warrant for the fact that he had a standard patter for library tours and laughed just as hard at his own jokes every time he told them to himself. He was an Archie Rice whose audience would never desert him; and the routine was opportunely captured for print, proving Barnum right when he said Merle Millers are born every minute.

It was an appropriate ending to an improbable tale. In the flush of our postwar euphoria, one man held more formal power over the rest of the world than was ever held before or since. The man was a weird mix of Archie Rice and Babbitt. The results? You see them all around you.

—*Esquire,* January 1976

AFTERWORD

As a result of this article, President Eisenhower's brother sent me material disproving Truman's boast that he ordered Eisenhower, by way of General Marshall, not to divorce his wife. Later, Dr. Chester Pach showed me that Dean Acheson helped create the myth of Truman's humiliation of Eisenhower. Acheson's memoirs recount a meeting at the White House to consider sending a naval force to the eastern Mediterranean in 1946:

> General Eisenhower asked me in a whisper whether I had made it sufficiently clear that the course we had recommended could lead to war. Before I could answer, the President asked whether the general had anything to add. I repeated his question to me. The President took from a drawer of his desk a large map of the Middle East and eastern Mediterranean and asked us to gather around behind him. He then gave us a brief lecture on the strategic importance of the area and the extent to which we must be prepared to go to keep it free from Soviet domination. When he

finished, none of us doubted he understood fully all the implications of our recommendation.

The President had to teach the whispering Eisenhower the facts of strategic life. But Dr. Pach, in the course of his research, came across a memorandum to General Eisenhower recounting the decisions reached at the meeting Acheson describes. Why would Eisenhower have to be briefed on a meeting he had attended? This puzzle led Pach to check Eisenhower's datebooks. He found that the general was not at the White House—not even in Washington—on the occasion Acheson recreates in such detail. Eisenhower was in Mexico City.

23

Eisenhower

Other Presidents are described in melodramatic terms as a mixture of the lion and the fox. But Eisenhower?—just like the Herblock drawings of him, sappy amiable pigeon, with a straggle of happenstance ivy in his beak. Even his equivalent of Air Force One was stiltedly called the Columbine. The man's rubberoid features, totally without menace, might have graced any TV puppet's body on a wacky kids' show of the fifties. Granted, some lightning of idiot good luck touched everything he tried—poker, soldiering, writing, investing, campaigning, reigning, grinning. (His *grin* was an investment.) But he was the last person the journalists of his day would blame for his successes. You just can't beat the gods at their most whimsical. If they wanted to lift up everyone's kid-grandpa to personify the Age of Geritol, who could thwart them?

It seems almost a blasphemy against these clown-gods from the machine to suggest, as Murray Kempton first did with any impact, that Ike carefully wound up all the machines ahead of time, and then feigned surprise when they went off, blowing little trumpets all around him. There could be nothing devious in him—apart, that is, from his meandering way with sentences before the camera. Who can picture Ike, with his homespun boring rectitude, mouthing the Crookback's boasts?

> I can add colours to the chameleon,
> Change shapes with Proteus for advantages,
> And set the murderous Machiavel to school.

But, with the help of Herbert Parmet's thorough research on Eisenhower's eight years as President, consider:

It is 1956. Ike has already tried, four or five times, to push Nixon out of his job one rung down the succession to a cabinet post. Journalists, catching hints of such push and tug between the two, ask Eisenhower if Nixon will run with him again. They are given an absent shrug: "I have never talked to him under any circumstance as to what his future is to be." *Why, I can smile, and murder whiles I smile.* As time goes on, and others try to dump Nixon, it is suggested that Ike's silence is a blessing on their effort. He calculatingly blows up: "If anyone ever has the effrontery to come in and urge me to dump somebody that I respect as I do Vice-President Nixon, there will be more commotion around my office than you have noticed yet." *And wet my cheeks with artificial tears.* Then, having indulged one of his theatrical angers ("Never lose your temper except on purpose" was a favorite maxim), he circles back to give Nixon the heaviest blow of that season, suggesting he chart his own course. After the storm cloud, smiles—the murdering kind.

Then there was the Bricker amendment controversy. (The amendment would have required congressional approval of international agreements and imposed controls on the President's foreign-relations activities.) Eisenhower studied the problem, called judges and law professors to brief him, read up on Constitutional history, spent three and a half hours trying to draft an alternative with John Foster Dulles, put steel in Dulles when he thought of compromising at mere delay. After all this, he pretended at a press conference that he was willing to go along if more learned men approved—he was not a lawyer, he reminded us—but admitted that "as analyzed for me by the Secretary of State, [it] would, as I understand it, in certain ways restrict the authority that the President must have." Poor Dulles, as often happened, got blamed for holding a tough position where the general had stationed him, before skipping off to a safer distance himself. Ike, while mocking the amendment to friends as an attempt "to save the United States from Eleanor

Roosevelt," was so gracious to Bricker that the poor guy thought he had White House backing. *And cry content to that which grieves my heart.*

After saying he would not get down into the gutter with Joe McCarthy, Ike dispatched Nixon thither, as a man not above such sullying; using one odious fellow to dispose of another. This was particularly subtle of him, for it made Nixon kick at his own constituency, using up prior credit in what looked, from the Oval Office, like a job of mutual destruction. *And, like a Sinon, take another Troy.*

Meanwhile, Ike himself would appeal to Nixon's right-wing constituency with more solidly based anti-Communist charges. He gave his Attorney General, Herbert Brownell, the go-ahead on accusations that Truman violated security procedures when he appointed Harry Dexter White to the International Monetary Fund. But once again, at a press conference, he barely seemed to know what was up, and assured the nation that his cabinet members were free to follow their individual consciences. (Just let a lieutenant try that one on the general back at headquarters.) When Len Hall, the party chairman, said this matter would help Republicans come election time, Ike serenely hoped in public that divisive issues such as this would not affect our politics. *And frame my face to all occasions.*

Yet for all this evidence, large and little, it is still impossible to think of him as a "Machiavel":

> Can I do this, and cannot get a crown?
> Tut, were it farther off, I'll pluck it down.

Tut—it is an Eisenhower word; and he got every crown he reached for; but we cannot think he had the conscious duplicity of Richard III, however washed off and turned to beneficent uses. He was so open and unblushing as he said now one thing, now another (as in the U-2 affair); he seemed above suspicion because below contrivance. We know, now, that he did contrive, more than his contemporaries thought; but always without seeming to, without adverting to his own best machinations. Duplicity of method, where the aim was clear, seemed nature's obvious blessing to him, no more odd than breathing through two nostrils. Kempton caught on, a full decade ago, when he realized with mixed admiration and horror that "Eisenhower has no sense of sin."

The truly perspicuous Machiavellian will not seem one at all. When Ike said religion was good for men, and he did not care what kind, the sheerly expedient came out sounding like the vapidly pious. He was duplicitous by instinct, not on principle—which put him one up on the author of *The Prince.* The Compleat Machiavellian not only will not confess deviousness, but will not think or theorize about it. That is why Machiavelli himself was a failed Machiavellian; he invented ways of succeeding after he had failed. Eisenhower could put this Machiavel to school, because he was not Machiavelli. He never wasted time reconciling what he did with strict or lenient ethical standards. He was too busy succeeding by his devious obliviousness. If that takes from him the self-conscious plotting air of Richard, perhaps the jollier Spanish schemer, amoral but serving his country, speaks for him in Middleton's *A Game at Chesse:*

> And what I have done, I have done facetiously;
> Abused all my believers with delight—
> They took a comfort to be cozened by me.

—New York Times, October 22, 1972

24

Johnson

Lyndon Johnson liked to have protégés around him. He had risen as the client of Sam Rayburn, and he seemed always to wish for as bright and loyal a disciple as he had been to "Mr. Sam." Some of his protégés became famous, one way or another—Bobby Baker, say, or Bill Moyers. Most did not. Yet he stayed on the lookout for young people to impress and promote—he made two White House Fellows his speechwriters: Tom Johnson and Charles Maguire. He also made it a practice, from early days, to woo or steal talent from his rivals. Even as a guest in some other politician's home, Johnson would openly try to hire the man's best aides away from him. It was often easier for the quarry to give in at the outset, since Johnson could wear down almost any personal resistance. The man who could persuade Arthur Goldberg to step down from the Supreme Court for a job like that of UN ambassador could obviously mesmerize us lesser folk into jumping off skyscrapers.

How did he do it? Booth Mooney, himself a speechwriter won over from the staff of Johnson's opponent in his first Senate race, describes one such campaign. Gerald Siegel was trying to leave Johnson's staff, and the senator offered him a job at the family TV station in Texas, to keep him "on the team":

> Siegel and his charming wife, Helene, went down to Austin on an inspection tour. They were accorded a typically Johnsonian grand reception, introduced to the Jewish community of Austin, and made to feel in general that the good life awaited them. As they were driven by car around the pleasant city, LBJ enthusiastically showed them the section where he knew they would want to live. He named clubs which they should join. He told them what kind of boat they would need for use on the several adjacent lakes.

(In fact Siegel got caught up in committee work, and stayed in Washington another year before going to the Harvard Business School.)

It is too often said that Johnson bullied people. He did, but by an unremitting blast of personal attention rather than by threats or surliness. Indeed, one of his most effective tactics was an abject profession of personal need. He could not go *on* without the services of the targeted individual. His very *life* depended on it—along with other things like national security and the fate of Western civilization. The object of his attentions was pummeled with hyperbole. James Rowe was recruited by the lavish shedding of real tears. Journalists were promised, "I'll leak to you like a dog on a hydrant." When one thing did not work, Johnson would try ten others—appeals to greed, duty, vanity, God. He knew there must be *something* that would work, and he would go through his whole repertoire and then start over again, if he had to. Few had the energy to outlast him.

Some, of course, did not put up much of a struggle; and Doris Kearns seems one of those. He *needed* her in Austin, to help write his memoirs. It would further her career. When she objected that she had to study, do social work, meet different people, he promised her hot-and-cold running scholars, poor people, and millionaire bachelors at the turn of a spigot. It was the old routine; but Ms. Kearns presents it as something peculiar to her and her situation. Did she really not know that was the standard Johnson pitch?

Take, for instance, the morning conferences in her bedroom—Johnson knocking shyly at her door, then crawling under the covers as she retired to a windowsill. The relation suggested sounds odd, if not positively loony. Kearns interprets her role as that of a surrogate mother, fending off dirty-minded psychologizing with the simple-minded kind. But Johnson grappled people to his body heat in multiple real and symbolic ways. He liked to talk to people while sitting "on the can," lying in bed, on his rubbing table, skinny-dipping. He undressed and got under the covers while arguing with Kay Graham in the presidential bedroom. Long before he showed off his gallbladder incision to photographers—in 1958, to be exact—he dropped his pajama pants before journalists (including a woman) to show how much weight he had lost after his heart attack. The Johnson family gatherings under bedclothes became something of a joke in the sixties, Papa Bear bundling with all his "Birds."

He wanted aides at his bedside the minute he awoke, and trailed people up with him at nap time, to be dismissed only after he was bedded and getting drowsy. Aides fumbled toilet paper between state documents. Ambassadors were received between towelings. The Johnson who needed three television sets and two tickers going all the time felt he should dictate the nation's course while defecating, or win a congressman's vote while he took a shower. It is a measure of his vast energy and ego, and of his fear that he would lose contact with the world if he lost its attention even for an instant.

The Mooney tale of Gerald Siegel's wooing contains one very revealing episode. Johnson was at the wheel of his car, ready to drive the Siegels on the customary breakneck tour of his ranch, but Mrs. Siegel had not appeared. Johnson fidgeted, honked repeatedly, let impatience undermine his diplomatic efforts. Finally Siegel explained that his wife was in the bathroom. "Well, hell, if she's not ready, we'll go without her," said Johnson, making the wheels kick dirt. Johnson's ever-open bathroom door, his state conferences on that shabby throne, seem intended to prevent anyone from leaving without him. If the door closed, and he were left alone, he might disappear down the hole.

Johnson practiced social intercourse as a kind of full-court press. He liked to hire both husband and wife, so there would be no family sanctum unreachable by him. His intrusion into the private lives of those associated with him had certain pleasant aspects. He showered them with gifts, large and small—a free operation or a gross or two of presidential cuff links. (He gave Ms. Kearns electric toothbrushes by the dozen.) But he also spied on people, went through their desks after they went home at night, opened letters not meant for him. George Reedy's press briefings were more ragged than they had to be because LBJ listened in on a remote speaker, and once even phoned the press secretary in mid-session to correct what had just been said.

Johnson aspired to total contact with any people he dealt with—knowing them, using them, writing happy endings to their sorrows, punishing them for the "disloyalty" of any response to him less urgent than his own assault. He would laugh, cry, dance, or destroy in order to keep their attention. His tales and "pitch" would change to fit their moods as well as his—what is "truth" in such a windstorm of flattery, bluff, ingratiation, intimidation? The *reality* was *contact,* and words had to serve that purpose alone. Having paid for attention with his own tears and abject pleas, he found ways to punish those who witnessed such a spectacle. He was randomly cruel—just another part of his act; it certainly caught people's attention. He mocked his wife before others, and subjected friends to a long obstacle course of hardships—were they *really* friends? He was always testing.

These well-established traits are manifested in what seems the improbable story of a President who recruited a young woman graduate student, met while she was a White House Fellow, to follow him into retirement and help write his memoirs. He tried to win Ms. Kearns as one of his opponents ("you Harvards"), to earn her sympathy, pity, admiration, devotion—*attention,* of *some* sort, goddammit! The pitch that seemed to work on her was one he saved for feminine targets—the tale of a mama's boy forced by his father to deny his own sensitivity for fear of being thought a sissy; Ferdinand the Bull, huffing and puffing on demand, but really in love with flowers. No doubt there was a strong element of truth in this tale, as in most of the routines he used for wooing others. What is astonishing is that he never had to use any other pitch to keep Ms. Kearns coming back for more and more notes on life behind his mother's skirt. The resulting huge book is written to a background of Johnson singing "Mammy" for hours and hours to little Doris.

The typescript of the book's prologue, sent out to reviewers, had no mention at all of Mrs. Johnson, of the strange desert flower that did not wilt under the lifelong blast of Johnson's love and cruelty. But handwritten late inserts scattered Lady Bird's name dutifully through the text, and are incorporated in the book. The changes may have been a precaution against prurient interpretation of the long hours Ms. Kearns spent alone with Johnson—we are now given the impression that she entered the family circle, sitting around and chatting with the daughters. But no doubt the first draft better reflects the reality. Johnson did not want a distracting audience present when he indulged in creative persuasion. Lady Bird had no doubt heard many variants of the Mammy song, used in more literal wooing contexts over the years. It always embarrassed Johnson to have witnesses from his past on hand while he was "remembering" that past—as when pointing to any handy log cabin near the Pedernales as his birthplace. Once his mother was standing by, and remarked indignantly that he was born in a *nice* house, not *that* thing. Real mothers inhibit the idealizations of the Mammy song—much as Presidents behind the arras can inhibit press secretaries.

Ms. Kearns admits, in a protective way, that Johnson took a fictive approach to his own past—calling up relatives who seem to have died like flies in every Texas battle since the Alamo, or expanding his own "war service" to fit different contexts. But Kearns guards her treasure trove of notes by saying that the particular *form* of invention he settled on was indicative. True enough. But indicative, among other things, of what he thought would impress her. Vast areas of the Johnson psyche are missing from her book—the shrewd and bluffing masculine side, obscene and voracious and game-playing—because he did not think that would "play" in Cambridge. Johnson gave of himself selectively, always expecting more in return than anything he had surrendered.

—*New York Review of Books,* June 24, 1976

25

Ford

For a while, it was enough not to be Nixon. We did not expect Ford to be perfect; just not to tell *unnecessary* lies. There was no economy to Nixon's lying. He added the pointless little excess—like falsifying his return date from Moscow, to suggest he missed the countdown toward a Democratic National Committee break-in. Nixon offended as much by being maladroit as by being malevolent. Ford's have been good workmanlike lies, the kind we allow politicians when telling the truth would embarrass them.

At his confirmation hearings for the vice-presidency, Ford had to squirm through long questioning under oath about his attempt to impeach Justice Douglas—yet he only told a couple of direct lies. It was a refreshing change from the Nixon performance. True, he said, "My action was totally independent of anything that happened in the Senate other than the coincidence that Justice Douglas had a somewhat similar arrangement with the Parvin Foundation to that of Justice Fortas with the Wolfson Family Foundation." He meant that his action was "totally independent" of the Senate's rejection of Clement Haynsworth for the Supreme Court—and that was one of his lies.

Ford rested his argument on chronology; which is what undid him. Ford said that he began his own investigation into the possibility of impeachment "sometime in the summer of 1969," after Justice Fortas resigned over the Wolfson fees: "it was sometime shortly after the Fortas matter." Since Fortas resigned on May 14, 1969, even an investigation launched in the dog days of a Washington summer was off to a sluggish start. Ford was just trying to establish that his work on impeachment began before September 3, 1969, when Haynsworth was nominated by President Nixon—and *well* before November 21, when Haynsworth was rejected. One cannot make the *post hoc, ergo propter hoc* argument if Ford began his effort *ante hoc*.

For most congressmen, "starting my own investigation" means one of two things: (1) a bit of information has been leaked or volunteered, which he wants to use, or (2) he wants to make a speech and has assigned an aide to start collecting information in order to draft it. Ford claimed that he began with (2), with an assignment to Robert Hartmann that he look into this subject, and only moved back to (1) when, shortly afterward, he asked the Justice Department for help in his independent investigative effort.

That, in itself, would be strange enough. Congress usually has an adversary relationship with the Justice Department. Why would an Attorney General leak raw FBI files to help one lowly congressman impugn the honor of a Supreme Court justice? But that is just what John Mitchell did in this case, according to Ford's sworn testimony before his fellows. Mitchell said he would look in the files and then suggest promising "areas" for Ford to investigate. In line with this, Mitchell sent Will Wilson (shortly afterward forced out of the department under shadow of a scandal back in Texas) with what Ford kept referring to as "blank papers," giving topic heads for Ford's staff to investigate. Again, the story is odd to the point of zaniness if one keeps trying to believe it. The Justice Department cared enough for the congressman, and so little for the Supreme Court justice, that it tried to help along an impeachment effort by hints and nudges; but, at the same time, it cared so little for the effort's success that it would not back up those hints with *information* gleaned from its own files.

But here is where chronology began to trip him up. Congressman Waldie of California asked Ford to check his appointments book and establish the date when Will Wilson brought him the "blank papers." Ford, promising candor after Nixon's resistance to subpoenas, came back with the information. Wilson had come to his office on December 12, 1969. Take a quick run-through of the dates again—May 14, Fortas resigns; sometime, vaguely in summer, Ford claims that he launched an independent investigation; September 3, 1969, Haynsworth is submitted to the Senate for confirmation; November 21, Haynsworth is rejected; December 12, three weeks later, Wilson arrives with the dirt on Douglas. The dates tilt and bunch around fall, not spring (when Fortas resigned), or summer.

Previously, Ford claimed that he called Mitchell at the *beginning* of his investigation. So, "shortly after" Fortas resigned, Ford asked Mitchell for a tip on "areas" he might look into. Mitchell promised his help. Then, while poor Ford waited, all the time wondering *what* to investigate, Mitchell sat on the request for six months. Even if we labor to stick to Ford's story, we must wonder why Ford didn't wonder at this sudden unearthing of material asked for half a year ago, in the aftermath of Haynsworth's rejection.

But we do not have to labor at Ford's original chronology—he was ready, by this time, to abandon it. He no longer called Mitchell "shortly after" the Fortas resignation. He did not even dawdle toward the telephone that "summer" (thus beating the September 3 nomination date). Now he thinks he called Mitchell in the fall. When he has to, he can change his story on the stand. Those who think Jerry Ford "too dumb to walk and fart at the same time" (Lyndon Johnson's unexpurgated phrase) need only read the confirmation hearings to see he can maneuver inch by inch to save his skin. While he could no longer say, very credibly, that he asked Mitchell in the summer for things he got in December, he keeps insisting on a date in the fall—i.e., before winter: "I would assume sometime in October." That is: a full two months before the Wilson material arrived (rather than six months earlier)—but still before Haynsworth was rejected. The canny old gum-chewer knows better than to multiply lies needlessly. So his tactic at this point was one of volunteered ignorance. He entertained confusion, hoping it would prove contagious:

Representative Waldie: So that there is no possibility that you could have in fact called Mr. Mitchell subsequent to November 21, 1969, the date that Mr. Haynsworth was rejected.

Ford: Would you repeat that again, please?

Waldie: There is no possibility that you could have called Mr. Mitchell subsequent to November 21, 1969, the date Mr. Haynsworth was rejected by the Senate?

Ford: It is possible that I did contact Mr. Mitchell prior to that time.

Waldie: No, no; I said subsequent to that time.

Ford: What I said was it is possible that I could have called Mr. Mitchell prior to that time, and—

Waldie: But what I asked, is it possible you could have called Mr. Mitchell subsequent to November 21, 1969, to set up the appointment with Mr. Wilson with you on December 12, 1969?

Ford: It is possible, but it is more likely that it was sometime in October, prior to November 21.

It takes an almost magic inoffensiveness to get away with that. Most committee members congratulated Ford on his candor at the end of these hearings. We approach the man's real genius.

Waldie, after brilliant questioning, concluded: "I believe the testimony confirms that in fact your call to your Attorney General was subsequent to November 21, 1969, the date the Senate rejected Mr. Haynsworth. There is a controversy in your statement and my belief. I believe that Mr. Mitchell was acting on behalf of the President and I believe further that you were aware of that, and that what was in fact occurring here was a political act on the part of the Department of Justice to assist in removal of a Supreme Court justice." In other words, Congressman Waldie thought that Congressman Ford had been lying under oath—and then Congressman Waldie voted to confirm Congressman Ford as Vice-President of the United States. I later asked him why. "Anything was better, at that point, than Richard Nixon in the White House." It was enough.

Return, for a moment, to those "blank papers" Congressman Ford got from the Justice Department. The papers were far from blank. One of them ran to two full pages of single-spaced typewritten summary, with facts and dates crowded together. Why "blank," then? *Because they lacked identifying letterhead or department stationery.* This was uppermost in Ford's mind, since he did not want to be put in the position of knowing he had leaked raw FBI files.

Ford had another line of defense that he clung to against the odds— that the many factual allegations in his speech against Justice Douglas were based on independent investigation by his office, and not on the "blank papers." He insisted on this with great urgency: "They gave me no factual information. I made some subsequent investigation." That was another lie. Under Waldie's questioning, Ford's office could turn up no record of investigative work beyond the "blank papers" themselves. Ford could name no "investigators" other than Robert Hartmann, who drafted the speech, "and myself." FBI stuff was incorporated almost verbatim in Hartmann's speech, errors and all, with no corrections or confirmation. The "blank papers," which Ford said gave him "no factual information," supplied all the apparent "facts" about the Parvin Foundation's ties to organized crime. Here is Hartmann's speech, as given by Ford.

In January 1963 the Albert Parvin Foundation decided to drop all its Latin American projects and to concentrate on the Dominican Republic. Douglas described President-elect Bosch as an old friend.

And here is the FBI source sheet:

> January, 1963. Albert Parvin Foundation decided to abandon all projects in Latin America not related to the Dominican Republic. . . . Douglas called Bosch an old friend.

It is not hard to write speeches—or, as Ford put it, "conduct investigations"—by that method. Here is another sample of Hartmann's work:

> With the change of political regimes the rich gambling concessions of the Dominican Republic were up for grabs. . . . This brought such known gambling figures as Parvin and Levinson, Angelo Bruno and John Simone, Joseph Sicarelli, Eugene Pozo, Santa Trafficante, Jr., Louis Levinson, Leslie Earl Kruse, and Sam Giancanno to the island in the spring of 1963.

And here is his source:

> There have been indications from time to time of other Mafia figures who have moved in and out of the gambling establishments in the Dominican Republic, including Joseph Sicareli of New Jersey, Eugene Pozo of Florida, and Santa Trafficante, Jr., of Florida. In the spring of 1963 Louis Levinson was negotiating with the Dominican Republic, as well as Leslie Earl Kruse of Chicago and also Sam Giancanna of Chicago.

Hartmann, compressing while copying his source, just strings out the names (Angelo Bruno was brought in from the immediately preceding sentence), not bothering to notice he has put Levinson in twice—and he adds Parvin just to keep the tenuous tie with Douglas. There have been many arguments against the leaking of raw FBI files to the public, and Ford's own leaking act confirms them all. Meanwhile, his doughty "investigator," Robert Hartmann, has become White House counsel, one of the principal advisers to the President of the United States.

Accompanying Ford to his hearing was a young lawyer named Benton L. Becker—he was seen conferring with Ford over several documents just before Ford's own testimony. Mr. Becker has had a wonderful history, hovering in the wings of several big stories, vivid yet peripheral, evasive even as he elbows his way in. He served three years in the Justice Department, and got to know about congressional investigations when he helped prepare a criminal case against Representative Adam Clayton Powell. In 1970, he left the executive branch, was hired as a staff member on Joe Waggonner's congressional staff, and also opened a private practice in Kensington, Maryland.

These activities were not enough to keep Mr. Becker busy—he also conferred extensively with Robert Hartmann on ways to keep the drive against Justice Douglas alive. Ford had given Hartmann's speech on

April 15, 1970, after Nixon's second Court nomination had been rejected. On September 3, 1970, Becker wrote a letter to the lawyer for Louis Wolfson, the convicted financier who had been associated with Albert Parvin. Becker, looking for Parvin material that might hurt Justice Douglas, asked for an interview with Wolfson, and promised "to assist him in any way available to me."

Becker claimed to be writing not only for himself but for "my clients, Congressmen Ford, Wyman, Waggonner, and others," who also expressed sympathy with Wolfson. Ford and Wyman both denied that Becker was their lawyer at the time; and working on Waggonner's staff did not seem to qualify him as Waggonner's lawyer, either. Representative Drinan put it best: "My legislative assistant does not write letters saying that he represents Congressman Drinan, and I do not think anybody on any staff of any congressman should say that they represent, that they have been retained; that is not really true. You were not retained by those congressmen. You were on the government federal payroll." By the time of Ford's hearings, Becker had called up Dr. Arnold Hutschnecker, the man involved in some of Robert Winter-Berger's allegations against Ford; and Becker again represented himself as Ford's lawyer, though he later admitted he was not. One would think Ford would learn, in Becker's case as in Winter-Berger's, to avoid those who misrepresent themselves as Ford's agents. But Ford showed up with Becker at the hearings, and had more business in store for him.

What kind of assistance did Becker's letter offer to Wolfson's lawyer, in the name of congressional "clients"? The lawyer, William O. Bittman, says that Becker, when he telephoned him out of the blue, offered certain information that would be of use. Asked to put his offer in writing, Becker composed the letter already referred to. In response to that, Bittman wrote: "When we spoke over the telephone, you indicated that your clients had some information and might well be in a position to help Mr. Wolfson in his present situation." Becker denied making any such offer, and said that he was calling merely to offer the "help" of an old friend from the Justice Department.

But Mr. Bittman does not remember knowing Becker in the department, and Congresswoman Holtzman pointed out that the eagerness to renew old ties is belied by the fact that he did not respond to Bittman's request by supplying him with any information. In an interview with Waldie, Becker claimed his offer was to donate his personal services to the Wolfson legal team, preparing briefs and the like (nothing at all can fill up Becker's busy schedule). Since Bittman was a member in a very large firm, it seemed unlikely that he would need a volunteer brief-writer. Mr. Becker lamely amended his suggestion to an offer "including but not limited to brief-writing or anything else that an attorney might do."

Floundering deeper all the time, Becker now thought he could become credible by admitting to selfish motives. He wanted to see if there was any business Bittman's firm could throw his way:

Representative George Danielson: Was it your motivation to assist Mr. Bittman *pro bono publico,* or did you expect to get something back for it?

Becker: What I expected, frankly, if anything, was simply to establish a relationship between a Washington firm and a Maryland firm.

Danielson: Where is Kensington, Maryland, how far from Washington?

Becker: It is in Montgomery County.

Danielson: Would it be over a hundred miles?

Becker: Oh, no, sir.

Danielson: Would it be two hundred?

Becker: Did you say over two hundred?

Danielson: Where is it, how far from Washington?

Becker: Maybe ten, maybe fifteen miles.

Danielson: Commuting distance?

Becker: Oh, yes.

Danielson: I note on the letterhead, which is kind of blurred, that Mr. Bittman's firm had sixty-nine lawyers as of the time this letter was written. What help do you suppose they would need in Kensington, Maryland?

Mr. Becker did not make a very good witness. He soon reached an advanced state of dither:

Well, wait, offering that to Mr. Bittman. What I was doing to Mr. Bittman, I felt, was simply—and this is, I should add, by way of reconstruction, when I think back this past weekend and looking at all these letters. What I think I was simply doing is simply letting Mr. Bittman know that a former colleague, what he was doing and where he was doing it, lest a possibility ever came up for some co-counsel relationship or something of that nature. I wasn't offering myself in that way.

But members of the committee could dismiss Becker, like Winter-Berger, as one who used Ford's name without his knowledge. Father Drinan seemed to speak for a number of those on the committee when he said: "I am afraid, sir, in my judgment, you seem to become more implausible at every moment. . . . I believe Mr. Bittman [about Becker's offer to give some government information or aid the Wolfson case], and I cannot accept or really believe your interpretation of your letter or the letter of Mr. Bittman."

After this unpleasant time with the committee, Becker continued to be controversial. He joined the law firm of ex-Representative William Cramer, an old Ford buddy, and became counsel for several figures prosecuted in Maryland scandals. One of these figures, Joel Kline, testified that he directed Mr. Becker to lie to federal prosecutors. He also said that Becker transmitted the false information to the prosecutors, and wanted Kline to pull the strings that would make him a Montgomery County judge. Becker denies the cooperation in perjury but admits that he asked Mr. Kline if he could help him secure a judgeship. At the same time, in 1972, Kline's partner Eric Baer admitted to perjury and said that Becker, acting as his lawyer, had known that he was committing perjury. By that time the prosecutors let journalists know they were interested in investigating Becker.

Meanwhile, what had happened to Becker's relationship with Gerald Ford? That interested Father Drinan at the time of the confirmation hearings.

Drinan: And you have had no transactions with Mr. Ford's office from the termination of the Douglas matter until his nomination as Vice-President?

Becker: No transactions. I have met him, I have seen him.

Drinan: You have not represented him?

Becker: No, sir.

Drinan: Would you say that you did say to the doctor that you represented him, that you were his lawyer, because we have a doctor here and he got the clear impression that you represented him.

Becker: Yes, I believe I testified to that.

Drinan: Represent him in what sense?

Becker: Mr. Cramer, Mr. Haber, and I are doing anything that we can to assist Mr. Ford with the confirmation proceedings.

It seems the kind of help Ford could do without. Yet when the news of Ford's secret transition team emerged, who but Benton Becker was among those planning for the takeover? Perhaps, one could add, Mr. Ford knew nothing of Becker's activity, since the transition team tried to leave Ford plenty of room for "deniability" about plans to become President while he was still Vice-President. But that will not explain the next, most surprising thing of all—that when Ford plotted the pardoning of Richard Nixon, working with only a few intimates, the man who commuted to San Clemente with the pardon in his briefcase was: Benton L. Becker! And Becker worked out the arrangement that would have given Nixon access to his papers.

Becker first shows up, in Jerald terHorst's biography of Ford, on page 229—in the last chapter, on the Nixon pardon. To judge from that book, you would think that Ford had no earlier dealings with Becker, no reason to wonder at his odd ways:

> On Saturday night I [Mr. terHorst] had been given a copy of a tentative draft of the Nixon statement that Benton Becker, a Washington attorney working on special assignment as Ford's emissary, had brought back from San Clemente the previous day. But we had no assurance that Nixon would follow that text to the letter, so I delayed reproducing it for the news media. It was a wise decision. Checking with Becker about thirty minutes before the President spoke, I learned that Nixon was doing what he had so often done with personal statements and speeches while in the White House: he was revising it up to the last moment. Becker advised me to expect some changes.

Perhaps the oddest aspect of the Nixon pardon was the use of this intermediary to conduct the delicate bargaining. The most interesting thing is that, if Becker is to be believed, Nixon told him, after concluding the negotiations: "You are a fine young man. I wish I'd had young men like you around me." Becker passes the Nixon test—which, if nothing else did, ought to give Gerald Ford second thoughts about his agent. Becker's relationship to Ford, through Robert Hartmann, seems to be that of a crony-of-a-crony. This suggests the real resemblance between Ford and that odd new idol for Republicans, Harry Truman. Hartmann stands in the noble line of Harry Vaughn, the embarrassing old pal who likes to speak for his boss. As Richard Reeves put it, "Bob Hartmann was the king of Gerald Ford's hill—and it was a pretty low hill."

—*New York Review of Books,* October 16, 1975

26

Carter

South Georgia is a mean and starved back corner of America. Union prisoners were shunted down to Sumter County, in 1864, to get them out of the way. At Andersonville, between 200 and 300 men died every week for over a year; they were lined up like dried giant peanuts in long ditches, the mass graves tended on neat lawns now by the Park Service.

Less than twenty miles from Andersonville, where two roads crossed, a cluster of huts was called, in Civil War times, "Plains"—after the Bible's Plain of Dura, where Shadrach, Meshach, and Abednego underwent their trial in the fiery furnace. It is a place that still bakes men in religion and other fires—though the town shuffled over about a mile at the end of the last century, adhering to the railroad track that finally came through.

It is a place of small moves and tenacious names. The village of Andersonville turns dustily on a marble axis—the monument native ladies put up to Captain Henry Wirz, the prison's second-in-command, who was "martyred" by Yankee execution after the war. "Ever been to South Georgia?" Governor Carter asked me in the spring of 1972. No, I answered—though I had told him I was born in Atlanta, my mother's hometown. "South Georgia is different," he said, with that amused air of measurement he assumes when showing outsiders the rigors and bias of his native turf. It is land oddly open yet stifling—a tropic sun without tropical foliage. It was lack of fresh running water that took the deepest toll of prisoners at Andersonville.

We were flying south from Atlanta in the governor's little plane, where Carter had to address a meeting of Georgia sheriffs. He was billed less as Shadrach at the Plain of Dura than as David entering the lions' den—but he went away to a standing ovation. On the flight back, Carter's eyes mocked and measured, with his smile's probing mildness, the "outsider" he was playing to. Did I get the point? He likes to show how he can dove his way in among serpents, back where native gods and blood count finally. Up in that tilt of little plane he expertly poised his city skills against country ones, aware of the contrast, using each as foil to the other. We swooped down for an Antaeus-touch at Plains itself, to pick up some mail. From the air the "town" looked like a Western movie set—one row of storefronts with a walk in front (just the kind for stunt men to whump down on, a whiff of lace curtain marking their takeoff point in the second-story window). It was only a glimpse; we took off right away, and talked again of Wallace, McGovern, and 1972.

I saw Carter next in Miami, at the Democratic convention, where he spoke for Henry Jackson's candidacy. He told me he liked what I had written about our Albany journey—all but my statement that he was obviously angling for the vice-presidency that year. "Well, aren't you?" I asked. He smiled his Dinah Shore smile, more enigmatic than the Gioconda's *because* it is so wide and open. Even then he could not tell a *direct* lie. When I saw him later, in Atlanta and at the 1974 Kansas City miniconvention, I was not surprised that he was aiming at bigger things than the vice-presidency. It was cheeky, but no more than Ol' Gene Talmadge's national barnstorming out of McRae in 1936. But I was astonished, like almost everyone else, when he rose so fast in the early primaries of 1976. This was explained, by Carter among others, as a sign of the New South's power. But there was a lot of the Old South at work in the phenomenon as well. So now I am going back to Plains, this time by car, in an effort to understand the Carter paradoxes—the ferocious tenderness, the detached intimacy, the cooing which nonetheless suggests a proximity of lions.

I began in Atlanta, talking with Hamilton Jordan, Carter's young campaign manager—he was even younger (twenty-five) when he ran Carter's successful bid for governor in 1970. Back in 1966, when Carter failed as a candidate for governor, Jordan was still in college, a Sumter County volunteer for Carter. Now he says: "It was good Carter lost in 1966. The country would not have been ready for him in 1972." There is the easy assumption—Carter's own—that the governorship was always just a step toward the presidency. What, I asked, made the country ready for Carter in 1976? Two things, he argued—the fading of sixties passions, and the new acceptability of Southerners.

In the car heading south to Plains, those words stick. Sheriff-types, as from Albany, are being interviewed on a talk show, and they talk about guns with those who call in. But there is less a feeling of strange country than in 1972. The nation *has* come around to the South in subtle ways. In the 1960s, black funkiness came out of the South—soul food, Afros, watermelons, all the things blacks were once supposed to be ashamed of. For some reason, white funkiness has become the fad of the 1970s— country music, junk food, soap opera, truckers' talk, evangelical "soul." If we are not careful, we'll all be eating hominy before long.

Of course there is still some cultural resistance to the funky side of Jimmy Carter. Take the name Jimmy itself. Some think it affected for a presidential candidate to use such an informal name. But "James" would be affected in Plains—not only for Carter's peers but for the still-deferential blacks, for employees in his peanut warehouse. South Georgia has a formal alternative to Jimmy; but it is not James or Mr. Carter. It is Mr. Jimmy—what slaves and servants called their masters. In Plains he is often Mr. Jimmy—just as his brother, in charge of the warehouse, is Mr. Billy, and his mother is Miz Lillian.

Another problem for some is the Jesus talk. One reads that Carter "raised the question" of his religion in North Carolina or elsewhere. But he could not avoid Jesus talk, even if he wanted to, where he came from. Gene Talmadge, whose religion barely went deeper than his sunburn, took every opportunity to preach from a Baptist pulpit on Sundays. Every twist of the dial finds a hymn or a sermon, this morning of my drive from Atlanta. Jesus talk is, at the least, a kind of static in the air—the very graffiti in the gas station toilet mix religious slogans with obscenity. It is a stickum or social glue, holding an angry countryside together, Bible talk among the guns, acting as buffer. One of the hymns on the car radio begins: "Jesus, I heard you have a big house . . . Jesus, I heard you have a big yard." It is the voice of respectability, dearly yearned toward. And one should remember that Baptists—considered a bit flaky in other areas of Protestant sensibility—form the prim religious establishment of the South, the quiet right wing of a religious spectrum that fades "left" into things like snake handling, tongues, and prophecy. Carter is amazed when people "up North" think his easy talk of Jesus implies mystical experience. That is like asking John Lindsay if he really bathes in lamb's blood because he is an Episcopalian.

Another twist of the dial brings in Albany—Georgia's Albany, pronounced All-BENN-ee—Hamilton Jordan's hometown. Just a week before this, Carter was on the televised League of Women Voters panel in

New York, discussing the city's problems. He said these problems were caused by the New York capital at All-BENN-ee. But it is typical that he sensed the audience's puzzlement, and worked All-b'nee into his very next sentence. He plays the native and the acquired against each other. Ol' Gene Talmadge, despite his Phi Beta Kappa key, would have surprised by coming up with the Northern pronunciation; Carter surprises by showing that the Southern one is instinctive still. But Merle Haggard is singing on the rental car radio: "The roots of my raisin' run deep."

Georgia clay spurts, with hot suddenness, lush foliage when it is given water—as in the short green tunnel that leads west into Americus, the county seat ten miles from Plains. There is a spelling bee on the Americus station, WISK, and a schoolmarm like the one Carter often pays tribute to still coaxes bits of alphabet from the quavery young voices. Frequent announcements remind the listener of that night's Miss Americus Pageant.

Americus is a great old Victorian town moldering away from the bypass to Albany—which is, in turn, a new town all hamburgery approaches and traffic lights, with an invisible core. But Plains is not really a town at all. Even the row of storefronts is only a husk. One end once held a grocery store run by Carter's Daddy, the other a general store and mule-auctioning center run by that Daddy's brother, Alton. Alton is still there, an antique in an antique building selling junk he calls "antiques." The general store is now in the middle of the row—it has the small items one does not drive to Americus for. The proprietor's wife is June Turner—she talks of Mr. Jimmy with matter-of-fact affection: "He loves the Lord and wants to bring the country back to what it was." It comes at the outsider like a mumbling of Georgia clay, the chewed dry stuff of a life lived at dirt-level. But it is the *relief* from that life she is really voicing, her cherished moments spent in the church across the tracks; a place good Plains Baptists visit four times on Sunday—for 10 A.M. Sunday school (taught by Carter when he is back home), 11 A.M. preaching, 6 P.M. training, and 7 P.M. worship. The church is an interesting bundle of planes folding over and around each other, with angled window facets—a small fan of stalls inside, womblike; but elbowed in strict board outside, and prickly with spire. It is a church well used. Carter's late business partner, Elliot Anderson, would not process peanuts on Sunday. That is schoolmarm day in Plains. It is civilization. Surcease. No wonder Carter liked its resonance of bookish power, of life shaped by words.

The principal structure in this little hamlet is the peanut-processing cluster run by Mr. Billy—who is up in Wisconsin today, observing the campaign. Hamilton Jordan says Billy used to be the family's closet red-neck—now, as Carter's slick national campaign leans back for strength on its native base, he is becoming something like the token red-neck. Carter's daughter is out of her school with measles—the school is made up of three prefab tunnels on a concrete base, with outdoor basketball courts dirt-surfaced, surfacing the kids with dirt.

Carter says he grew up poor—but as a member of the reigning family in "Archery," a clump of houses a mile down the track from Plains, where two dozen or so large families of blacks lived dependent on Carter's father. Carters, spread about the area, were dwarf lords of this drab world, whites in the predominant black population, held down close to black living conditions but retaining a psychic distance for that reason. Carter's mother, a nurse, was kinder to blacks than his father—but she kept herself insulated by subtler means, with diction, books, and knowledge. She let one black visit her by the front door, a signal favor still recalled as a prodigy—but that was a *college-educated* black. Such creatures frightened Carter's father, but confirmed the distinction of gentle Miz Lillian. Carter has clearly shed the racism of his crabbed youth, but the psychic distance he learned to keep with the swarm of black boys he grew up among has been retained for political use—it lets him nod, grin, and hug his way through crowds while keeping some detached zone of dignity untouched. He grew up a dirt-hole aristocrat, with schoolmarm power over sheriffs as well as blacks. Apparently that showed even in the intimate traffic jam of submarine life, where he was always the officer.

Though Plains is still a small pond, Carter has become a big frog—with a beautiful red-brick home he built in 1961 (like the second Talmadge home in McRae). He points out that his daughter goes to an integrated school, more than half of its students black. But this little town needs no full-time policeman, so strong are old ties of deference. The town repeats, on a somewhat larger scale, Carter *père*'s reign over shacks at Archery.

In Alton Carter's "antiques store," I talk with Alton's son, Hugh, who grew up with Jimmy and took over his seat in the state senate when Carter ran for governor. Hugh, like all the Carters, talks endlessly of Jimmy's work and of his own. All Carters work hard, and are fascinated by that fact. Hugh and Jimmy never had allowances as boys. They made their spending money by energy and ingenuity—collecting old newspapers for the two blacks who sold fish wrapped in the paper;

collecting scrap iron for the smithy; making ice cream and selling it on mule-auction Saturdays; breaking "pennywinkle" grubs out of tender branches to sell for bait. When they had enough money, they would hitchhike to Americus to see a Western movie.

"All Carters are hardheaded and hard workers." And they are all successful. The eighty-eight-year-old Alton boasts of his brothers and his two boys, as well as of his nephew Jimmy. Carters bustle in a world where others (which meant mainly "niggers" in Carter's youth) dawdle. Survival at the pennywinkle-catching stage was easy here. But rising was hard. The earth clung to one, broke one, dimmed one into its red color—as it still paints the cars and trucks of Sumter County. One had to be tough as flint to strike a few sparks of life from that hard land. Carter boasts of the hard work he can do as a campaigner. By enduring any mule-labor rigor, he proves that the other candidates are soft—do less; study less; indulge themselves. His confidence, which awes so many, is a tested reliance on his ability to do anything he calls on himself to do. He lifts and defines himself out of the dirt, white against the red and black clay, not slipping into the dread slide of po' whites down to practical niggerhood. It is commonly said that conventional Democratic politicians fear Carter because they have no claims on him, no ties or debts to organizations, urban machines, labor unions, old delegate clusters. But a deeper fear moves under that—the fear that he is not like the rest of us sinners, so little disciplined, so subject to his schoolmarm condescension.

Morris Dees, the best of Carter's workers, is a crusader against the death penalty. I asked how he could support a candidate who believes in capital punishment. "Only for some crimes. And I have not really gone to work on him yet. When I have, he'll change his mind." But changing his mind is not a skill Carter shows any familiarity with. He is not easily deflected from his course, once chosen. His self-confidence is a confidence in all his own opinions, once formed.

Even his religious confidence has a narrow base. He is a man who reads the Bible every day and has taught it all his adult life. I asked him if he ever entertained, in Sunday school classes or private study, the "form criticism" of the New Testament. "What is that?" he asked. A bit dumbfounded, I said, "The kind of textual analysis Rudolf Bultmann did." He knew that name but could not remember if he had read anything by Bultmann. For a bright and educated modern man, dealing with the thing he says matters most to him, he shows an extraordinarily reined-in curiosity. It suggests a kind of willed narrowness of mastery.

He moves quick and certain within a deliberately circumscribed territory. He sticks to his base, his certitudes. To Plains. That gives him his power.

There is a concern over Carter's religion that is not mere bigotry. It may seem unjust to punish real religion when we reward empty religiosity; but the thing makes sense. When Birch Bayh goes to his native hamlet and says he never feels closer to God than when he is there, we do not care because we know he does not mean it. When Nixon used Billy Graham to bless the billy clubs of the sixties, the insincerity gave us at least the relief of mockery. Kennedy's Catholicism was made up of gestures. But when a man *means* what he says in this awesome area, he drifts outside the ties and shared weaknesses that keep us in touch with each other. Carter says he will not be compromised. The claim has been dismissed as unkeepable. But I wonder. The scary thing is that he might have some way of keeping it.

Carter's second-birth conversion came in the aftermath of his one defeat, the gubernatorial race of 1966. Hamilton Jordan calls Carter "the world's worst loser." The inner peace denied him in loss was only renewed by fresh energies of determination to win. His power, briefly doubted, was reasserted—his mastery, the quiet spell he works on others, the authority given him by labor and self-testing. I thought of that as I drove back to Americus, at dusk, to watch the twenty-fifth Miss Americus Pageant. (Last year, Jimmy's niece Kim was a contestant.) The Jaycee organizer of the affair says it will bring out only half the crowd it used to—600 instead of 1,200—and his prediction is confirmed. Even so, drawing 600 people away from TV sets on a weeknight is not bad. Americus has few sources of pride these days—two of its three mobile-home assembly plants are closed, and the third working only part-time. Yet there is a picnic air of real enjoyment as not-so-slender young women parade in their Atlantic City bathing suits. This black belt of Georgia supplies no black beauty to the contest. The announcer makes an act of blind faith when he opines that tonight's Miss Americus might go on to become this year's Miss America. It is as unlikely as Jimmy Carter's rise, his hope to become this year's Mr. America. His power over his own is linked to their pride in him, to the kind of adoration Ol' Gene got from the wool-hats.

Outside the Americus High School Gymnatorium, one can almost touch the tropic sky of stars unobstructed by buildings, hills, or foliage. It is a fierce sky, not distanced by Argus myths of pagan levity—a sky of older Americans, physically starved and spiritually overfed. God stares with a thousand eyes on the isolated souls of Sumter County—which is enough to make one crazed and desperate; or cowlike in resignation; or President of the United States.

—*The Atlantic,* June 1976

AFTERWORD

The journalist Theo ("Ted") Lippman, who comes from Georgia, wrote me about this piece, wondering how anyone who mentions "form criticism" got out of South Georgia alive. Others thought it odder for me to ask about Bultmann than for Carter to be ignorant of biblical scholarship. But Carter's intelligence always promised more breadth than he managed to acquire. Precisely because he takes religion and books seriously, I was surprised at the narrow and repetitive intensity of his thought about religion. It was one key to the personal narrowness that remains one's lasting impression of him in the presidency,

27

Reagan

I

Some imp of history seems to be urging Reagan on. He is the serene beneficiary of turmoil, the amateur foil of professionals, the sane relayer of craziness. There is no way to have at him. His very flaws promote him, and weakness prevails. He wins campaigns by being a noncandidate. His lapses rivet his followers closer to him. There is deep magic here.

But he is *old?* Sure; and that reassures. He is walking evidence that the past perdures—it is out there, ambulatory, defying time. Representative Philip Crane's 1980 campaign was supposed to establish that Reagan's unsound mind could exist in a sound body: as Reagan decomposed along the way, there would still be a handsome head swiveling on a mobile body, saying the same decrepit things. How was Crane to know that a modest decomposition was part of the charm? Time softens Reagan, taking sharp edges off. The older he gets, the better he looks, enacting the truth that old values are still viable. Creaky, perhaps, in this joint or that; rambling, not marching. But even the ramble soothes, where marching would disquiet.

So rivals find that age is Reagan's tease, the booby trap he sets for them. Attack his age, and you seem ungallant, the kind of person who would suggest that Mae West's yips of sexual ecstasy have been feigned for the last half-century or so. Who wants to believe *that,* true or not?

Wild inaccuracy is another of the pitfalls he seems to be digging for himself—as he watches others fall into them. But why should his friends care about a fact here, a fact there, when he so obviously does not? Facts come and go; statistics are symbols or persuaders, when they are not lies; reality adjusts to belief, not vice versa. There is a sophisticated calculus to Reagan's naiveté.

Not that he sees all the understructure of his own appeal, girder by girder. That would not fit the very genius of his appeal. He floats above the quibbling because it never touches his self-esteem. He does not, like most politicians, have to defend his arguments. It is his *attitude* that is right, and arguments cannot reach that. His right-wing attitude was unruffled even when he enacted rather left-wing laws in California (such as a permissive abortion program). He rather observed his own administration than became its creature, or creator.

Like most politicians, he lacks pride. But, unlike them, he also lacks the vanity that makes a claim, at least, to discursive consistency on issues. He looks unchallengeable, even in error, because he feels secure. He shrugs off attacks as so clearly benighted they scarcely deserve attention, and certainly not anger. He brings to the Right the thing it always needed most—relaxation. Consider how rare has been sheer friendliness on the right wing of our political spectrum. Robert Taft was loved only by a hard effort of those conservatives who threw their hearts into the arctic blast of his number-citing rectitude. Reagan has a shrugging semirectitude that will not stand on ceremony or split hairs: he tugs hearts out before they can be thrown.

Richard Nixon, unlovable himself, was warmly supported for the enemies he made. A hired gun, his very nastiness could be put to an exterminator's use. But Reagan is a *good* guy willing to take on the baddies; and after the shoot-out, he will not rifle the till, like Nixon; he will not save the town only as a way of lining his own pocket.

Reagan's deepest appeal is to those who cheered George Wallace on but who felt a bit shabby and soiled after he worked them over, massaged their hate glands, made them queasy with acrid emotions sweated out of them. Reagan croons, in love accents, his permission to indulge a functional hatred of poor people and blacks. Nothing personal about it. It is really an act of patriotism not to let the hardworking middle class be dragged down to *their* level. Imagine what a godsend this is for rightwingers with some small claim left them to fastidiousness. Poor William Buckley had to fashion a nine-foot pole for dealing with all those Spiro Agnew types a gentleman should not touch with a ten-foot pole. But Ronnie he can walk right up to and hug. Even the caricaturists have a hard time putting Joe McCarthy's scowl or Agnew's dopey viciousness on Reagan's face.

But Reagan would not be important if he just made the right wing giddy with relief at finding a nontawdry spokesman for its tainted views. Reagan has stronger historical tides running in his favor, tides that affect our whole society and not just the right wing. He is the legitimate-looking heir to our government's illegitimacy. Carter is limited by the sectional appeal that made him important to the Democratic party, resolving the anomaly of our postwar politics—the fact that Democrats could outregister Republicans two to one or more, yet have not won the presidency without the South. The very thing that gives Carter his marginal purchase on the South—his loyalty to a clutch of Dogpatchers like Bert Lance and Hamilton Jordan—weakens his links to the everyday experience of Americans less regionally marked.

Reagan, by contrast, has *been* our everyday experience through years of popcorn matinees, through decades of relaxed success as a campaigner. He is by now our neighbor as the orator, the hometown boy as a success. Not only the baseball player and soldier and football hero he actually played on the screen, he has become the inheritor of roles played by his coevals—Jimmy Stewart as Mr. Smith, Gary Cooper as Mr. Deeds, even Judy Holliday as Billie Dawn. He is the innocent as celebrity, and the celebrity as the real authority behind all screens and shams of officialdom. He is the first serious *counter*authority with an *air* of authority. He is the perfect denier, the double negative that comes out, somehow, as a positive. The disinherited air of his predecessors made us let them into the American psyche through some trapdoor (like Joe McCarthy and Nixon) or side entrance (like Agnew and Carter). But Reagan strides confidently in through the front door.

That, of course, is the scary thing about Reagan: that he does not scare us. He so obviously means well that it is gauche (like mentioning his age) to notice that he means nothing. He represents that stage of our government's inanition of authority where it is no longer a wild claim, but mild dogma, to say that the rule of tradition and reason has ended in Washington—that everything depends, now, on one man's personal qualities. Facts and evidence are not important in the time of a private vision publicly yearned for. In that situation, to claim authority is to profess illegitimacy, to be one of "them," of "the gummint" that must be got off "our" backs.

This process—ridding American backs of encumbrance—is entirely negative in logic; but Reagan's approach is not obviously denigrative. He does not mean the meanness of his views. He does not scowl racism at us, like Wallace, nor leer hatred of Ivy League types, like McCarthy or Nixon. He does an actor's walk-through of those men's lines without losing his unruffled air of meaning well—just as he walked through the animosities of both sides in the Hollywood Red-hunt days, coming out an ambiguous half-hero (or at least nonvillain) to all sides. He uses ideology without being trapped inside it. The very thing that frees him of authority keeps him clear of what others take so seriously. If the authority of facts and argument matters so little, how should ideology itself confine its wielder? With Reagan, we get Vietnam defended without Goldwater's bluster, blacks put in their place without Bull Connor's dogs, patriotism hymned without Lyndon Johnson's oleaginous defensiveness. Reagan is so patently unmalicious as he speaks for war and divisiveness that he may, indeed, kill us with kindness. He is the wholesome hometown sort who can drop the bomb without a second thought, your basic American Harry Truman.

In a world being emptied of authority, Reagan has some of authority's characteristics. Continuity, for instance. His age does not bother people; rather, it suggests a rootedness in essentially deracinated views. He is unmenacing because he was always around, so much a part of us (and how could *we* be menacing?). He is both Henry Aldrich and Grandpa Walton, our remembered and our present selves, our fantasy of afternoons with popcorn and the "real" world of TV politics. Where so little is stable, the emptiness at the center looks eternal.

Reagan is the calm eye of history's hurricane; and we hope, by moving with it, never to slip toward the edges and to chaos. But we will.

—Esquire, August 1980

II

When Carter austerity gave way to Reagan splendor, a certain moralism was driven from the White House. President Carter had not been good but goody-goody, we were told, in his wish to deal only with nations devoted to human rights. The man was a touch monkish with his cardigan sweater and talk of limits. He was doing penance over the Panama Canal. He expected the CIA and FBI to reform. He scourged Americans for their malaise.

Reagan, by contrast, came to pep Americans up, to make them feel good again. Gloomy "moralism" is not the style for this candidate backed by the Moral Majority. The successive presidencies represent, in fact, two different moralities—almost, one might say, two theologies of America. From the Reagan point of view, Jimmy Carter was not just inept; he was a heretic. He seemed to admit, if not proclaim, the fallibility of America. To counter that claim, Reagan comes to vindicate not an imperial but a papal presidency. We are to believe in America, and in Reagan as its symbolic embodiment. Carter was not only unbelievable as a leader; he also seemed to have lost his own faith in the national destiny. Popes do not wear cardigans.

The two moralities clashed most obviously on the question of human rights abroad. Carter asked that other nations be moral. Reagan asks that they be on our side, and that will *make* them moral. If the overriding moral issue of our time is the victory of freedom, and America stands for freedom, then almost anything that helps America prevail is by definition a force working for freedom. Unfree regimes that become our allies are unfree regimes that promote freedom. Working from that logic, Reagan tried to make Ernest Lefever our human rights spokesman, on the understanding that he would not look at any human rights issue but the right to be aligned with us in our crusade for freedom.

It is popular to say, now, that the Reagan presidency attempts a counterrevolution—a return to Herbert Hooverism after the Keynesian revolution. But that covers only the economic side of Reagan's administration, and only one aspect of it. In this administration's economic breviary, George Gilder's *Wealth and Poverty,* even the supply-side doctrine is at root theological: "Economists who distrust religion will always fail to comprehend the modes of worship by which progress is achieved." As David Stockman, in his *Atlantic* interviews, said of supply-side theory: "The whole thing is premised on faith." (Actually, Gilder got Arthur Laffer's supply-side theory a little mixed, leaving out his insistence on a gold standard. I guess Laffer kept saying "gold" and Gilder kept hearing "God"—presumably because Laffer would use the same hushed tone of awe for either word.)

The Reagan effort is not so much a counterrevolution as a counter-reformation—a return to the full claims of America upon its citizens' belief; like the reaction of the papacy when, in the sixteenth century, it called the Council of Trent to reassert the Catholic faith against the Protestant Reformation. The effort to "make America great again" is directed at those gloomy critics who said America could not recapture world leadership. The failure in Vietnam, they said, had taught us that America cannot be "the world's policeman"; gas lines had taught us that we are vulnerable to Arab regimes; Watergate had shaken confidence in our leadership. In a time of dwindling resources, they said, America, as the world's principal consumer, must retain friends by surrendering part of our affluence. Reformers in America—some of them expressly theological, like Richard Barnet—said it was a moral duty to cut back on America's expectations, demanding less from the rest of the world because we claimed less for our own nation. There was a call to an age of limits and institutional humility.

This was the heresy against which the Reagan counterreformation meant to throw all its forces. According to Gilder: "It is said we must abandon economic freedom because our frontier is closed; because our biosphere is strained; because our resources are running out; because our technology is perverse; because our population rises; because our horizons are closing in." These evil doctrines, he suggested, amount to "a sociology of despair," to a "calculus of impossibility," against which the old faith must be reasserted.

Ronald Reagan was not elected to wrestle with the reformers' problems but to deny that they *are* problems, that America has anything to surrender or regret—except, perhaps, any timidity in asserting our claims. Vietnam, in his words, was not a failure but a "noble cause." There is no oil shortage, just a lack of will to get the oil we want. Watergate was not the real scandal of Washington; bureaucracy is the scandal. America will benefit others by producing more, not less—as we help our poor, internally, by making rich people richer. In David Stockman's words, Reagan offered a "happy vision of this world of growth and no inflation without pain."

Reagan thus, in one leap, not only changed the ground of the argument but also lifted his program above the realm of argument. His appeal is not to reason but to faith, to his vision of America as an infallible mother church. His theologians have worked out a doctrine of the trinity that confounds reason but attained, at first, a mystical acceptance —the triune program of monetarism, fiscal restraint, and supply-side incentives. Fusing these incompatibles has made the Reagan economic team seem more like bishops at the Council of Trent than like secular economists. The program will work if only we *believe* it will.

And, of course, once heresy is hunted, it will be found in the Vatican itself—as when certain bishops were forced to toe the Tridentine line. Even before David Stockman confessed to heresy in public, supply-side inquisitors like Rowland Evans and Robert Novak had sniffed a lapse from orthodoxy. Stockman's recantation was a doctrinal submission to presidential infallibility: it humbled mere reason, much as Galileo's submission had.

If people do not recognize that the basis of Reagan's program is *faith,* they must be puzzled to this day that facts or arguments matter so little to him or his admirers. If one accepts, as part of a *creed,* that America produces wealth by consumption, then environmentalists are so heretical on the face of it that their carping about details is beside the point. Thus Reagan could mock any concern about industrial waste, could say that trees pollute more than cars do, that oil slicks purify the air, that seeing one redwood should suffice the most rabid aesthete, that there is more oil under our feet than we have taken out of the American soil in the last century. A challenge to the factual basis of these statements receives no answer, because fact was never at issue—only faith. As one believes, in the economic sphere, that three are one, so one must believe in the incarnation of American consumerism as the savior of mankind.

The principal environmental office went, therefore, to a theological appointee, to James Watt. The reformers had looked at the supplies of clean air and water, of food and fuels, and said—like the cartoon character with the poster sign—The End Is Near. James Watt, who has given the Second Coming deeper study, assures us that we cannot be sure about the only End that matters; and, besides, the Lord clearly wants Americans to use up things just as fast as possible, since that is the way to enrich others.

No counterreformation catechism, of course, would be complete without a theology of the devil. Diabolic motivation was ascribed to the original Reformers. Papal literature portrayed Luther, at his death, being taken to hell after doing the devil's work on earth. A similar insight informs the champions of America's counterreformation. The Lutheran reforms began among restive monks. The modern reform movement has for its spearhead some radical nuns. Mrs. Jeane Kirkpatrick was the first to identify the threat from Maryknoll sisters; but it was Secretary of State Alexander Haig who saw how wondrously the devil works through these ecclesiastical cat's-paws. When nuns were killed, the San Salvador regime was not to be blamed. The devil had sacrificed some of his own in order to discredit the forces of good. The proof of diabolic intervention, according to Haig, was the way the nuns managed to be raped and then shot in the back of the head while running a roadblock. That scene comes, clearly, from some counterreformation woodcut mocking the Reformers. It does not strain the credulity of believers to find the grotesque mixed with the miraculous—how else would devils operate? If Maryknoll sisters are the modern Luthers, the modern Satan has been Leonid Brezhnev—just as surely as Soviet Russia is the modern hell. The original sin of Marxism has tainted mankind and made us subject to the devil's will. He leers in through our window of vulnerability. He haunts our sleep. He corrupts our friends. He assumes many disguises. He infects the very air around us, lurks in apparently orthodox breasts. He has planted "disinformers" in our midst. He not only has bigger missiles than we do—he has the Institute for Policy Studies!

Only orthodoxy can break the chains this devil throws around us; and it is always a near thing, a last-minute rescue from the jaws of death. Exorcism may drive out the possessed nuns; Admiral Denton's new Inquisition may alert us to the devil's disguise. But we must work out our salvation in fear and trembling; we must put on all the armor of the Lord. That means not only selling indulgences to nations with sinful records, or catechizing the public on our economic trinity, or punishing the preachers of America's fallibility. It also means restoring the country's religious orders to their maximum effectiveness.

The Protestant Reformation was largely energized by a critique of the religious orders, of their corruption and departure from the Gospel. The bishops at the Council of Trent responded by admitting to certain abuses—the occasional greedy friar or lecherous nun—but proclaiming that the Reformers' real target was not abuses in the orders, but the orders themselves. Luther did not want monks to be more chaste. He wanted no more monks.

America's reformers have criticized the CIA—its clerical privilege to spy on others, to be secretly funded, to depart from the Constitution, to live by separate doctrines. But Reagan's people know these attacks are not on the abuses of the CIA, but on the CIA. The reformers do not want the CIA to stop making powders that will shave Fidel Castro against his will; they want the CIA to stop bothering Castro at all. So, as Tridentine reforms strengthened the orders, Reagan's counterreformation means to license, once again, secrecy and covert action in the CIA, and remove freedom-of-information restraints from the FBI. It would be folly to let some *abuses* make us forswear our armies of the night in the long battle with the Prince of Darkness.

The sixteenth-century papacy not only strengthened the old orders in their monastic seclusion—the supernatural weaponry of the praying convents—but also created new orders to go out into the world and teach Tridentine doctrine. Foremost among these were the Jesuits, who adopted some of the Reformers' techniques. And it is no wonder that a new teaching order should arise, in our modern counterreformation, a new team of Jesuits to develop sophisticated arguments for indulgencing "authoritarian" regimes and excommunicating "totalitarian" ones; to preach the economic trinity; to debate heretics and denounce radical nuns. As "neoconservatives," the American Enterprise Institute priesthood retains a certain rhetoric of its democratic past, using reformers' weapons against them. These new Jesuits, like the first ones, also try to discipline their own side, correct its sillier spokesmen, and rescue orthodoxy from embarrassment—the blocking of Melvin Bradford's appointment to the National Endowment for the Humanities is but one instance of this function.

Presiding over this great effort of faith, Ronald Reagan has performed creditably, as popes go. Like the best pontiffs, he combines a personal humility with strict guardianship of his infallible gifts. He excommunicates more in sorrow than in anger. He stresses the sunny side of faith—while not forgetting that Moscow is hell's center. Like the Counter-Reformation popes, he puts on a good show—papal splendor in the White House, business executives flanking him like a college of cardinals. The baroque style enjoyed the blessing of Counter-Reformation popes. We have a California baroque—$1,000 boots instead of silk slippers, expensive turkey shoots instead of deer parks.

And it never hurts for a pope to work miracles. Our aging President survived an assassination attempt with preternatural ease. (He did rather better, in fact, than his counterpart in Rome.) Reagan's very election seemed to show he is the darling of the gods. The spirit, as they say, moved over the electoral conclave. For a time Reagan exacted from Congress a profession of faith in the economic trinity. If some members of his curia have stumbled or fallen since, that does not seem to besmirch his robes. The pope's own faith is beyond question, and many believe in his belief.

But can this counterreformation succeed? The first one, after all, did not suppress heresy or make the Catholic church universal again. And it had the advantage of being *expressly* religious rather than *implicitly* theological. As a Catholic, I have no difficulty accepting the leadership of the pope over my church; but I find it a little burdensome to accept a pope for my country as well. Gilbert Chesterton said that very few people really lose their faith—they just transfer it. When they can no longer believe in a church, they believe in a country. Disappointed by theology, they take up politics—which often means that they mix the two without realizing it.

How long can Reagan float, on faith, above inconvenient errors and fudged facts, above an economic performance that does not match the mystical formulas? The world is polycentric now, not just a struggle between two superpowers. Communism is not a monolith (witness China), nor anticommunism a guarantee of our best interests (witness Taiwan). There *are* limits to what America can do in the world. Ignoring or denying these realities may make for a neat system of catechizable doctrines. But reciting the Trinitarian creed will not reduce the conflict between lowered taxes and expanded defense spending. The mystical trinity of the supply-siders is producing, in the order of fact, a messy trio of high inflation, high unemployment, and high interest rates. Keeping the faith is getting more difficult for all but the devoutest members of the White House staff itself. The country's faith is slipping far more drastically, as every poll confirms.

At some point, the naive faith of Reagan—which inspired so many hopes for him—may become his greatest weakness if he continues to believe after the rest of the country has learned to doubt. At that stage, he will be a pope without a church, and the country must seek a President again, a man with programs instead of a creed, compromises instead of doctrines, and arguments instead of faith. But if we want that kind of President, we must wait for Ronald Reagan to leave office.

—*Soho News,* January 19, 1982

III

President Reagan has kept his principal (if implied) campaign promise —not to be another Jimmy Carter. At a post-election symposium held by the Kennedy School of Government, managers for both Reagan and Carter agreed that the main determinant of the 1980 contest was the fact that no one wanted the Wimp in the White House anymore. Between an economy out of control and an image of weakness abroad, the President seemed a mean-eyed do-gooder who was giving decency a bad name.

Reagan could not offer a more dramatic contrast. He seems a likable gunslinger, able to give even toughness a good name. No more sweet talk and self-doubting. Reagan would tell the world what America wants, and then make sure we got it. Carter hid a strong mind within soft manners—which convinced some that strong minds are not necessary equipment for the presidency. Carter knew, in a way, too much; he was overburdened, paralyzed by his preoccupations. That would not be a problem with Ronald Reagan, who is wittily called "underburdened" by his biographer, Lou Cannon.

Reagan hides a weak mind within his strong manner—and that, for a while, was enough. Style can matter more than what a man is saying. Reagan, while thundering against the Russians, can give them grain and still look like he just won a shoot-out. Carter, though skeptical of bellicose rhetoric, could cancel the Moscow Olympics and still look accommodating toward communism. No one cared that Reagan invented whole new sciences during his campaign, teaching us that trees pollute and oil slicks purify, that car exhausts are less noxious than Mount St. Helens, that evolution is devolving. The President is not expected to be a biology teacher. Besides, Reagan has a natural grace in assertion; and, more important, an eerie inability to be embarrassed while having to recover—often—from mistaken assertions.

Americans quickly acquire a way of thinking about their Presidents, and facts get sieved through that thought-screen, sorting themselves out in recurrent patterns. After that, the mental picture is hard to shake or disturb. Kennedy was grace under pressure, as Truman was guts under pressure. Johnson, as David Levine showed us, kept looking at his personal scars and seeing Vietnam. Ford was as clumsy at doing good as Nixon was deft at mischief. Carter was so nice it put one's teeth on edge, till he tried being mean, and that was worse.

Reagan was fixed in people's minds, even before assuming office, as the man who could say and do hard things without having a hard heart. During the campaign, Democrats and Republicans debated just how hard Reagan's talk was—would it lead to war, would it scare allies? No one suspected that the real problem would not be the tough rhetoric but the tender heart. Even opponents had to assume that Reagan is "a nice guy," and try to get around that obstacle to their success by pointing out that even nice guys can blow up the world today. Supporters never doubted that their man's niceness was their greatest asset. If he made mistakes, or sounded bellicose, at least people had to admit he was likable about it. Thus the one thing that would trip Reagan up was almost entirely neglected—his way of pleasing others.

Almost immediately after the election, Reagan started demonstrating that, for all the shoot-out oratory, he is a pushover. He does not like to say no to anyone. It was assumed, during the campaign, that Reagan had to accept all the contradictory components of his economic program in order to get elected. After he was securely in office, he would have to *choose*. You cannot have supply-side tax cuts, monetary controls, fiscal balance, and defense boosts all at once. The programs undercut each other. In the final campaign debate, Carter tried twice to make Reagan address this problem. He failed; but economic reality, it was assumed, would succeed where Carter failed.

Yet economic reality proved to have as little purchase on Reagan as his opponent did. He smiled and ignored the one, as he had the other. Conservative Republican economists have been divided, in the past, between congressional budget-balancers and the Milton Friedman money-controllers. Reagan, instead of joining one of those camps, threw his arms around both and then added a third and a fourth school—the supply-side tax cutters and the cold war defense boosters. They would all be one happy family. But the family was never happy, except in Reagan's presence. Each group was supporting Reagan to get its own way—which meant the others could not get theirs. David Stockman was soon complaining to *The Atlantic* that he could not advance his supply-side program for investment incentive; the interest rates were stifling that initiative. He was trying to fight inflation with new productivity while the monetarists were fighting it with measures that blocked production. Ronald Reagan could not adjudicate the quarrel because he resolutely refused to see it. It would be too disturbing, in his rosy mental world, to think all those nice fellows were not really on the same side.

The first major surprise of Reagan's presidency came just after the

1981 tax cut. Reagan was supposed to be the businessman's President, and Wall Street was expected to do little leaps and frisks of joy about his success with the Congress. It did nothing of the sort. It glumly settled into a stall. The President, puzzled, said that Wall Street just didn't have faith enough—it must learn from Main Street, where he was still getting cheers. But Wall Street would not respond till Reagan's faith underwent adjustments, and he raised taxes rather than cut them. Typically, Reagan said he had not reversed himself—he just could not say no to a *fifth* inconsistent item in his program, Senator Dole's "revenue enhancement."

Wall Street had never bought the whole Reagan package. It was in for its penny, not in for Reagan's pound. The same was true of other sectors of Reagan support. No one could believe he *really* meant he would stick with all the parts of his campaign oratory about the economy.

In foreign affairs, Reagan was just as unable to say no to supporters with different approaches, which meant that for a while we had five foreign policies, emanating from Secretary Haig's State Department, Secretary Weinberger's Defense Department, Presidential Assistant Allen's Security Council, Ambassador Kirkpatrick's UN post, and Senator Helms's Senate Office. Again, the President thought they were all one big happy family, even as they were cutting each other up in various combinations. The anomaly should have been apparent during the transition period. Reagan appointed a transition team for foreign policy that was very hawkish, pleasing to Richard Allen, who had been Reagan's main campaign adviser on foreign policy. But Richard Allen had run into trouble over his finances during the campaign; there was never any question of making him Secretary of State, which would have involved full-scale confirmation hearings. So Reagan listened to other friends and appointed Alexander Haig to the State Department—a protégé of Henry Kissinger, who was Richard Allen's pet hate. Haig airily dismissed the transition team, which sought refuge in Senator Helms's office. Helms responded by blocking as many Haig appointments as possible and sniping at the ones who were passed.

Haig at first tried to ingratiate himself with the hawks by adopting Jeane Kirkpatrick's opposition to the human rights policy of Carter, as that was symbolized in El Salvador. This brought him fire, from within, as too bellicose—from the same people who would soon call him too pacific on arms control, on relations with China and Russia; or, alternately, too hard-line in support of Israel. The foreign policy "team"

was pulling itself apart, and Reagan did not notice. He thought the five
or more voices were all speaking his own message, and how could that
be inconsistent with itself?

Reagan's desire to please made a mess of his appointments. He came
into office with a declaration that he would dismantle large parts of
the government, including Carter's human rights policy, the Department
of Education and the Department of Energy, the SALT II treaty, government support for abortion and busing, and government opposition to
school prayer and religious education. Instead of attacking these departments and programs outright, he appointed people to administer them
who did not truly support them. They were supposed to phase themselves out of their jobs. But in order to be confirmed, some of these
people had to make themselves minimally acceptable to the constituents
of the relevant department or program. The results were ludicrous.
Ernest Lefever, appointed to the Human Rights post at the State Department because he did not think such a role belonged in our foreign policy, tried to get confirmation by telling the Senate he really *did*
believe in human rights policy, of a sort. He never got that story straight
enough to win confirmation.

A man who believes in book banning was put in charge of the Department of Education. A dentist was put in charge of the Department
of Energy. Neither phased his department out. Eugene Rostow was put
in charge of disarmament policy because he did not believe in disarmament; but, again, he had to pretend to believe in it during his hearings.
To combat abortion, Reagan appointed as Surgeon General a man over
the statutory age limit, one who had toured America showing a film in
which Supreme Court justices beat up people. He appointed a devout
foe of abortion to the Office of Personnel Management, one who tried
to get abortion money taken from the insurance of federal employees.
On busing and segregation, he appointed an Attorney General, his old
financial adviser, who tried to eliminate by executive order a court-en-
joined refusal of tax exemption to segregated religious schools. The
common note in all these efforts is that they were indirect, done by people Reagan appointed because he liked them, or his friends did, and not
as the result of policy decisions announced and carried out with White
House support. Reagan avoided open confrontation. He wanted to
eliminate large parts of government without being personally "negative." Reagan had to back off on the federal insurance and tax exemption issues, saying he did not know these were matters that courts had

spoken on. (He also said he just did not know there was any segregation left in existence.)

During the 1980 campaign, Ronald Reagan said the answer to teen-age promiscuity was simple; young women would have to learn, again, how to say no. But over and over during his time in office, Reagan has been that character from the *Oklahoma!* song: "I'm just a girl who cain't say no." He could not say no to Paul Laxalt on the MX basing system. He could not say no to the auto workers on protection from Japan, though he campaigned as a free-trader. He could not say no to the sugar growers, though it meant undercutting his Caribbean Basin plan. He could not say no to the farmers on the grain embargo. He could not say no to China or to the Taiwan lobby, leaving everyone confused on our most important strategic relationship, the one that keeps Russia patrolling four thousand miles of hostile border.

He is omnidirectionally agreeable. He could not say no to Israel—or to Saudi Arabia. He has sent confusing signals to allies, while claiming that relations with them have never been better. He cannot say no to arms talks, but neither can he say a final no to those who want to end the observance of SALT II. He cannot say no to détente, but neither can he say no to those who still argue for it (as Secretary Haig did). When he "loses patience" with Israel, he does it by denying Defense Minister Yitzhak Shamir a *smiling* "photo opportunity." When he does say no to Europe, he claims it is not really a no, but the implementation of agreements he thought existed all along. As he did not think there was any segregation left at home, so he thinks there is no discord in NATO.

Ronald Reagan, as he himself says, has been an extraordinarily lucky man. Things have gone his way, and he showed malice toward none. Not even his denunciations of things like welfare workers have any note of personal hostility. He cannot imagine they will be taken as applying to real people. He has got along by going along, by living off his considerable personal charm, and he sincerely does not see why others cannot do the same thing—avoid disturbing confrontations, say yes to everyone, promote amity by overlooking differences. He told people they just had to believe enough in his program, and that would make it work. He thinks things will go along merrily if everyone is just as preternaturally confident as he has managed to be for most of his life. In Reagan's most important screen test, Pat O'Brien, playing Knute Rockne, asked George Gipp if he could run the ball. Reagan glanced at him as if he were crazy to ask such a silly question, and said: "How far?"

But in the movie George Gipp runs up against intractable reality—the most intractable of all, in fact: death, from pneumonia, before he could demonstrate his potential as a football star. Forty-two years later, O'Brien and Reagan went back to Rockne's and the Gipper's school, where the President was giving Notre Dame seniors their graduation address; and, surprisingly, Reagan showed he understood his old movie better than his current assignment in the White House. This is what he said:

> Now, today I hear very often, "Win for the Gipper," spoken in a humorous vein. Lately, I've been hearing it by congressmen who are supportive of the program that I've introduced. But let's look at the significance of that story. Rockne could have used Gipp's dying words to win a game at any time. But eight years went by following the death of George Gipp before Rock revealed those dying words, his deathbed wish.

Rockne knew how to wait; knew how to say no to the impulse to please everybody all the time; knew how to deny himself the satisfaction of a quick response to each difficult game. If Ronald Reagan could learn that, his team players might finally become a team; but the clock is running. He has won too many, too easily, too soon, by being too nice, by agreeing, by not saying no. And all that early winning just makes it easier to lose, in the second half.

—Family Weekly, October 31, 1982

VII
Religion

If the business of America is business, so is the religion of America religion. Everyone should attend the church of his or her choice—or should at least devoutly tend that platitude. I have devoted a good deal of my journalistic time to religion, because I think it is the least studied aspect of our national life, and perhaps the most important aspect. When Jimmy Carter focused new attention on America's religion, Bob Silvers asked me where all this new stuff was coming from. "It's not new," I told him. "It's always been out there." I think he found that a scary thought; and so do I. But I also find comfort in it.

28

Born-Again
Watergater

Is Charles Colson, famous dirty trickster and convert, a feigning Christian? I don't think so. I believe he may be the truest American Christian since Andrew Carnegie—the man Mark Twain took as the perfect specimen. If I had any doubts, they were dispelled at a Washington prayer breakfast for Christian law professors and students. The air was stickier than the buns, full of sweet-Jesus billings and cooings. A steady diet of this would try the nerves of anyone less entirely converted than Colson —who not only billed with the best, and cooed biblically, but seeks out such gatherings with increasing, not lessening, devotion. It is an obstacle course to try anyone's skills at feigning.

Colson's sincerity has been carefully examined by those best able to judge it, by responsible members of the evangelical community. That community suffered a great psychic blow from the Watergate revelations. Richard Nixon had achieved a rare empathy with fundamentalists. It went beyond his friendship with Billy Graham—though that was a factor, of course. More to the point, that friendship had itself been based on Nixon's confession of the evangelical faith, his account of teenage conversion at a revival. Many evangelicals worked hard for Nixon in 1960, trying to save America from a Catholic President. The most famous of these, Norman Vincent Peale, had to back off from his public attacks on John Kennedy; but others worked in less public circumstances, convinced that Nixon was more evangelical than Quaker in his outlook. Nixon remained the beneficiary of that presumption—and of evangelicals' willingness to separate the good Christian from his political views, to stand by a brother despite disagreement, to comfort him under attack, growing more loyal as the siege mounted.

But the support given Nixon was betrayed. It was hard to find the "good Christian" speaking on the White House tapes. Wes Michaelson, an editor of the neoevangelical journal *Sojourners,* says, "Nixon embodied many evangelical hopes. Watergate would remain a huge disappointment for them, except for Colson. He redeemed the experience in their eyes." Colson's book, *Born Again,* has sold several hundred thousand copies in circles where reading is an exercise in prayer. He is the current star of the prayer-breakfast circuit, a friend-brother of major religious spokesmen, a broker of unity among fundamentalist factions. Watergate, which might have threatened evangelical belief, he turns into a miraculous confirmation of that belief. It was all worth it, to bring this man to God, who is bringing so many others with him. I looked at two days of his busy schedule—1,200 at a prayer breakfast in Omaha, 600 at a dinner in Lincoln, lunch in Fort Wayne with Associated Seminarians, dinner with 500 businessmen that night. The Lord loves a cheerful speaker. Evangelicals can claim with some justification that theirs is the characteristic American religion, the kind you get when you "get religion." In terms of doctrine, evangelicals tend to be fundamentalists— though they are more willing to call themselves that if their politics are right wing (Colson says he is a fundamentalist). In terms of ritual, they favor a revivalist fervor, though often in a traditional framework of mainline "services." In terms of organization, they are individualist, each major preacher wanting his own crusade or campaign. In politics, they range from a populist left wing (now quite hospitable to ex-leftists of the "Jesus People" sort), through Billy Graham's soft center, out toward a hard Right that thinks the devil is trying to take over America for socialism. But all branches stress the private emotional experience of conversion. The convert is as central to this religion as the pope is to the more hierarchical belief of Catholics.

But the conversion must be sincere. That is why Colson, after his rebirth, had to be thoroughly probed and tested. He must be the genuine article to be useful to them. Short of that, he would be a menace, threatening them with ridicule, repeating the Watergate experience, not redeeming it. The last thing they need or want is a recycled dirty-trickster. So he was "vetted" by the movement's "stars," like former Senator Harold Hughes. He was even exposed to Mark Hatfield, who flirts with the heretical "social gospel"—but who, for that reason, would be quicker to suspect sham if Colson were playing some game.

He isn't. He's the real article. Which turns all his potential menace to literal godsend. A religion based on conversion tends to measure the height of a man's rise by the depth of his fall. He who has sinned greatly gives the most impressive testimony to the Spirit's redeeming power. The whole drama of the Bible is summed up in the sinner's return. And the returning sinner is bound to bear witness to the power he has felt.

Some have doubted Colson's motives precisely because he spoke out so soon. Even while calling himself a baby in the Christian life, he has taken to preaching it. It would have been more seemly, in some people's eyes, for him to have gone off and learned to live his new life, to study its teachings before publishing a book and lecturing, alternating prayer meetings with the major talk shows. But the evangelical tradition has always held that sinners are the greatest preachers of the Gospel. If divine rescue is the central fact of the salvation drama, what better witness to this than the recently rescued, the shipwrecked man still dripping from his scrape with spiritual death? Therefore, for millions of Americans, Watergate proves to have been the heaven-sent whale. Colson is Jonah, the man charioted toward life by that monstrous intrusion into the normal order of our politics. He is the crippled man who walks off the faith healer's stage, holy and passing on holiness, touched by the Spirit and touching others. He is the town drunk "witnessing" at the town revival. He is what it is all about.

Colson understands that since his conversion a great deal rides on him, and he works very hard at his new vocation. Many Watergate figures came out of jail calling for reform. He is doing something about it. In his genuine concern for prisoners, the desire to convert and the desire to help coincide. This repeats his own experience of being helped and converted simultaneously. His brothers in prayer helped him through his ordeal—they even offered to serve part of his time, and found a legal basis for it. Such love is humbling.

At the Washington prayer breakfast I attended, a small affair of fifty or so in a church basement, my neighbor at table was a law professor. Our conversation turned to separation of Church and State, on which he is as fundamentalist as Madalyn Murray O'Hair could desire. He rattled off court cases and precedents. But when Colson spoke, very movingly, of Christian prisoners he had met, my professor seized on one young fellow's plight, crying out for him in the Spirit. When the group, about to scatter, joined hands in a lopsided in-and-out-of-tables circle, the professor's hand in mine was wet with fervor and he prayed aloud, formulating an appeal more important than any phrased for lower courts. These people are not pretending.

They share a dense culture of their own—and Colson is now a central part of it. For them, the exact moment when he broke into tears, sitting one night in his car, has entered the history of *tolle, lege* ("take up and read") great conversions. (He had just left the home of millionaire Tom Phillips, who advised him to accept a "personal relationship" with Christ. The tears came as Colson was driving away, so he pulled to the side of the road and "prayed my first real prayer.") In his talk, he could refer to his book, confident that everyone there had read it or had its conversion lesson preached to him. No one had noticed when the teenaged Richard Nixon came down the aisle at that California revival meeting he later made into a part of his life story, but Colson's is a "witnessing" no concerned evangelical could miss. The crush around Colson after his talk was both affable and reverential. Law students shook his hand shyly, as if communing with the Spirit through His handiwork. Jesus is where Jesus works.

Colson told them in his speech that there is much Jesus work to be done in jail. He asked for volunteers to study legal reforms and the confused statistics of penology. He told them what he and Harold Hughes have been doing while waiting for more basic prison reforms. The two went to the nation's highest corrections officer and asked to take twelve prisoners out of confinement for two-week discipling sessions at Fellowship House, in Washington. It was a daring request, since the prisoners chosen—by Colson's group, setting its own norms—would not be guarded during their two weeks on the outside. "As Harold talked, I prayed; and as I talked, he prayed." The request was granted— miraculously, Colson believes. Three groups have so far been cycled out and back in without mishap. Colson describes the good effect the prisoners have had, on those outside as well as on the inside. "I took them to the Senate and a friend asked me where my prisoners were. When I pointed them out, she said she had mistaken them for a tourist party. I told her, 'Put prisoners in street clothes and they are just like the rest of us.'" The piety of the prisoners has converted the prison chaplain. This brought laughs—a staple of evangelical humor is the need of so-called Christians to be converted, to become real Christians. Colson said the choosing of prisoners for Bible study and prayer outside is an easy matter. "Take no one recommended by the chaplain or the warden." There is a refreshing animus against authorities among those who follow the Spirit, combined with a yearning for real authority.

BORN-AGAIN WATERGATER / 279

How does Colson choose his twelve disciples for brief respite from
the pokey? "You have to get someone who knows the prison and can
move around in it. One way is to go into the mess and see who bows his
head in prayer before the meal." Heads bowed in agreement all around
me. This is a world of semiprivate signals, a culture maintained amid
aliens—somewhat like the world of principled gays: Jesus Lib. A great
many of those present wore crosses or some other Christian sign like
the fish—but unobtrusively. There had been almost invisible nods, a
spark transmitted instantly, when Colson described how prison authorities stripped him naked for search when he arrived, but did not take the
cross from around his neck. Even the "screws" of old movies live by the
inhibition that movies placed on Bela Lugosi.

In our later conversation, when I gave him some examples of what I
consider an anti-intellectual treatment of the Bible by fundamentalists,
he asked, "Are you a Bible scholar?" No, I answered, but I read the
New Testament every day, since I am a Christian. "You are?" (His
eyes lit up.) "I had no idea!" (I had given none of the signs.) Yes, I
said, I'm a Catholic. "Oh . . ." (Lights out.) He thought I meant a real
Christian.

Though he ended his speech with the appeal for legal help on prison
reform, Colson's emphasis throughout was not so much on releasing
prisoners as converting them, sending them back to form prayer cells,
new Christian "catacombs," changing society by changing men's hearts,
one by one. Indeed, Colson is a bit shuffling and apologetic over his
reform efforts, since much of the evangelical movement opposes the social gospel. Yet, as I told him later, some of his protest against dehumanizing prison conditions reminded me of things I have heard from
my friend Philip Berrigan. He laughed at the comparison, and said,
"Yes, and at a book-and-author luncheon in Pittsburgh where I appeared with Jerry Rubin recently, I said there was a time when I looked
forward to reading Jerry Rubin's prison memoirs. It is one of the ironies
of the era." But Colson came back to the Berrigan comparison, to deny
it. He still opposes clerical involvement in social activism, all those
"men of the cloth out parading and demonstrating."

The resistance to public formulation of a just order (other than conversion of each man to God) is one important tradition of evangelism.
Yet I pointed out that Colson was working with legislators on prison
reform; he hoped to discuss the problem, next week, with Governor
Exon of Nebraska. "Yes, but I approach all these people at the level of
individual concern. Besides, I am not a man of the cloth." Would you

not work for prison reform if you were a minister? "If I were wearing a collar, I would not—no, I guess I would be doing the same thing. Maybe it isn't as great a distinction as I thought. But I do think it wrong for some Catholics to be siding with the Teamsters and other Catholics to be working against them." (The Teamsters, you remember, became Colson's clients when he left the White House.) "That just divides the community of believers." I said that an unwillingness to discuss any issues that might divide Christians could well make the Church irrelevant. "No, the Church is the most relevant thing, in the world. It looks to the only thing that matters, to the spiritual salvation of the individual." The overriding importance of this consideration makes it wrong to separate brother from brother on any less urgent matter—which explains the support Colson instantly won from the liberal Democrat Harold Hughes.

The evangelical faith works from a whole series of paradoxes. A public order built on the most private imaginable experience. Stability based on radical conversion. Separation of Church and State that nonetheless allows for secret signals between Christians in political life. Colson's centrality, after an overnight change, is a symbol of these paradoxes. His private experience restores his public status; in fact, enhances it. Those who do not grasp the paradox of his new religion might be surprised that Colson still hobnobs with governors, pleads with high government officials, takes "his" prisoners to convert senators, and works out of the posh Fellowship mansion. Precisely because evangelists do not share an expressly social vision, they depend on the social *status* of converts or potential converts. The private experience of conversion is what matters. But more people will notice a private individual if he is in the public eye. It is the man who counts, not his views —but certain men are more watched than others. When Campus Crusade takes its mission to the colleges, it seeks out the team quarterback, the paper's editor, the student government's president. One conversion made at that level will raise the question of conversion for many others. In the opening of Sinclair Lewis's novel, the charismatic evangelist Jud Roberts seeks out the football hero "Hellcat" Gantry—whose conversion and instant preaching before the school work wonders. I asked Colson if he had read *Elmer Gantry* since his conversion. He said, "No, but I get the thrust of your question. There are fakes in all professions." My point was not that Elmer's conversion was fake. It was real. I wanted to discuss the need evangelists have to go "head-hunting" among celebrities. Colson himself would not have come to the attention of Harold Hughes if he had not been a former White House aide.

Famous athletes serve a twofold purpose in the evangelical scheme of things. Affected as it was by the philistine milieu of the nineteenth century, American evangelism had to fight the sneer that religion is a "sissy" concern of Milquetoasts and women. The result was the muscular Christianity that rejoices in "he-man" ministers and churchfolk. At times, the whole evangelical movement seems to be turning into one Fellowship of Christian Athletes. When the sermons preached at Nixon's White House were published as a book, sports references and background were one of the dominant themes. One minister quoted a coaching friend. Nixon introduced a pastor from Ohio with praise for the Ohio State team. He introduced another man of God as the fellow who "won nine varsity letters." Among those called in to preach was a college baseball coach, Bobby Richardson. Billy Graham, of course, carries on a locker-room ministry at all the better golf clubs. That allows him to say things like this: "The first time I became suspicious that the United States might become involved in Indochina was in January 1961. I had just played golf with President-elect John Kennedy, the then Senator George Smathers of Florida, and Billy Reynolds, president of Reynolds Aluminum." One of those who visited Colson in prison is described this way in *Born Again:* "Jim Hiskey, who had a ministry to professional golfers throughout the country." Colson writes that his first prayer circle behind walls was made up of "three former tough, pragmatic marines"—the authentic Jud Roberts style.

Listening to Colson address the Washington prayer breakfast, I was amazed at how many Christians could be described in a single talk as "burly." Colson referred to himself as an ex-marine, and talked of doing some things "on the back stroke, as golfers would say." His prize anecdote was about the prisoner sent out for two weeks of "discipling" with Colson's group. He was predictably "burly"—a 280-pound black man whose nickname was "Soul." Colson described with glee how Soul went into a cellblock to spread the Gospel, though its inhabitants had told him to stay out. Close up, his size made them change their mind. "Thank God he's on our side," Colson said, to appreciative laughter.

Of course, other skills than athletic ones are admired by evangelists. Tom Phillips, the man who converted Colson, is president of Raytheon Company, the largest employer in New England. Colson found a testament to the Gospel in Raytheon's stock report: "With any other man, the notion of relying on God would have seemed to me pure Pollyanna. Yet I had to be impressed with the way this man ran his company in the equally competitive world of business: ignoring his enemies, trying to follow God's ways. Since his conversion, Raytheon had never done better, sales and profits soaring." Evangelical talk often turns on the accomplishment, even the net worth, of prominent Christians. By their own favored paradox, worldliness is made the guarantor of spiritual values. The prayer-breakfast movement boasts that such busy and important men take time out for the Bible.

Colson sees his ministry to crowds as a religious duty. "I never liked to give speeches. I was not an up-front guy—that's why I never saw you people of the press. I was a back-room guy. I enjoyed things like getting Ed Brooks [sic] into politics." If he exaggerates his importance prior to his conversion, that is in the tradition—Senator Brooke told me he had been running for various offices for ten years before he met Colson. Yet Colson is so traditional he grants himself hypothetical achievements in the past: "If I had been in the Senate at the time of the Tonkin Gulf Resolution, I would have joined the two who voted against it." Indeed, it was only his antiwar feelings that kept him from being appointed Assistant Secretary of State: "Elliot Richardson asked me in 1969 to replace Bill Macomber." But Colson could not in conscience work for a policy connected with the Vietnam war. (It is odd that he felt he could work with Nixon and not be involved with the war—he admits proposing to Nixon that Haiphong harbor be mined, explaining that he was against the war but felt that if we were going to make a fight, it should be a good one.)

Watergate, redeemed for others by his experience, is left rather shadowy in Colson's book. I asked why his confessions are so nonspecific. He volunteered a "spiritual," almost a fictive, guilt for savaging Daniel Ellsberg's name, since he did mean to do it but did not participate in the break-in that fulfilled his desire. He says he had no part in most of the things he was blamed for (e.g., the White House enemies list). Some things he did—like giving Life magazine writer Bill Lambert material to use against Senator Joseph Tydings—were based on what he thought was truthful information. In his book he expresses most regret for sins of omission—that he did not impart a Christian atmosphere to the White House. He admits to only one and a half sins of commission (counting the virtual smearing of Ellsberg as a half-crime at most). The effect is as much exculpatory as confessional, and I asked about that. He answered: "Does it have to be in the book to have the penitence? Do you have to catalog your sins in a book in order to cleanse yourself?" St. Augustine apparently thought so. "Then that's the way *he* felt he could cleanse himself. I didn't feel the need to do that."

I asked what specific sin of commission he most regretted now, and he mentioned again the one deed of sheer malice confessed to in the book—floating the false rumor that Arthur Burns was asking for a pay raise at the same time he was working for price controls. "I regret the Burns thing because I saw the hurt that I had done to him on his face that day when we had the prayer breakfast." At that nondenominational White House prayer breakfast (the one that broke the news that Colson had been converted), he was reconciled with Burns, who was chairing the meeting. Later, at Burns's Watergate apartment, the two men prayed together. When columnist Tom Braden called Burns to get information on Colson's dirty tricks, Burns vouched for the sincerity of Colson's religion. So Colson turns his one admitted crime into a scene of reconciliation with a very important man, who then gives him a character reference, repeated in Colson's own book.

The book describes how other Christian politicians forgive each other and forget their differences—e.g., how Harold Hughes, who opposed Earl Butz's appointment as Secretary of Agriculture, apologized at a prayer breakfast and said, "If I'd known you then as I do now, brother, I might have led your cheering section." This picture of men agreeing in love, without regard for political views, is the evangelical ideal. It animated Abraham Vereide when he launched the prayer-breakfast movement in the fifties. And it takes on new force now, when there is special urgency to put Christians in high office after Watergate and other scandals. These episodes in our history are seen not as failures of policy, but as departures from personal morality—for which the obvious remedy is the installation of men whose personal morality one can trust.

Bill Bright, the influential leader of Campus Crusade for Christ, has launched a campaign to "turn America around," saying that the country will be taken over by God's foes unless a sufficient number of His friends reach positions of influence. *Sojourners,* the Left evangelical journal that keeps a wary eye on Right evangelicals, claims that businessmen have already pledged large sums of money to a future effort to make evangelical Congressman John Conlan the President of the United States. Conlan is a spellbinding orator, and Colson told me, "Everywhere I go I am asked about Conlan, whether he would not make the ideal President." Conlan, a Republican in his mid-forties, is the son of major-league umpire Jocko Conlan; he was a Fulbright scholar in Cologne after graduating from Harvard Law School. He claims to have been converted to Christ from socialism, and has combined a vivid oratorical style with a right-wing voting record as representative from Arizona's Fourth District. He is competing now for Paul Fannin's seat in the Senate.

Colson does not support specific candidates, but his practiced eye sees great political potential in Conlan. So does the appraising eye of Richard Viguerie, the direct-mail entrepreneur who raised funds for George Wallace's 1976 campaign. He says, "The next real major area of growth for the conservative ideology and philosophy is among evangelical people." He expects to do direct-mail work for the movement and for candidates it favors, including Conlan—he calls him the man who can mobilize national opinion (Conlan is on the board of the Committee for the Survival of a Free Congress, one of Viguerie's accounts). So much for the division of "converted" Church issues from the State.

When Colson read the *Sojourners* description of Bill Bright's organizational genius on the right wing of the movement, his immediate response was to set up, for Bright, a meeting with Mark Hatfield, the best-known spokesman for leftist evangelicals. Brothers should be united, not divided, over politics. At the meeting, Bright told Hatfield and others that his effort to bring America back to God was not political in any partisan sense. He is for the election of real Christians, whether Republican or Democrat, conservative or liberal. Hatfield found his distinctions elusive, but he admits that the meeting was a bit disjointed, with vote-calls on the floor of the Senate. Another meeting is being arranged by Colson. The "back-room guy" is now an "up-front" preacher, but the man who likes to think he launched Ed Brooke's political career may play some role in the future of men like John Conlan. Colson's conversion, undoubtedly sincere, has luckily—providentially, some would say—put him right at the center of the religious action.

—New York *Times,* August 1, 1976

29

Stained-Glass Watergate

Local patriotism has long made me defend Baltimore as America's place of nicest scandal. I know other towns can beat us in terms of organized corruption. They monopolize crime. We are free enterprisers of the petty cheat. Spiro Agnew, bought for comparative pennies, captures our style to perfection. We recently had three of the area's top county officials in jail—no record, to be sure. But one was indicted, though not convicted, for "carnal bribery," a crime from some minor biblical epic in Technicolor.

More recently we have added to our gallery of scandal a little group of priests called Pallottines. They have computerized the wheedling arts of Chaucer's Friar with such success that some of Maryland's better-known financial types have become involved with the reverend fathers in matters of hard cash. But success turned recently to scandal, indeed, to two scandals: one involved the way Pallottine money was raised, and the other, how it was spent. Two separate investigations—one of Maryland charities and one of Maryland's governor—converged unexpectedly on this obscure order of millionaire operators, and two reporters from the Baltimore *Sun* became the Woodward and Bernstein of a stained-glass Watergate story, prying details week by publicized week from behind a stone wall.

The Pallottines are named after a nineteenth-century priest, St. Vincent Pallotti, who offered sentimental Italian piety as a form of "social action"; in America, the order's priests served poor immigrants. They have had a rather undefined purpose since those immigrants became educated and active in American life. But the order also has some missions in foreign lands, and it has parlayed distant poverty into wide American holdings. The Pallottines' Eastern headquarters used to be in Baltimore, where their fund-raising operation is still run out of the Mission Center and a vast inaccessible warehouse. But its very presence was a secret to ordinary citizens like me, since the priests, with an animal fastidiousness, did few mailings near their nest. Shortly before the scandal broke, the provincial headquarters was moved to New Jersey. Archbishop William Borders of Baltimore tries to give the impression that he is one of the few important figures in Baltimore who has not dealt much, even indirectly, with the Pallottines—and that may be the case. These priests show a marked preference for pols over prelates.

On one occasion, the friend of many of the order's friends, Governor Marvin Mandel of Maryland, needed $54,000 to maneuver his way through a politically risky divorce. That sum was lent, in three packets, by the Pallottines to one C. Dennis Webster. (Webster's uncle was Governor Mandel's campaign treasurer in 1974 as well as a campaign accountant for Spiro Agnew; he is the Pallottines' accountant and financial adviser.) Webster lent the same amount to the governor for his divorce. Then, too, the Pallottines became financially involved with one of Governor Mandel's fellow defendants in an ongoing case of alleged racketeering. (This defendant is Dale Hess, whose firm, Tidewater Insurance Associates, insures the vast Pallottine holdings.) And the Pallottines also put money in a modular-classroom deal that led to a fifty-one-count indictment (hefty even for Maryland) of the state school-construction chief. The order had roughly a third of a million dollars in the firm that allegedly got preferential bidding.

Other scandals go outside politics, the Florida motel, for example, owned by the order, that had a broker problem. But the order sold that motel, at great profit it is thought, to Dennis Webster. The order's three other Florida motels apparently have been as innocent of venality as of any mission work, other than keeping Gideons in every room.

Hard digging by the *Sun*'s young reporters—Mark Reutter, age twenty-five, and Steven Luxenberg, twenty-three—turned up an astonishing series of Pallottine investments in real estate and direct-mail operations in Maryland and Florida and North Carolina. And a preference for lawyers and others with political connections—with Spiro Agnew's former lawyer and his campaign accountant, with campaigners for Edmund Muskie, with ex-Senator Joseph Tydings, with many friends and supporters (and codefendants) of Governor Mandel. And a series of questionable practices: having religious tax exemptions for activities and quarters devoted to secular affairs. Archbishop Borders, too, has concentrated on the uses of Pallottine funds, calling for a public audit of what was invested and delaying his own moral commentary until such time as the audit is published.

But the Pallottines' way of raising money is even more interesting than how they invested it. There were millions to invest because of a very sophisticated begging operation. The *Sun* reporters uncovered this by getting postal records for charity mailings at the Baltimore post office. The records, released in November of 1975, were hastily studied; two days later, the Pallottines held a press conference on the Mandel loan; it was to be the first and last press conference in the months leading up to release of the audit. Reporters Reutter and Luxenberg worked long into the night with a calculator and ended up going off the calculator's capacity—which meant they were dealing with mailings in the triple-digit millions. They found that, by 1974, the order was mailing over one hundred million begging letters, greeting cards, and sweepstakes flyers per year. It had dispensed up to five million pieces of mail (all "individualized" by computers) in a single day, making the Pallottines the second-largest private bulk mailer in Baltimore. In return, the post office delivered to Pallottine headquarters two to ten pouches of mail every day—each pouch containing about a thousand letters and most letters containing money to be spent on the needs of baptizable pagan infants.

The plight of these infants was touchingly depicted, in letters mass-produced to look personal. The letters had a poor feel about them, though they were expensively produced. Machinery scrawled convincingly handwritten messages of great urgency: "*You,* dear friend, were much on my mind about fifteen minutes ago. I was on my knees in prayer at our Mission Chapel. I was praying for the poor in our Missions of South America, India, Australia, and Africa. I was also praying for *you.*" Most such letters included pictures—infants of odd hue in desperate condition. Mission activities seemed to be devoted almost entirely to infants—nothing wrings the hearts of poor old women like even poorer young children. The letters suggested poverty in the writer as well as in the writer's flock: "Please return this photo so I can send it to one of our friends. Thank you." That guaranteed a return (with cash). But the returned photos were flipped into wastebaskets along with the returners' letters. Only the money counted—and counted up, fast. Twenty or so women were employed to open letters, sorting checks and cash into a dozen pigeonholes at each desk; each handled about a thousand dollars per day. Meanwhile, at the Pallottines' warehouse, modern equipment slipped new photos into computer-keyed repeat appeals: "If you break down the figures Reverend Felix wrote me, you can see that $20 will provide books and tuition for Jettu Paul for another year." (A picture of Jettu Paul, appropriately rickety, was sent along.)

Other gimmicks used were the charity sweepstakes, with cars and other things awarded in the interests of Jettu Paul (or whoever was being saved with millions of returnable pictures that month). "Free" Christmas cards were sent out to prod the consciences of their recipients into a donation. And St. Jude pens.

It is true that the Baltimore Pallottines sent some money to their foreign missions—about a quarter of a million dollars in 1974, less than they put into the modular-classroom project and a mere fraction of what they had to invest in other things, from condominiums to golf courses. All the invested money was ultimately intended, we are assured, for the order's work. But the women responding with five or ten badly needed dollars to help a child pictured as just minutes from starvation could not dream their money was going to wend its way, at a leisurely percentage rate, through investment channels. Besides, the Pallottines showed the true Baltimore style of petty corner-cutting when they listed in their comparatively exiguous outlay on the missions the cash value of medicine donated by drug firms.

Nor could donors dream that the mastermind behind those tear-jerking letters, the Very Reverend Guido Carcich, was doing some of his mission work in Las Vegas, rendering to Caesars Palace what belonged to Caesars Palace. (He was there with Maryland connections, one of whom rented a VIP suite while attending a direct-mail convention.) The *Sun*'s Mark Reutter traced the Florida comings and goings of a "Mr. John Carcich," who maintained an apartment and an air-conditioned car in Fort Lauderdale to oversee the multiple holdings of the Pallottine order in that area.

Father Carcich used to be seen dining at restaurants favored by Baltimore politicians. But since the scandal broke, he has become a ghost, spooking in and out of town, untraceable. Neither of the *Sun* reporters has ever seen him, despite strenuous efforts—Steve Luxenberg once chased a man into a shoe store in the forlorn hope that it was he. Mark Reutter heard his voice on the phone, some time before the scandal broke, telling him the *Sun* had no right to investigate Pallottine affairs.

Father Carcich was not always so shy. Once he required the employees who opened money mail in the Mission Center to buy a posh book of photos celebrating his rise from poor Yugoslav immigrant to entrepreneur and spiritual big shot. At the time of Vincent Pallotti's canonization in 1963, Carcich was publishing a magazine called *P.A.* (for *Pallottine Apostolate*—though Father Carcich referred to it, jokingly with his pol friends, as "Pain in the Ass"). The issue devoted to St. Vincent's canonization by Pope John contained four pictures of Father Carcich, the magazine's editor. At that time, some devotees of St. Vincent invited Father Carcich to address them in Pennsylvania, to tell them what they could do to spread devotion to the saint. They were disturbed when his advice all boiled down to different ways of sending him cash. One of those at the conference, a descendant of the Pallotti family, told me that Father Carcich showed contempt for the poor Italians with whom he had grown up and gone to the seminary in New Jersey.

The women who worked for Father Carcich, opening the mail at the Mission Center, had special reasons for disliking him. He monitored the time they spent in the lavatory, and cut off office conversation. One woman, Lillian Goralski, told the *Sun* reporters: "If you took your sweater off [at the coatrack] when you came in and then got cold, you weren't allowed to put it back on until the midmorning break." If she had been a starving black baby from some exotic clime, Father Carcich might have felt obliged to clothe her—but only after taking her picture for inclusion in the appeals.

The thing that bothered these women most was the callous attitude shown toward contributors whose letters, most begging for prayers, were discarded—all but those with big contributions. Several ex-employees told the *Sun* that letters requesting masses were discarded unless they contained at least ten dollars. Other letters asked that a candle be lighted for the donor at the Pallottines' garish and vulgarly promoted St. Jude's Shrine in Baltimore. Mrs. Goralski said she, too, gave money to have a candle lighted for her son in Vietnam. "But after a while I realized that they only lit twenty candles no matter how much money was sent in, so why bother?" Once she asked Father Carcich what he would answer if a contributor asked him where his or her candle was. "He told me he would just point to any of the twenty and say, 'That one.'"

Archbishop Borders has been curiously in and out of this affair from the start. He requested the public audit of the Pallottines but then told the press he could do no more than request it. I wondered why. Canon law says that the "local ordinary" (i.e., spiritual ruler of the diocese) must see that Church law is obeyed, to the exclusion of superstition and "anything in conflict with ecclesiastical tradition or savoring of sordid profiteering [Canon 1261]." Good for Church lawyers. They know religion can be used for profiteering, and the savor of it is all around the Pallottine affair.

I went to ask the archbishop about this and told him I came both as a member of his flock and as a local journalist (I had met him recently for the first time when he administered the sacrament of confirmation to my daughter). The archbishop repeated that he had no power over the Pallottines since they are a religious order with their own superiors, all ultimately answerable to a superior general in Rome. True, *but:*

1. "Exempt" orders are those that have taken the three solemn vows of poverty, chastity, and obedience. Congregations that take simple vows (like the Pallottines) are separate from the exempt orders (Canons 615 and 618), despite some "extension" of privileges by analogy with the exempt orders.

2. Even exempt orders are subject to the local ordinary insofar as their parishes are administered within the local ordinary's jurisdiction (Canon 631), and the Pallottines have two parishes in Baltimore, including the one that houses St. Jude's Shrine. Since that shrine is cited as a beneficiary in Pallottine mailings, a right to oversee any superstition or profiteering arising from its use is clearly within the archbishop's power.

3. Even exempt orders are subject to visitation by the ordinary if he has "reason to suspect that there exists a failure to comply with his regulations" on the heads of superstition and profiteering (Canon 1261).

4. Ordinaries of dioceses *other* than the one in the order's place of residence can stop or control any "systematic" alms-gathering, even by exempt religious orders (Canon 621). Thus, all Archbishop Borders had to do, if he wanted, was to ask some other bishop to look into the matter and commission him to stop the alms-gathering racket.

5. The Pallottines are an apostolic order with lay membership in charitable institutes, which is to say that laymen, by contributing money, can join with the clergy in a "spiritual partnership." This is a situation over which local ordinaries have special power, especially "in the event that abuses are reported [Canon 1491]."

Besides, this is all a dance of legal gamesmanship. It is commonly admitted that the bishop of a diocese can control a religious order's activities inside his boundaries if he really wants to. Thus, Cardinal Spellman encouraged New York Jesuits to exile Daniel Berrigan to Latin America in the early sixties. The very week I went to see Archbishop Borders, the president-headmaster of Baltimore's Jesuit high school denied Philip Berrigan the right to speak to students, and gave as one of his reasons that he did not want to get into trouble with the local ordinary—though the Jesuits are an exempt order with far better ecclesiastical connections than the Pallottines have ever aspired to. Archbishop Borders pretended to powerlessness not because he was indeed powerless but because he wanted to appear powerless.

The archbishop, a nice man slightly less ineffectual than his predecessor, was obviously perturbed by the scandal; but the center of his perturbation seemed to be the publicity it had raised and the effect it might have on other fund-gathering operations. The "common good," he said, demanded no "recriminations—I am not investigating the Pallottines." What was needed was sincere desire to amend. But amend what? How can one know, until one investigates, what needs correction? Well, he was going to find that out.

But the audit can only show certain things: how much money came in and from where; how much went out and to where. It cannot show, for instance, anything about the morality of the personal letters. And morality is supposed to be the realm of the archbishop, not high finance. The audit can say nothing about prayers asked for and not given, masses paid for and not said, or candles not lighted. The archbishop said he would get to the truth about any possible simony, which is the Church offense. How? By investigation (he said he was not *doing* that). When? "In time." Besides, he was pretty sure simony had not occurred. Why? Only one woman had said so, to the newspapers; and the archbishop does not like people who talk to journalists. But in fact, the *Sun* has worked on the confirmation-by-two rule of investigative journalism, and its reporters quote one woman "and other mail sorters" on the subject of mass requests being discarded. Had the archbishop made any effort to talk to these women, themselves Catholics and members of his flock? Did he intend to? No. Then how would he discover whether they were telling the truth? By asking the Pallottines! "They are not entirely without conscience." But according to the women, the ordinary Pallottine priests had no way of knowing about mass requests discarded the minute they came in. Only Father Carcich, who gave the women their instructions, and perhaps Father Peter Sticco, who oversaw their work, needed to know about those instructions. In other words, the archbishop would ask the *accused* if he did any thing wrong; but he would not ask the accusers, whose report he had already told me he disbelieved, on no other grounds than that they had given it to the newspapers. (And besides, they were mere laywomen, not priests.) The archbishop became most eloquent when describing the dangers of "prejudging" the Pallottines—though minutes before he had prejudged laywomen who talked to reporters, and shortly afterward he expressed the harshest secondhand opinions about a priest sincerely trying to live the Gospel in his diocese, Father Philip Berrigan. (The archbishop admitted he had never met or tried to meet Father Berrigan.)

Yet the archbishop assured me he would rectify things. He had a panel that was drawing up guidelines for charitable donations. I thought he had no power over orders engaged in that work. Oh, yes, he said, these will merely be suggestions—but "*strong* suggestions." All right. Who made up this panel? Good people. Could he give some names? No. It is best that we trust our rulers, working for our good in secret. (The *Sun* reporters later got wind of the panel's existence and persuaded a couple of its members to reveal the names of the others.)

When I asked the archbishop if he had visited the Pallottines' warehouse and he said he had no power to go there, he added that he might be admitted "as a private citizen." How, as a private Christian, would he express himself on the way the Pallottines had conducted themselves in this affair? Did he remember that *parrhesia,* confident openness, was described as a defining Christian attribute in the Acts of the Apostles? For that matter, did he remember St. Peter's very strong and slangy Greek expression when Simon Magus approached him to discuss spiritual powers merchandisable for silver? I quoted the passage, which is best rendered: "*Damn* your silver, and you with it!" (Acts 8:20). Aren't there times when that is a better Christian reaction than his strongest commitment so far, expressing "disagreement with the basic fund-raising philosophy" of the Pallottines? He remarked mildly that that was just *my* interpretation of the Bible passage. Oh, yes, and private interpretation is only for Protestants, isn't it?

The photography in mission literature gives the impression that Christian infants never grow up in darkest Africa. The archbishop seems to think the same is true in darkest America.

—Esquire, July 1976

30

Pope John Paul II

Novelists have toyed with the scandal of a Marxist on the throne of Peter. But has fact already outrun fiction? We might think so when reading Karol Wojtyla on the alienating effect of industrial labor:

The Car Factory Worker

> Smart new models from under my fingers
> whirring already in distant streets.
> I am not with them at the controls
> on sleek motorways; the policeman's in charge.
> They stole my voice; it's the cars that speak.

During the conclave that elected him Pope John Paul II, Wojtyla was seen reading a Marxist journal, and he is the first pope to give the *abbraccio* to Communist mayors of Italian cities. Then why do the anti-Communists love him so?

It would be an odd man in any circle who could write this:

While the somatic dynamism and indirectly the psycho-emotive dynamism have their source in the body-matter, this source is neither sufficient nor adequate for the action in its essential feature of transcendence.

And then write this:

> The moon became a tambourine
> thrumming deep in their eyes
> in their hearts, deep.

Odder still for that man to become pope.

Athlete, linguist, theologian, musician, poet—this Wojtyla pope seems too good to be true, and probably is. He is, for one thing, ominously likable. His Wednesday audiences in Rome had to be moved out of the new audience hall into St. Peter's itself, to handle the crowds, and then into the piazza of Bernini. The crowds grew, early last summer, to 100,000, tying up Roman traffic even after the hour was moved back to placate city officials. The pope outdraws rock stars, and his name sells disco records. *Time* magazine, in a cover story on the dearth of leadership, found him the only natural leader of obvious stature on the world scene. A writer in the Jesuit magazine *America* thinks he may fulfill Plato's dream of the philosopher-king. There has been no such elation around a new pope since the election of Pius IX in 1846. Nightly torchlight processions turned out all Rome for the first days of that reign, and the world press rejoiced that an enlightened ruler had arrived to reconcile Rome with progress. Yet *Pio Nono* came, in fact, to make that reconciliation impossible for the next century and more. His liberal political views fell victim to his theological conservatism.

Now John Paul is cheered by the ecclesiastical right and left, by cold-warriors and peaceful coexisters, by the Communist mayor of Rome and a Ukrainian dissident from Russia. He cannot, one would think, please all these antithetical types, yet he seems to. Most of the world thinks of him as open and flexible, but the most intransigent bureaucrats of the Vatican Curia think he is really on their side. Who is wrong? Or can this miracle worker square the circle, make dogmatist lions lie down with ecumenical lambs?

One solution to the mystery is to see in Wojtyla a consummate diplomat and compromiser, one who distributes his favors so evenhandedly, left and right, that he does not get trapped in a sterile one-sidedness. That reading would make the pope succeed at what Jerry Brown calls "canoe politics"—paddle a bit on this side, then on that. But such flexibility in a Brown looks wishy-washy, and that is the last thing one would say of Wojtyla, this chunkily handsome man with the blunt stare and the John Wayne lumbering walk.

Still, his record is curiously ambiguous for one who seems to radiate decisiveness. During the Nazi occupation of Poland, he neither collaborated nor defied—he withdrew to a seminary in Cardinal Sapieha's house. (Some romanticizers of his youth claim he went into hiding there because he was on a Nazi "wanted" list for helping Jews, but the pope's own friend Mieczyslaw Malinski makes it clear that all seminarians had to live indoors because they could not get the German work permit that gave access to the streets.) At the time when Cardinal Wyszynski was imprisoned, Wojtyla managed to avoid his superior's fate, yet went farther than he in stressing freedom of religion. In Poland he championed liturgical reform and ecumenism, but in Rome he stressed unchanging doctrine, like his rigidly Thomist mentor, Father Reginald Garrigou-Lagrange. On the tenth anniversary of *Humanae Vitae,* Paul VI's encyclical condemning contraceptives, Cardinal Wojtyla was the principal speaker at a Milan conference celebrating the encyclical.

Since becoming pope he has emitted contrary signals. Two of his early acts were a moratorium on the laicization of priests (which lets the discontented leave their ministry without forcing them out of the Church) and a strengthening of an "apostolic constitution" on doctrinal uniformity in papal seminaries. He went to the Puebla conference in Mexico, where the "hot" issue was liberation theology, and avoided using the term all the time he was on Mexican soil. He has, apparently, given the cardinal's hat to a Lithuanian, Steponavicius of Vilna, encouraging one of the most resistant ethnic groups in the Soviet Union, yet he did it *in petto,* secretly, so as not to offend the Russians. In the same way, he granted a private audience to the Ukrainian dissident Valentyn Moroz but removed his name from the official list of those received. Unlike his part-namesake, Paul VI, John Paul stayed out of the Italian elections last June—which only helped the Christian Democrats (papal intervention in electoral politics regularly backfires in Italy). One leftist member of Parliament I talked to in Rome said she feared John Paul will make reaction respectable, while most people say he is making liberal attitudes blossom in that oddest of settings, the pope's great Renaissance palace.

How can John Paul send so many signals in so many directions without getting tangled in his own footwork? Far from looking dithery, his every gesture comes off as magisterial. He combines an easy friendliness with an equally easy certitude. As people plop all kinds of silly hats on his head, he never looks silly. A journalist who went with him to both Mexico and Poland told me John Paul never seemed, like most visiting officials, to be puzzled by his surroundings. Entering a strange room, he knows where all the doors and obstacles and people to be greeted are. His authority is accepted because he does not have to assert it—he, like everyone else, just assumes it is there.

I was thinking of this as I waited in the rain for one of his Wednesday audiences last June. I stood, by the rail he would pass, with a tiny Italian nun who shared her umbrella with me and a tall Chilean priest who had shrewd and disillusioned things to say about governments and the Church. When the pope came along the rail, they both lunged and grappled like any groupie at a rock concert. I tried to lean back and watch, despite the frenzy, with some journalistic detachment. I certainly felt no emanation of excitement from the pope. He keeps a distance of his own, dipping in but moving on, his eyes taking in the whole scene while his hand is seized and released. I had heard he brought an American style of politics to the Vatican—"working the fence" as deftly as Lyndon Johnson ever did. But American pols try to summon a quick effusiveness, feign momentary intimacy, as they shake each hand. A President, asking for votes, is a suitor. Leading a democracy, he must be ostentatiously egalitarian despite the trappings of Air Force One gleaming behind him. John Paul, with Bernini's columns swirled around him, feels no such need. When his sweeping eyes meet those of another person leaning back, their expression is just short of a wink. His calibrated friendliness is paternal, not democratic. A favorer of the *abbraccio* (especially with children, as Amy Carter learned), he welcomes the world to his arms—but to give warmth, not take it. And though he encourages directness, one sees at a glance that he would not tolerate "liberties" with his person. He is even less a Mick Jagger than a Lyndon Johnson.

The pope's principal scholarly work, *The Acting Person,* is an attempt to reconcile the anthropology of Max Scheler with Aristotle and Aquinas—a tricky maneuver, since Scheler tried to reconcile Plato and phenomenology. The doctrine of Scheler that Wojtyla stresses, even in his poetry, is the way man's grasp of particulars is both an a priori knowledge of essences *and* a renewed act of self-possession:

> Only the one who has possession of himself and is simultaneously his own sole and exclusive possession can be a person.
> Love and move inwards, discover your will.

The strong will of this self-possessed man, all of whose moves are so controlled and sure, creates a mystery: Why, with his sure grasp of himself, has he remained so elusive to others? Even his appearance fluctuates oddly: He manages to look both lean and lumpy, handsome and feral, refined and vulgar. In the Vatican, no one knows the first thing Roman gossips want to know about any pope: Who has his ear? He is everywhere and nowhere—the first pope to visit all the agencies of the modern Vatican, to begin the systematic journey to every parish in Rome. He likes people around him—at his daily mass in the Vatican gardens, at meals, in the planes he takes to far places. It is as if his years of travel through the Iron Curtain have given him a Frostian dislike of walls. He has not defied the Curia, just gone around it, ignored it, been off on his travels.

All this energy is free to play because the pope shows no doubt about his goals and beliefs—as Chesterton said, the hands are free because the heart is fixed. Wojtyla seems to have escaped the soul-searching that other Christians have gone through in recent decades. A symbol of that is the way his play, *The Goldsmith's Shop,* reflects the poetic drama of Paul Claudel. Like the early Maritain—or, for that matter, Pound and Eliot in the same period—John Paul is ready for all kinds of technical experiment because his faith is very old-fashioned. This shows up in his sophisticated use of folk piety, especially to the Virgin. In Poland, a Catholicism familiar in the thirties and forties has been preserved as in a time capsule—and not surprisingly. Persecution tends to call up heroism, willpower, determination, the fixed heart—and tends, as well, to put doctrinal questioning aside "for the duration." The noble folly of defiance is thrown, as a halo, around the mere caution of old dogmas. In a persecution men die for beliefs they have not time, in the crush of heroic events, to discuss. Even accidental symbols of the faith became banners of resistance—the pope told priests in Rome to resume their clerical dress as a proud uniform. John Paul's words have the noble simplicity of a voice from the catacombs.

But that voice, bracing and almost swashbuckling when it sounds from the catacombs, can be disturbing when it issues from a throne. John Paul said, in Mexico, that priests should administer the sacraments and let the laity deal with politics. That division reflects a subtle and old-fashioned clericalism, separating (and ranking) spheres of the natural and the supernatural, but it also reflects, no doubt, John Paul's experience in Poland, where simply administering sacraments *is* a political statement, a defiance of official atheism. John Paul's attitude toward the state is tinctured with theocratic traditions from Poland's past—which makes him well disposed toward the "integralism" of Spain's right-wing lay order, Opus Dei. (A collection of John Paul's addresses to Opus Dei members in Rome has been published in Italian as *The Church's Faith.*)

In theology, John Paul is even more doggedly "orthodox" than Paul VI was. Asked to preach a series of sermons to the papal household in 1978, he devoted one discourse to praise of Paul's reassertion of the reality of the devil, brought Mariology into every discussion, and attacked the ideal of scientific progress that has given us nuclear weapons—and the means of contraception. Some have accused John Paul, in this period, of politicking among Italians for the pope's chair (he said he sat with Italian-speaking bishops at synod conferences because they were having a problem with the Communists), but there can be no doubt that his faith is sincere and unquestioning.

Well, what's the matter with that? Perhaps the Church needs a return to certitude about its own doctrines; that may make it deal more confidently with the world's problems. Besides, what does the world at large care about internal theological disputes of the Catholic church, so long as the pope uses his moral authority for such causes as peace and human rights? The trouble with this analysis is that Pope Paul VI (like *Pio Nono*) demonstrated how sterile can be the combination of political liberalism and theological conservatism. No one could have been more sincere or eloquent than Paul in arguing the cause of the poor and the oppressed. But he shot down his own troops before they could carry his enlightened social program into action—driving priests and nuns and laity away from the Church with his views on contraception, on the role of women in the Church, on the celibate priesthood. The flow of missionaries, money, priests, and nuns to implement the pope's humane programs dried up. And this new pope is even less ardent on social programs than Paul was and more rigid on doctrine.

To understand the crisis of the Catholic church, one must recognize that the Church's distrust of the modern world was effectively challenged by John XXIII and the Second Vatican Council. But the distrust once reserved for "outsiders" has now been transferred *inside* the Church, from such niceties of doctrine as the pope's infallibility (defended against Protestants) to questions of internal morale and manners. I was shocked, while at dinner with an urbane monsignor high in a Vatican agency, to see him turn ferocious with glee over John Paul's moratorium on the laicizing of priests. Many in the Curia hated Paul VI's comparative leniency on that count. (Paul thought that if celibacy was to be maintained, it must be a free choice.) "Now," said the monsignor, "we don't have to pretend that we believe Father So-and-So when he tells us he found Christ by screwing Sister Mary Sue." I asked what possible good could come of forcing those soured by the priesthood to stay on—or, more likely, to leave both the priesthood and the Church. "It is the morale of those who stay that we have to consider." Apparently, those who freely choose celibacy can enjoy their choice only if others are made to suffer for disagreeing with it.

John Paul is the pope least likely to change those self-defeating attitudes in the Church. That is why we should take very seriously the confidence that Curial conservatives have in their new man. They have been there a long time and know what works to their advantage in the long run. The stronger-willed the new pope, the more will he impose his Polish church's views on the role of priests and nuns, on women and the Virgin, on celibacy and discipline. In fact, there may be something anachronistic in the wish for a "strong" pope. What does that mean? The larger the man, the smaller looks the toy castle in which he has to stand, so that his best efforts undermine his own work. That was true even of John XXIII, who asserted his will by creating the commission on contraception that has largely emptied Catholic churches. It was especially true of Paul VI, who went forward with reforms while seeing how they backfired. A Marxist politician in Rome said, "Paul at least had some doubts about his own role in the world—a tragic sense that I found appealing. This new man has no doubts at all. That's what's scary."

Perhaps the modern papacy is innately tragic, a historical freak, trying to combine the eternal Gospel with a passing Renaissance ideal of *The Prince*. What would St. Peter make of its modern trappings—he who came to Rome as a missionary to the Jewish community that had embraced Christianity, undoubtedly unable to speak Latin, probably accompanied by his wife, and who was killed before he ever had a chance to become "bishop of Rome"? He was not a man self-possessed, but "carried where he did not wish to go." What would he make of the media superstar who rides as high in the world's opinion now as Pius IX did in 1846? Would he find in him an apostle or The Prince? That question might give pause even to the formidable John Paul II, if he were given much to pausing or to doubts.

—*New York,* October 1, 1979

31

Dorothy Day

In an earlier time, pious legends would have been woven round the life of Dorothy Day long before she died. Giotto could have painted her meeting with Peter Maurin the way sacred artists painted the meeting of St. Francis with St. Clare. And that might, in fact, be the best way of rendering the event.

It occurred in 1932. Dorothy had been a Catholic, but a rather aimless one, for five years. Trying to find again the ardor of her radical pre-Catholic days, she had gone south to write about a hunger march in Washington. The experience was unsettling: "Is there no choice but that between communism and industrial capitalism? Is Christianity so old that it has become stale, and is communism the brave new torch that is setting the world afire?" The hunger march led her to anguished prayer in a Washington church that is much criticized on aesthetic grounds, a prayer that would have made the place a holy spot in medieval times, just because of Dorothy's ordeal there: "I went to the National Shrine at the Catholic University on the feast of the Immaculate Conception. There I offered up a special prayer, a prayer which came with tears and with anguish, that some way would open up for me to use what talents I possessed for my fellow workers, for the poor."

When she got back to her New York apartment, where friends were caring for her young daughter, Dorothy found a shabby stranger sitting there; she fed him; she turned him out. For one thing, he smelled, and Dorothy was fastidious. But Peter was persistent; he turned up again the next day, and the next day. She remembered later: "No matter what it was that I had to do, housework, shopping, ironing, mending, cooking, Peter followed me around, not only interpreting daily events in the light of history, but also urging a program of action." It took her some time to realize that he was the unlikely-looking answer to her prayer in the shrine. Was there nothing to choose between capitalism and communism? To Peter Maurin, the philosophizing French hobo, that question would never have occurred, so obvious was it that his brand of Christian "personalism," worker solidarity, and agrarian "distributism" was the only respectable historical option for a Catholic.

Peter's narrow dogmatism, the plodding peasant certitudes of a self-educated man, paradoxically opened up a new world for Dorothy. It began the union of essentially conservative thinking with radical concern that marked all her activities. She would henceforth confound, equally, those who wanted her to be a rebel and those who wanted her to be submissive. She scared the orthodox by taking their doctrines seriously and the radicals by taking religion seriously. She was a "difficult" person because she found no difficulty in being dutiful toward Cardinal Spellman and admiring of Fidel Castro.

She met Peter just before Christmas, and by May Day she produced the first issue of the *Catholic Worker*—four people passed out the first 2,000 copies in Union Square. By September, twenty thousand copies were being produced. In the fall, a Workers School was opened. In the first Christmas issue of the *Worker*, she wrote: "It was just one year ago that I spent the morning at the National Shrine of the Immaculate Conception in Washington." The story is too good to be true, and too similar to a thousand hagiographical tales. Peter hounded Dorothy the way Ignatius hounded Francis Xavier through Parisian streets, breaking down his resistance to grace. Indeed, the greatest story of conversion is repeated here—Saul wandering blind until Ananias came to him: " 'Saul, my brother, the Lord Jesus, who appeared to you on your way here, has sent me to you so that you may recover your sight, and be filled with the Holy Spirit.' And immediately it seemed that scales fell from his eyes. . . ." (Acts 9:17–18). In more credulous days, the events in that New York apartment, between one December and another, would have been recognized as providential—and what better explanation can we offer?

The story is made more intriguing by the fact that its earlier parts read less like the embrace of Francis and Clare than like Lillian Hellman's affair with Dashiell Hammett. Rayna Simons, the radical beauty who died in Moscow, played Julia to Dorothy's Lillian. In those early years, Dorothy seemed to have a knack for being in the right place at the right time. Her father's raffish racetrack career took him to Califor-

nia in time for the San Francisco earthquake. Then he moved to Chicago, where Dorothy won a scholarship at the University of Illinois,
which led to her meeting with Rayna. Then it was on to New York,
where she helped edit *The Masses;* to Washington, where she went to
jail with Alice Paul's army of suffragists; to Staten Island, for summers
with Village literati; to California, as a screenwriter (again the Hellman
resemblance); to strikes and demonstrations, as a radical journalist.
Dorothy's best friend, who would die in a Catholic Worker house, was
married to Malcolm Cowley before she took on herself the thankless
job of beating young sailors away from Hart Crane. Cowley, meanwhile, fell in love with Dorothy's sister, Della; and Eugene O'Neill seems
to have been interested in Dorothy because she looked like the departed
Louise Bryant. Dorothy believed, with good reason, that she was one of
the models for Josie in *Moon for the Misbegotten.*

Dorothy, who would later be censorious about the sex and drugs of
young people who drifted through Catholic Worker houses in the
1960s, flushed the heroin down a toilet stool after Louis Holladay died
in her arms from an overdose. She had one thwarted love affair (which
led to an abortion and a suicide attempt), a perfunctory marriage
(quickly terminated), and one stable affair that left her with a daughter
(Tamar). Malcolm Cowley claimed, in *Exile's Return,* that "gangsters
admired Dorothy Day because she could drink them under the table;
but they felt more at home with Eugene O'Neill, who listened to their
troubles and never criticized."

What, in one woman's life, tied these two worlds together, the worlds
of Giotto and Jack Reed? Again, the only answer seems to be a mystery
that other ages at least had a name for: grace. Dorothy grew up with a
simultaneous and unbidden attraction to what seemed, in her culture,
opposite arenas, to the Catholic church and to radical concern for the
poor. She noticed, of course, that there were more poor people in the
immigrant churches than in others she entered, on impulse, all through
her youth. But Catholic officials were often at odds with the most ardent
champions of the poor. She saw a connection, nonetheless, that was too
obvious for her to explicate.

Not that she was very good at explication—a failing that hagiog-
raphers would also call providential. She found her true work because
she failed at her first (and lingering) ambition, to be a writer—a novel-
ist, she hoped; or, short of that, a journalist. She was neither. Her one
novel, *The Eleventh Virgin,* is important only for the autobiographical
details it supplies us. Although she continued to write her columns for

the *Catholic Worker,* they were valuable only for what they tell us of her activities. They were a kind of extended correspondence for the *Catholic Worker*'s scattered family. She was not an original thinker or a convincing stylist. What was she? What would a job applicant trying to fill her shoes choose as his or her title? The only answer imaginable is: saint. Yet one hesitates to use that diminishing description. Dorothy reminds us that it *is* a job description, not a vague tribute to a person's supposed perfection. In many ways Dorothy was not "saintly." She could hold a grudge, explode with anger (one group of sixties radicals complained of the way she "stomped" them), resent others' curiosity about her (though she kept a public diary in the pages of the *Worker*), alternately smother and strand her daughter: Dorothy Day was not a saint because she was good, but because she got dragooned into the job.

What is a saint's job? To be in all circumstances so focused on the divine presence that the world gets lit through from the other side by the intensity of one's gaze at and beyond immediate surroundings. It is a thankless job, since real saints scare people—who try to "tame" them by reducing them to "saintliness." Saintliness is, in that sense, private and irrelevant. Being a saint is the ultimate challenge to others, as God's prophets realize when the divine press gang comes round. There was nothing about going to jail that appealed to Dorothy (the lesbianism, in fact, appalled her). Some of those attracted to the *Worker* movement— Ammon Hennessy is the most obvious example—liked the push and shove of confrontation politics. Dorothy, essentially reticent and prim, did not. Her real respect for priests made it a torture for her to see how many of those who came to her Houses were helpless alcoholics. When those at Hospitality Houses stole her beloved books, she barely contained her rage. Her work was not one that permits any satisfying measure of success, or even progress—the thousandth hungry person fed is not much different from the first. She could, by struggle, keep loving the poor, but she also kept hating the lice they brought her. St. Francis was, no doubt, much more hospitable to "Brother Flea." Dorothy died with her allies and followers at an acute stage of their perpetual inner squabblings (a fate that she shared with Francis). What, then, did she accomplish? Nothing but what the saint always does and must. She bore witness; she brought the message; she came from God.

—*Sojourners,* September 1982

32

The Pope in America

A Washington editor told me, "Haynes Johnson went belly up on this one." Catholic theologian David Tracy said, "I can't *believe* Frank Reynolds's bad imitation of Bing Crosby as Father O'Malley." John Chancellor called the pope "His Majesty." Mary McGrory made up an even grander title, "pastor of the planet." The Chicago *Sun-Times* headlined its October 6 coverage A CITY NESTLES IN THE HANDS OF A GENTLE PILGRIM. My own hero, Murray Kempton, wrote that the pope "saw the children singing and his face was a bath of light." Some adoring commentators gave the pope implicitly divine status, but Jimmy Breslin more modestly concluded that he had a monopoly on humanity: "He is, I believe, the first real man I ever have seen on television."

Oddly enough, just when the Catholic church is forswearing the imperial claims known as "triumphalism," the secular press brimmed with imagery of war, of the pope's conquests during his October processional. A DAY OF TRIUMPH was the New York *Post*'s headline to its Wednesday story on his New York visit. The pope's seven-day, Monday-to-Sunday U.S. journey was "all-conquering," according to M. W. Newman in the *Sun-Times*. Some varied Caesar's formula. Charles W. Bell of the New York *Daily News* filed this lead from Dublin: "He came, he smiled, he conquered." And conquer he did—hearts and heads, and standards. As Gail Harris, a Boston TV reporter, told *Newsweek:* "It was one story where it was OK to drop the reportorial objectivity." No, not okay. Mandatory. The giddy stuff from the columnists was expectable—James Reston bent his presbyterian knee. But was it true that a real pro like Haynes Johnson of the Washington *Post* threw all pretense of reportorial objectivity to the winds?

Alas, it was embarrassingly true. His series of perfervid dispatches finally boiled over, on Friday of the week's visit, in this declaration:

> He preaches love and mercy and justice. When he says, in his heavy accent, "dear B-r-r-others and Sisters, I love you," this old American WASP, baptized an Episcopalian, confirmed a Congregationalist, and lapsed from any church in recent years, has to respond, along with many of his fellow citizens, "I love you too."

It was not enough, on this trip, to report excesses—Peter Hebblethwaite, Vatican affairs writer for the *National Catholic Reporter,* grimaced at the story of a woman who said she was going to catch the pope and ask him to hear her confession for the sin of having fallen in love with him. One had to *commit* excesses, and swoon in headlines. The New York *Post* ran this statistical paroxysm across the top of its Thursday souvenir story: "For 29 hours and 34 minutes this Pope, this Polish priest, this common man sent shimmers of magic through cathedral and arenas and along 58 miles of New York streets."

Mayors and senators tried to bask in the pope's presence—Senator Moynihan of New York, writing what purported to be a "think piece" for the Washington *Post*'s thirty-one-page supplement on the pope, declared: "I have given him my heart." (When it came to the pope's strictures on capitalism, the senator had to stipulate that he kept his head to himself.) Papolatry, of which Catholics were once accused, was now celebrated and imitated. In the Washington *Post,* Father Timothy S. Healy, the president of Georgetown University, even applied to the pope St. Paul's words about Christ's own divine love: "Anyone who thinks he comprehends the breadth and depth and length and height of this man's love had best think again."

Thinking again was not an activity much in favor during the week of the pope's visit. There was a contest to see who could be more awed, and Gilbert A. Lewthwaite, of the Baltimore *Sun,* lapsed into a kind of holy stuttering: "For once, officials of awesome temporal authority found themselves looking in awe at an awe-inspiring spiritual visitor. It was that sort of memorable moment." Charles T. Powers of the Los Angeles *Times* interviewed a twenty-one-year-old secretary, sophisticated lady in tough New York, and concluded: "It was hard to define, this feeling, but to say that she adored him, standing there as the rain started again, would not have been putting it too strongly." On the basis of this conclusion about a single person, Powers's editors headlined his story NEW YORK SHOWS ADORATION OF POPE.

It took an army of reporters to assemble these "facts"—and to write story after story on the popemobile, the foods prepared for the pope, the names of his various capes and caps. The Boston *Globe* printed an AP story from Maine on the brand of raincoat he wore. The Philadelphia *Inquirer,* in a caricature of investigative reporting, managed to find out the brand name of the sheets he would sleep on in Washington. But the popemobile was obviously John Paul's most fascinating attribute of office. Pictures of it were shown in various cities before his arrival, sometimes with him in it, sometimes not. Since the truck was a modified Ford Bronco, the Chicago *Sun-Times* could not resist the headline POPE TO MOUNT A BRONCO. But the Washington *Star* did the definitive piece on papal locomotion, leading its automotive section with pictures of five cars he has been carried in (the Rome popemobile is a Toyota, which he was tactful enough to leave at home, even though he was not visiting Detroit). Needless to say, the triumphal note rang throughout this celebration of what were called, in the headline, THE PAPAL PROCESSIONAL CHARIOTS. The serious historical approach was signaled by beginning with the chariot St. Peter presumably rode into Rome.

Manning the froth machine involved the nation's papers and networks in the logistics of a journalistic D-day, mobilizing "the largest press corps ever assembled," according to Christopher Dickey in the Washington *Post.* Church officials said fourteen thousand journalists had been given credentials and cleared by the Secret Service. Two hundred and seventy of these flew on the three papal planes. The New York *Daily News* assigned over sixty of its staff to cover the New York visit. The *Sun-Times* fielded seventy. The *Globe* topped that by assigning one hundred twenty-eight to John Paul's American arrival.

What light was shed, what knowledge diffused, by this massive effort? Or is that the wrong question to ask? After all, there was an air of celebration to this visit, of simple pageantry, that made people react as to the Bicentennial festivities. The pope was this year's Tall Ships, fun to watch. But the press claimed to be covering more than a spectacle. The pope was here to deliver a serious speech at the United Nations. Most papers felt obliged to print as many of his sixty-nine addresses as they could find space for. Columnists kept admonishing us to listen to his words while they kept swooning at his manliness, his "sexiness" (Andrew Greeley), his "broad hands" and "broad back" (Mary McGrory).

What were those thousands of journalists doing? What kind of story was this? Some dodged the question by calling the visit unique or unprecedented. No matter what the pope did, some journalist was bound to write that no one had done it before. For Breslin, it was carrying a human face onto the television screen. For Haynes Johnson, it was talking about peace and poverty: "In range and context these were extraordinary things for a religious leader to address," he wrote of the UN speech. But Paul VI had said the same things in the same place. Better-informed religion editors, like Kenneth Briggs of the New York *Times,* found nothing added to the social message of recent papal encyclicals. Columnist Colman McCarthy made a good case, in the *National Catholic Reporter,* that John Paul had subtly softened the plea of recent popes for nuclear disarmament. Besides, "religious leaders" (Johnson's term) have been speaking out on social issues for years—not only the National and World Councils of Churches, but all the Protestant sects, even evangelical ones, and various Jewish organizations, as well. But little if any hyperbole would get challenged in this week of breathless inaccuracies. The adjective "historic" got stuck to the word "visit," not to come unglued the whole time John Paul was on American soil. After all, you cannot apply standards to an event that is outside all comparison.

There was, of course, one sense in which John Paul's visit was historic. It was the first time a pope had been formally received by an American President at the White House. (Lyndon Johnson met Paul VI informally in New York when that pope addressed the UN.) As such, it was a landmark in the disappearance of nativist suspicions about the loyalty of American Catholics. It was presented, along with John Kennedy's election of 1960, as a kind of one-two punch to defeat prejudice. We were told, over and over, that this could not have happened even a short time ago.

All that was true. But an even more interesting point was nowhere alluded to. In a way, all the worst predictions of the nativists had come true: the pope was treated almost as a divinity, or at least as a unique man. Temporal leaders did genuflect to him. TV journalists called him a king. His more dubious titles (for example, successor to Peter as bishop of Rome, for which there is no historical evidence) were recognized and solemnly repeated. Policemen in two cities referred to the visiting dignitary as "our commander in chief." Stalin's comment on the pope's divisions was endlessly quoted by an army of journalists acting as his troops. The President even requested that the pope bless his White House guests, Catholic and non-Catholic. Not only would that have horrified nativists of the 1950s, but American Catholics, too, would have denied that any such thing could happen, or should happen. We

Catholics were brought up on a careful language: adoration (*latria*) for God, special reverence (*hyperdulia*) for the Blessed Virgin, reverence (*dulia*) for saints, and respect (*honor*) for the clergy. Back in the 1950s the president of Georgetown University would not have applied scriptural language for divinity to the pope. So we witnessed not only acceptance of the Catholics, or toleration, but indulgence of their wider claims and wilder language. Why was that? Because we have become so tolerant (the self-gratulatory message that was conveyed in most discussion of this matter)? Or because religious language has lost all technical meaning? Did we babble these things lightly because they no longer matter? Interesting questions, never raised in the love feast, for lack of historical imagination or investigation.

Legitimate questions about the separation of Church and State were not raised in their own right. The matter, when discussed at all, was reduced to the sums various cities might have to pay for police protection or construction of outdoor facilities. The politicians wanted no part in hindering this popular visit, so the objectors were covered only insofar as their suits reached, or threatened to reach, the courts. The priest-columnist Andrew Greeley wrote in his syndicated column that those bringing suit, principally the ACLU and Madalyn Murray O'Hair, just wanted "to get some cheap self-advertising out of the papal pilgrimage." Those who once thought Catholics would like to silence all speech against their claims could take some comfort in the way Greeley fulfilled their prophecies.

Most papers solved the Church-State question by saying the pope's visit "transcended" both categories—it was unique, and could submit to no standards. This was not a religious story or a political story, but something beyond either. The hard news was that there was no hard news: what mattered was his "humanity"; not what he said, but his "simple goodness." Yet transcending the religious and political categories did not mean that either was canceled. It was *also* a religious story, *also* a political story. And though the transcendent note makes an application of standards impossible, we may still ask about the vestigial components of the story. What, precisely, was this week-long outpouring of print and of viewing time, considered as a religious story?

The first instinct of journalism when a religious story arises is to seek the "objective" ground of sociology. So we had the endless potted histories of recent Catholic developments and divisions—Pope John XXIII was neat, Paul VI constipated, John Paul I incipiently sweet, and John Paul II expansive. We were told that Catholics are divided on sexual matters—contraception, abortion, premarital sex, masturbation, celibate clergy, sacerdotal females; but few journalists asked why sex was the node of all this dissent, or what that meant in terms of Catholic politics in American history—the experience of immigrants coping with Irish standards of sexual mores, the odd blend of Hibernian and Mediterranean stresses this created (seen in the difficulty of forming a celibate clergy among Hispanics), the special problems posed a church with those strains in the midst of a sexual revolution. Sociology was kept antiseptic in the newspaper treatment of religion.

The New York *Times*, objective to the point of irrelevance on Wednesday, October 3, exhausted its sociological expertise in giving us statistical maps of how many Catholics live in the New York area (over five and a half million) and where (everywhere, though somewhat fewer, and with fewer priests, in Nassau and Suffolk counties). The next day the paper refined these mapped statistics to show Catholic, Protestant, and Jewish population borough by borough. That was interesting, of course—especially for those who think of New York as an Eastern Establishment WASP nest, since Catholics outnumber the others in all boroughs—but this procedure is so standard that a newspaper can do it in its sleep.

I came across only one example of the method used in a way that stimulated (a crime, where stimulation can mean disagreement)—in the *National Catholic Reporter*. Discussing a map in VATICAN ON THE POTOMAC, Patty Edmonds not only identified Catholic real estate holdings in the District of Columbia, but also showed that the holdings are greater than those of any tax-exempt organization except for the federal and district governments, and greater than all other religious holdings put together. Presumably only a Catholic paper would print such statistics, since others would be accused of giving aid and comfort to Ms. O'Hair.

What approach, other than the sociological, is left for journalists covering a religious story? What kind of story is it? One might look for historical parallels—though they suggest that the story is comparable to something else, which would take away the journalistic protection afforded by the visit's "unprecedented" and "unique" character. The visit was called "historic" because, paradoxically, it was seen as occurring outside history.

I found only two attempts to place the pope's visit in some context of American history—not as the end of nativism, but as a parallel to some earlier event. Russell Chandler of the Los Angeles *Times*, writing in the

National Catholic Reporter, said, "I think something of this nature has not happened since the great revivals of the eighteenth and nineteenth centuries." He did not elaborate, and the comparison hardly bears elaboration.

True, it was a religious visitor's arrival that helped spark the first Great Awakening—when the famous preacher George Whitefield came for his second visit in 1739. But the Great Awakening worked in apparently opposite directions. It showed some of the characteristics of mass hysteria—faintings and "quakings"—but it aimed at individual conversion. It was theologically reactionary—calling for a return to the community of "visible saints"—but politically progressive, cutting across denominational lines and suggesting the possibility of a national consciousness. Any superficial similarities with John Paul's visit merely reinforce deeper differences. The pope might be called theologically conservative and politically liberal, but his embattled old doctrines are not being heeded even by American Catholics, whose attitudes on contraception and abortion approximate the national and not the papal norm. The mood the pope elicited was not one of individual repentance and conversion but of celebration. We were asked to recognize how good we really are, not how sinful.

The same facts militate against Kenneth Briggs's more extended attempt to compare the pope's visit to an evangelical revival. American Catholics have had their own form of revivalism, as Jay P. Dolan has argued. The parish "mission," preached by itinerant priests, led to a hellfire sermon and the individual repentance signaled by lining up for the confessional box. But John Paul's speeches had no such shape or intent. The signal he sent was to "rejoice," not to "repent." Even his UN speech calling for a redistribution of the world's goods was largely congratulatory toward the UN gathering itself and the nations represented.

Given the futility of these attempts to place the visit in a serious historical context, the press seemed to settle for the category: Religion Equals Superstition. The New York *Post* editor, Steve Dunleavy, who is supposed to have told a reporter to "Find me a miracle," was in good company. Finding miracles became the journalistic game. Charles Nicodemus, in the Chicago *Sun-Times,* marveled that no one died in the crush of people around John Paul's visit to Chicago—though 341 suffered minor injuries. A paramedic was found to conclude, reverently: "It seems like kind of a minor miracle that in a crowd of this size, with all the excitement, there were no fatal heart attacks or serious incidents." The *Sun-Times's* kicker on a story about the absence of a commuter jam in Chicago's subways was TRANSIT 'MIRACLE.' T. R. Reid, in the Washington *Post,* reported on the speed with which the platform for the pope's Boston mass was assembled, and concluded in his own words: "One could almost call it a miracle. But then, hundreds of thousands of Boston Catholics are willing to place this entire day in that category." When a palsied girl had trouble getting to the pope's Philadelphia mass, but was helped to reach it "just in time," the *Bulletin* ran a headline across the entire page: MIRACLE OF GOODNESS: PALSIED GIRL REACHES MASS. The perfect modern miracle, which made the pope a technological faith healer at long distance, appeared in the Washington *Post:* POPE, THROUGH TELEVISION, BRINGS INNER PEACE TO DISABLED PAINTER, ran the head on Art Harris's story about a mentally disturbed man whose shouting seizures disappeared when he saw the pope smile in black and white.

But most reporters turned to their favorite superstition, searching for signs in the heavens. Indulging a meteorological triumphalism, the press made even God a Catholic. As Murray Kempton, back in form, put it: "Respectful journalism worked earnestly to suggest that the sun came out as though by decree of the heavens whenever the Pope appeared." But few journalists could summon that skepticism amid the believing crowds. Henry Tanner, in Dublin for the New York *Times* to report the pope's stopover, checked his miracle with the weather bureau. "Even the weather seemed to be on the Pope's side. It was sunny and warm, the warmest September 29 in 30 years, according to the Weather Bureau. In Phoenix Park, a bearded young man in shirtsleeves said: 'We had rain all year. It stopped two days ago and now look—it renewed my faith.'"

But what if it rained on the pope, instead of shone? That could not defeat the Misses Lonelyhearts who had taken over editorial desks this week. The New York *Daily News* excelled at turning raw weather into a blessing. HEAVENS WASH BOSTON FOR THE POPE'S ARRIVAL ran an October 1 head. Or, if the sun failed, the pope would *become* the sun. Charles W. Bell, writing the next day in the same paper, began: "In a brilliant red robe and beaming with a smile that cut through a gray mist, Pope John . . ." The head on another *News* story the same day developed the idea further: PONTIFF IS BEACON IN BOSTON FOG. Rain or shine, heaven was on the pope's side.

It was a charmed time, and reporters were sending season's greetings, not facts. Bob Herguth, of the *Sun-Times,* sought out a Dickensian blind street musician named Bob Holdt and assured us that "Bob said it all: 'It's like Christmas in the summer.'" Santa Claus had come to town, as we could tell by the "twinkling" eyes (Haynes Johnson) in John Paul's ruddy face. Like all Christmases, it was a time for the children, for trying to believe in their belief in Santa Claus. Watching the children open their presents is always fun. T. R. Reid of the Washington *Post* watched a little girl watching the pope go by: "She buried her face in her tiny hands and cried—as she had known all along she would —for the sheer joy of it." It's all in the "tiny."

Full of blah and hokiness, the press did not often strike a realistic note while the pope was here, and it was resented when it did. The papal coverage was one long letter saying "Yes, Virginia, there is a pope." But this supposedly reverential air is implicitly condescending to religion, reducing it to fairy-tale status. That reduction made it harder to take anything about the pope's visit very seriously.

> Miss Lonelyhearts did not answer. He was thinking of how Shrike had accelerated his sickness by teaching him to handle his one escape, Christ, with a thick glove of words.

Maintaining an air of make-believe involved the press in a complicity with the national mood, one reaching the level of self-censorship. Inflated crowd estimates were published (one explanation of the "miracle" of uncrowded Boston subways was a smaller turnout than had been publicized). Yet when the Chicago *Tribune,* two weeks after the pope had gone, questioned its Grant Park crowd estimate of one million, a radio commentator on the press, John Madigan of WBBM, criticized this effort "to try and take some of the gloss off of the papal visit."

Anyone who printed dissenting views about the pope could expect angry letters. (I wrote a column defending ordination for women, and got a letter from a pious Catholic calling Sister Theresa Kane—who had criticized the pope for his stand against ordination and whom I did not even mention—a "bitch.") Even paraphrases of the pope himself were criticized if they broke the mood of geniality. Darrell Holland of the Cleveland *Plain Dealer* told the *National Catholic Reporter:* "Our headline said 'Pope ends tour scolding U.S. for self-indulgence' and people called us to complain. They are caught up in reaction to the man. But they are not listening to what he was saying."

The pope's trip was arranged to avoid questioning and criticism. Journalists were not allowed to ask him questions at any point—and, far from protesting this arrangement, some blamed it on themselves. When the pope tried to come back and say hello on the crowded plane from

Rome to Dublin, cameramen almost mobbed him, and subsequent accounts suggested that the press had scared him off. That would not have happened at a formal press conference; but he fielded no questions on this trip. Comparison with the Mexican and Polish trips makes this seem more than an accident. In fact, a few were worried about the pope's exposure to a free press in a pluralistic society, something of which he had no prior experience. Priests defensively said beforehand that he was not used to such circumstances, and might not understand open questioning of his views on contraception or divorce. Andrew Greeley, the sociologist and columnist, wrote a hysterical column, on the eve of the pope's visit, advising him to cancel it because "poisoned mobs" awaited him. The papacy was not ready for the mass media, the putative poisoners: "He ought to try again later when the American church has learned better how to deal with 'libertarian' mass media tactics." I presume "libertarian mass media tactics" is a sneering way of referring to a free press. The ironic thing is that this pope was praised, just after his election, for his openness to questioning. That did not show anywhere in America.

Father Greeley need not have worried. The press went belly up. The mood of celebration muted the protests of loyal Catholics who demonstrated for women's rights and other matters. Even the minimal coverage this guaranteed was attacked in letters to papers and in the columns of Father Greeley himself. Before and during the visit, this priest was the most voluble commentator on the journey, increasing his column releases to a daily basis and appearing on various television shows as an expert on the Catholic church. Of the nuns who stood in silent appeal, this charitable expert concluded that "they merely want a piece of the publicity action for themselves." He dismissed them this way, in the name of the pope's true Christian "charisma": "Would you want to be trapped in a stalled elevator with any of them?"

The one time the pope heard any challenge at all, it came by accident. Sister Theresa Kane was chosen to greet him because she seemed safe, a respected head of a religious order, elected by her peers to head the congregation of women's orders. The TV coverage of the pope's Washington appearance picked up Sister Kane's address, of course—it could not do otherwise. But the papers that reported it heard in their letters columns that they should not have. It was the wrong time and the wrong place—which meant there should be *no* time for the pope to hear the voice of loyal Catholic dissent. She was trying to steal his limelight, the papers were told—which meant only mayors, senators, and President Carter should bask in that.

This trip belonged to the public celebrators and journalistic adulators. Phil Donahue was invited to the papal audience in Washington, not Phil Berrigan. The pope shook hands with Barbara Walters and Shirley MacLaine, not with Dorothy Day. He chatted with Zbigniew Brzezinski, not Cesar Chavez. Real religious questions could not be discussed, not even with Catholics—and not with Protestants or Jews either. He took no questions from those he addressed in an "ecumenical" meeting which, Kenneth Woodward, *Newsweek*'s religion editor, told the *National Catholic Reporter,* "set the ecumenical movement back a hundred years." That address widened the differences caused by doctrine to include, for the first time, moral and ethical matters—this was taken, naturally, to refer to abortion and divorce. In his UN speech, the pope offended Jews by repeating the Vatican's unnuanced demand that Jerusalem be an open city.

The repression of frank coverage reached its climax when not only dissent from the pope but reporting of his own words was attacked. The nerviest ploy of all was Father Greeley's, when he resented in print the reporting of the pope's remarks on contraception during his address to bishops in Chicago. Greeley said that attention to the pope's comment was out of place because it was just a "one-sentence reaffirmation of the church's traditional birth-control teaching . . . which hardly anyone noticed . . ."; it was a "private" remark to bishops; and, besides, one should not pay attention to two pages on sexuality in a seventeen-page address which had not, Greeley felt, absorbed the pope's attention. "One senses," confided the father, "a ghostwriter working in the bowels of the Vatican."

All this is hogwash. The pope had saved his first explicit and very formal reassertion of the controversial doctrine of *Humanae Vitae* for America. He reaffirmed it in Chicago in the most solemn manner: "I myself today, with the same conviction of Paul VI, ratify the teaching of this encyclical . . . 'by virtue of the mandate entrusted to us by Christ.'" That would have been news if it had been said in Rome. That he saved it for America made it doubly newsworthy.

Father Greeley need not have worried. The press restrained itself. It needed no help from him or the angry letter writers. This is made clear in answers to a *National Catholic Reporter* poll of ten religion editors and reporters after the visit. Most were critical of the pope's visit on one or more counts—his UN speech, his lack of openness and responsiveness, his ecumenical and sexual views, the shallowness of the crowds' reactions. But these judgments, which should have influenced the way the event was covered, apparently remained private and closely held.

The self-censorship was most evident in the matter of the Pauline monks. Just a month before the pope's arrival, the Gannett News Service ran a brilliant series of articles on the misuse of millions of dollars collected mainly from Polish-American Catholics to build a shrine to Our Lady of Czestochowa in this country. The Vatican had investigated the priest responsible and recommended his disciplining—but that priest is Polish and a protégé of Cardinal Wyszynski, Karol Wojtyla's own patron. So, seventeen days after taking office, Pope John Paul II quashed the procedure against him. (Six government agencies in America are still pursuing the matter, and an important legal concession by the order took place the day after the pope left America.)

At a time when the press was trying to turn up new things to say about the pope, right down to what sheets he slept on, I asked writers who called me (David Butler from *Time*, Kenneth Woodward from *Newsweek*) why they did not discuss the pope's protection of a man who bilked Polish-Americans of millions of dollars. Their voices trailed off as they said, "Oh yes, that might . . ." They obviously had no intention of spoiling the mood. So far as I can find out, there was not a single story on this matter while the pope was in the country.

Considering the reaction to such harmless things as televising Sister Kane's appeal or reporting the contraception ban, it may not be surprising that editors did not want to brave the storm of protest which discussion of the Paulines would predictably have called forth. (Even before the pope's arrival, the Gannett papers were attacked for anti-Catholicism.) But the Watergate reportage raised similar protests. Is the test of what we print its inoffensiveness? Only, it seems, where the pope is concerned.

Perhaps the pope was not challenged because a religious story is not often a serious story. One would get that impression from the thinness of religious coverage in general, and the tendency to take "serious" stories—like the Ayatollah Khomeini's crusade—out of that category. William Buckley, for example, has protested the treatment of Martin Luther King's civil rights campaign and Jim Jones's cult as religious stories. At any rate, the "transcendent" nature of the pope's trip guaranteed that it would be more than a political story, as well as more than a religious story, without preventing it from being *both* along the way.

And our heaviest political analysts weighed in with their thoughts on the trip's political significance.

If, as a religious figure, the pope became Santa Claus, as a political figure he was quickly reduced to Dr. Feelgood. Wilfrid Sheed accomplished the reduction as intelligently as possible in the Washington *Post*'s papal supplement: "If a priest can be cheerful behind the Iron Curtain, there's not much excuse for anyone else not to be. In a quaint reversal, an old man (by fairy-tale standards) from the Old World has lightened a load from a bunch of those traditionally optimistic Americans." Andrew Rooney on CBS-TV put it rather more sappily: the pope convinced us that "we may not be as rotten as we thought we were, after all." Much was written in the "Thanks, I needed that" vein. We needed a little Christmas, right this very minute. James Reston began with this thought (if that's what it was) on Sunday: "Pope John Paul II arrived in Washington just when we needed him. In recent days, this city has been down in the dumps."

What about the pope cheered us up? Well, for Breslin it was having his first glimpse of a human face. The political answer often suggested the rhetoric of John Connally's campaign aides. Here was a leader, a take-charge man, one who knows his own mind. It sounded, as well, like Allard Lowenstein's recommendation of Senator Kennedy as one who can "make us feel better about ourselves." Most commentators who took this line went out of their way to say they did not agree with all the pope was saying but they were so glad to see that he *meant* it, whatever it was.

Others took the popularity of the pope as a gauge of America's resurgent conservatism, a kind of counter-Woodstock, a "spiritual be-in" (M. W. Newman, in the Chicago *Sun-Times*). On this reading, his reception showed a yearning for old-time values as supported by old-time religion. But there was no sign that even Catholics were tempted by the pope's old-time sexual morality, and it is doubtful that any poor country in the world was the slightest bit better off for his speech at the UN. What kind of leadership is it that elicits cheers but no followers? The kind that does not make one do anything except feel good about oneself (and curse those who don't join in the festivities)?

The most blatant attempt to find a political use for John Paul came from the right wing, which has decided that he and Solzhenitsyn are symbols of tough resistance to communism. This meant they had to supply him with words he did not say (there were no denunciations of communism as such) while dismissing the words he did speak against the hedonism of wealthy countries (Senator Moynihan suggested the pope needed instruction in the blessings of capitalism from columnist Michael Novak).

There was one resemblance to Solzhenitsyn which, however, was not mentioned by anyone I read. The Russian attacked our free press in his Harvard speech two years ago, saying the public has a "right not to know." The pope, in his Washington address to theologians, delivered the old defense of theological censorship in these words: "It is the right of the faithful not to be troubled by theories and hypotheses that they are not expert in judging or that are easily simplified or manipulated by public opinion for ends that are alien to the truth." The man who says that will not be quick to take questions from a press that might "simplify" or distort the answers.

What kind of story was the pope's visit? To answer that question would take a historical consciousness that the press seems unwilling or unable to acquire. There has been a growing recognition that matters like the law and the economy and medicine demand better training of the journalist than simple apprenticeship with the police blotter. Lawyers are being hired by newspapers and networks, economics is being taught to journalists. The coverage of the swine flu scare led some to recognize that the press should know something about medicine. But religion and history are not thought of as hard disciplines needed for covering a story like the pope's visit, which was a big story, but soft news.

Yet *did* the pope's visit have "no precedent" (Haynes Johnson)? America's most famous visitor of the nineteenth century was the Marquis de Lafayette. American museums are full of the 1824 equivalent of those T-shirts and buttons hawked wherever the pope appeared. The nation turned Lafayette's visit into a national celebration, which its principal modern historian (Edgar Ewing Brandon) could call, in 1944, without parallel in our history.

All parts of society turned out to greet Lafayette in a series of parades and banquets. Criticism of the guest was not brooked. When he spoke of something embarrassing, like the abolition of slavery, people politely averted their gaze rather than break the spell. Contemporary press accounts were adoring: "There is something of imposing grandeur in this visit of General Lafayette to the United States. In moral sublimity there is nothing like it in the history of man." The young were told to look up to him. *The Carolina Observer* called the time holy, "a pe-

riod in which none but the nobler feelings of the heart were exhibited."
Faction was suspended, leading James Fenimore Cooper to write:

> The newspapers, which, the evening before, were furiously combating for
> their favorite candidate, now closed their long columns on all party dis-
> putes, and only gave admission to the unanimous expression of the public
> joy and national gratitude. At the public dinners, instead of caustic toasts,
> intended to throw ridicule and odium on some potent adversary, none
> were heard but healths to the guest of the nation, around whom were ami-
> cably grouped the most violent of both parties.

The devotion poured out on the visitor has puzzled some historians.
Attempting to explain it in *Emblem of Liberty*, Anne C. Loveland
sounds at times like James Reston. The nation was down in the dumps.
The "era of good feeling" had come to an end in the blatant par-
tisanship of Jackson's supporters. The old revolutionaries were dead or
dying. Political life seemed sordid. Lafayette, who was only nineteen
when he shed his blood for America at Brandywine, had come back as
a living reminder of the Revolution. He was outside time—he had never
departed from the first pose he struck on behalf of the American Republic. As a spiritual son of Washington, he was greeted like the very
soul of the nation returned to its semianimate body.

The virtue most often praised in Lafayette was "disinterestedness"—
the republican virtue to people brought up, like Washington, on Ad-
dison's *Cato*. The American civil religion was one of classical devotion
to the Republic, presented in religious guise by Parson Weems and
others, but kept entirely separate from sectarian Christianity. Where the
sects divided, cult of the republican gods united. As Lord Acton often
argued, religious tolerance is the side effect of sectarian bigotry—where
reconciliation is impossible, we agree to exclude discussion. America's
Enlightenment ideals of brotherhood and equality kept a deliberate distance from Christianity and Judaism, even when they were theistic (like
Voltaire's).

By contrast, the talk of brotherhood in our scientific age comes more
naturally from the churches, which have lost much of their dogmatic
fervor and now stress ethical community, the role of republican religion
in Lafayette's time. Where political ideology divides, the religious denominations now cooperate in public service. The civil religion, once at
odds with sectarian beliefs, now seeks its home in them—in the "brotherhood of man under the fatherhood of God" with which Nelson
Rockefeller closed his ecumenical campaign homilies. A celebration of
unity that came naturally when Lafayette, a Catholic, arrived in

America and people agreed to *overlook* his Catholicism, was now possible for a pope precisely *because* of his Catholicism. We were celebrating ourselves, our tolerance and hospitality, in a liturgy of national unification resembling our national rites when Lafayette arrived.

There are specific differences, too, and troubling questions that arise. If our civil religion is now at home in the religious denominations, why is this happening at a time when our principal civic institution, the Supreme Court, is removing sectarian references from public life? Maybe the muted voices that demanded separation of Church and State asked the right question, though no one wanted to listen: isn't it odd, not to say schizophrenic, to celebrate with all our national resources a papal Christmas in October while cutting back on religious symbols in December?

I do not offer this comparison with Lafayette as the best or only one available; but it does show there *are* some comparisons to be made, some events that might set a standard for coping intelligently with this odd outpouring of time and money and attention around one man. But the press did not choose to explore the event, to reflect it and reflect on it; it became an unthinking part of the event, joining in all moods rather than deepening them, trivializing with empty acclaim.

—*Columbia Journalism Review,* January–February 1980

AFTERWORD

The adverse mail that greeted this article was so emotional that it shook the board of directors for the *Columbia Journalism Review,* and was used against Robert Manoff, the enterprising editor whose independence led, shortly after, to his being fired. No criticism of the pope's visit was going to be permitted in that silly season. When Manoff went to the *Soho News* as its editor, he asked me to contribute, and I did—just before the journal ran out of money and folded. I asked Manoff if he noticed that every time I wrote for him he lost his job. He had; but, not being a superstitious man, he hoped we would work together on still another magazine—and I'm sure we will.

33

The Devil

"Which side are you on, Marlowe?"
"I don't even know which teams are playing."
—Raymond Chandler

There have been times when it seemed impossible to make a movie without a priest. C. B. De Mille thought it up—a God's Advocate to take the curse off Claudette Colbert's thighs. Bing Crosby tinkled bells at us, as Father O'Malley, with the blessing of staff censors. But this churching of cinema reached a new level last year, at Georgetown University, when a vestmented priest came out to bless the set of *The Exorcist*. Better safe than sorry—the devil might have heard that the author of *The Exorcist,* William Blatty, means to give him a bad press. Priests not only advised on the set, but play walk-on parts. Two Jesuits have speaking roles, one of them quite large. A real Father O'Malley—not the Bing Crosby character—plays the part of Father Joe Dyer, a straight man to the movie's hero.

Some will, no doubt, think the devil has not been given his due in *The Exorcist*. A generation that acquired freak-rock sympathy with the devil may resent his demotion to a villain in this shabby little tale. It is getting ridiculously easy to shock the impious. Not long ago, when I lectured on "Witchcraft in Shakespeare" at Notre Dame, a young woman came up to me before the talk and said, "It's about time someone began taking us witches seriously." She looked hurt when, during the lecture, I said the only Renaissance witches to be taken seriously were the ones who *did* try to compass mankind's harm.

The devil has a thousand advocates today, and has good cause to be embarrassed by them—a rock star unable to escape his groupies. They are a grubby sort, for the most part, looking for a higher (or lower) palmistry, using the devil for self-promotion. Better to be a cosmic rebel, reasons Anton LaVey, than a carny barker—and he works out a style combining the two. Given this circus kind of Old Nick cult, some priests feel *The Exorcist* cheapens true religion—that God should really keep better company than this Adversary foisted off on him. A fellow Jesuit, Raymond Schroth, argued against Father O'Malley, the movie's actor-apologist, in *Commonweal,* a Catholic publication. *The Exorcist,* he pointed out, is good for cheap thrills, but it leads to no revelation of God, even in the story's tawdry terms. The trade-off is two healthy men for one sick girl, and the result of the whole action is to make the possessed girl's mother believe in Satan, but not in God.

Father O'Malley, who says he believes in God but is still checking Satan out, thinks that believing in a devil must, anyhow, be a step in the right direction—better believe in Nothing than in nothing. This keeps one "susceptible to transcendence," breaks open man's enclosure in "the antiseptic and spookless streets of 1984." Putting spooks in the street has become, for him, an apostolic endeavor. If the price of the supernatural is Satan, he is willing to buy it on those terms—or, more properly, let others do so. He feels more at home, as a theist, talking to a Satanist than to a secularist. He told me why: "People find it easier to believe in spooks than in God. I put on *The Crucible* in the high school where I teach, and the actors were behaving very strangely after rehearsals—especially those who were not faith people. When I heard they were going to go to the cemetery and try to raise a ghost, I hit the ceiling. 'You are supposed to be acting these parts, not becoming them,' I said."

This touch of sorcerer's apprenticeship did not make Father O'Malley give up the summoning of spirits (which may or may not be there). He hopes that people will be led to ponder devils, if not deviltry, as a result of *The Exorcist.* The attitude seems to be that Christianity saves men from devils (even if there are no devils to be saved from), so people had better believe in devils, if only as an excuse to spread Christianity. To Father Schroth, this argument is a way of telling lies for God.

It is not a new temptation—using the devil to drum up business for God. A great many of the old "hellfire" sermons were celebrations of the Evil One. Ishmael leaves Melville's chapel more impressed by storms and whales than by God—or, rather, impressed by God as the force behind such evils. Forensic artistry, if nothing else, drew preachers toward such promising material. Heaven is a dull, unimaginable place, compared to hell. Scripture spends a great deal of time trying to describe a final peace in complex symbols; but the scattered hints of hellfire are easier to fan into bonfires for the mind. Compared to heaven, hell is home—accessible and familiar as each of our own private nightmares. Michelangelo's Charon is far more energetic and compelling than the muscle-bound introvert who judges him on the Sistine wall.

The temptation to preach hell can return to God's ministers in a thousand plausible forms. In the summer of 1972, priests and other clergy in Spanish Harlem conducted a series of unauthorized "exorcisms" at known centers of heroin trade. Using the old tools of holy water, candles, and incense, they tried to pray "the devil" out of corner bars and neighborhood halls. Some saw this as a profound updating of an ancient symbol. But one could just as well consider it a tickling of the Puerto Ricans' ancient superstitions to effect their good. A touchy point of conscience. If you use the devil, are you guilty of the ancient crime of summoning him? Is such "exorcism" a form of sorcery? People who smile benignly on the exorcism of drug centers are forced to a second thought or two when Ted Patrick, known as "Black Lightning," captures Children of God and "deprograms" them. Mr. Patrick thinks he is exorcising the devil from such people, and can call on scriptural accounts of demoniacs bound and carried to their exorcist. In his deprogramming, fire is used to drive out fire—one fundamentalism to drive out another. And his carelessness of human rights reminds us how dangerous fundamentalism can be. Those who believe they can command the devil are not going to worry about putting snakes under their power —and so men still die religiously of rattler venom.

On the other hand, if preachers banish the devil entirely, they might end up with a world as sappy as Franco Zeffirelli's, in the movie *Brother Sun, Sister Moon.* That movie, also, had priest advisers and actors—a Franciscan historian, liturgical directors, and an American priest, Raymond Bluett, who traveled about getting permission to use churches and church property for the film. (I was reminded of all those gutsy leatherneck pictures that sign off with deepest appreciation to the U.S. Marine Corps for doing their propaganda with taxpayers' help on the ordnance.) Father Bluett, describing his movie apostolate on the "Today" show, said he was assigned to the film by the Catholic Communications Foundation, and reminded viewers that the Church had always been a patron of religious art. Maybe so. But Charles Reich would have been more at home on that movie's set than a Catholic priest. The

St. Francis portrayed is a Bird Man of Assisi, strolling around with aimless grins, learning everything from the birds and the bees except sex. The poor fellow has a very short attention span—after going all the way to Rome to get an answer, he forgets to ask the question. But that may be because he is not at home except in fields—where he never has to worry about Brother Wolf. No creature more aggressive than a doe or butterfly is allowed into Zeffirelli's enchanted forest, which makes the animated world of *Bambi* look like stark realism. Here it is that pretty Clare and handsome Francis wander about in a baby state of love that never thinks of intercourse. There is, in fact, only one evil thing in the movie—clothes. "Papa" Bernardone and the pope are cut off from nature in beetle-shiny shells of jeweled fabrics—which cause war, greed, and the Generation Gap. The movie Francis undergoes his conversion in the sweatshop run by his merchant father on proceeds from a war. Leaving the movie, I thought it needed a devil or two more important than the clothiers' guild.

When the Francis of Giotto enters Arezzo, the sky is dark with angry departing spirits. The real Francis did not preach the devil, but he fought him. There was one animal he would always beat mercilessly—Brother Ass: his own body. And the stigmata, after all, did not appear on Francis after a parental whipping. Yes, Francis blessed the birds—and even that implies the power to bless. (The movie Francis has nothing to tell them; they preach to him.) But he also blessed Sister Death—he stretched out his hands toward the darkness; and the darkness pierced them. To tell the story of Francis and leave out the supernatural is far worse than playing Hamlet with no reference to the Danish throne —it is telling the story of Jesus and neglecting to mention the Roman instrument of execution. In the records of St. Francis, I do not see the devil; but I think I see an angel who sees him.

If there are devils, I doubt that they get their kicks torturing little girls on M Street. Conflict with them goes on at a level of the spirit most of us do not attain. Where a holy man is, there will the devils gather—around St. Anthony in the desert, or St. Francis; around Jesus or the Baalshem; near high crazed visionaries or those stooping to the lowliness of saints —near Blake, or Ahab, or Kit Smart. Such men live in a blaze of light that seems haloed with the darkness driven out of them.

Defenders of *The Exorcist* can claim, on New Testament authority, that devils do delight in mean little tortures of the innocent. Father Schroth would answer that superstition once blamed all kinds of natural or nervous disorders on possession—the fact that Jesus cured these disorders does not mean he was endorsing the beliefs that helped cause them. Both sides of this debate seem to miss the point of the main Gospel anecdotes. The most striking part of those exorcisms is the debate devils hold with Jesus, fencing over the heads of the possessed. This struggle has little to do with the agony of afflicted individuals. It centers on the mission of Jesus. The possession is just an excuse to get at the holy man, a continuation of the tempting in the desert. These emissaries of Beelzebub want Jesus to give up his secret, change his calling, league himself with their powers, betray a weakness. It is all an assault on mystery—devils gather in the halt or cripple or crazed as a way of thronging toward a dangerous saint, tempting him toward miracles, to all the evil that can come of doing good. But this man cannot be diverted from his mission, even to live in a constant miracle of local cures. He is tempted to feed the poor, to be a philanthropist, to rule the world—as Dostoevski realized. The main act in this spiritual drama is not an exertion of power but a restraint from it. Jesus not only casts out spirits, he silences them. When he says Beelzebub's house is not divided against itself, he does not mean that devil will not cast out other devils to snare men; magic is, after all, the devil's game. But Jesus goes beyond casting out—he also discredits. The point is not so much the loosing of the possessed, as the binding of the possessors. He first defeats the devils, and then—more important—mocks them, making them oink their way to swinish deaths in Gerasa, surrendering their claims along with their victims.

If devils gather around holy men, it is not because they were summoned. The saints dispel and scatter demons; and the act of Jesus—consummated in his last struggle with "the power of darkness"—was to exorcize the world, to deliver nature from its thralldom to the Powers and Principalities. That is the point of that majestic ending to the Lord's Prayer, so pallidly undertranslated in the past: "And spare us earth's Torture [*ho Peirasmos*]; rescue us from its Torturer [*ho Poneros*]." The cosmic harrowing, or *Peirasmos,* is the Apocalypse; and the Lord of Evil is the opponent conquered there.

The Christ of the Gospels does for the whole world what Francis did to Arezzo—fills the air with the wings and the bat screams of *departing* evil. Various "spiritualists" work themselves up to a worthiness for séances and other interviews with the unworldly. But the saints refuse to entertain strange spirits, breaking up the permanent unscheduled séance in which they move. Medieval saints used tongs on the devil's nose, as Jesus used pigs on the devil's troops. Whatever the demythologizers do to scriptural history or the saints, many saints had earlier done to the devils. Laughter, that great demythologizer, is also your best exorcist.

The early Christian fathers applied this scriptural treatment of devils to all pagan gods and idols. Father O'Malley suggests a kind of ecumenical tolerance toward all forms of the supernatural—like clerical courtesy between the priests of differing religions. Heaven and hell have a stake in each other's reputation, and should form a protective league. But early Christians were felt to be *anti*religious, opposed to the syncretic "openness" of their time, when men hedged their bets about gods who could hurt or help them with an omnidirectional deference. The patristic era is far from deferential—it is scalding and satirical in its debunking of magic, astronomy, and demons' promises. The pagan gods were laughed to death, and St. Augustine played Mencken to their bumblings. These devils only went out with scorn and deriding.

The Gospel opposition to idols and false gods, drawing on the Torah's high monotheistic discipline, extends even to magic in its lesser forms. The tale of the Magi is read by some scholars as St. Ignatius of Antioch read it in the second century: the astrologers have come to admit defeat, surrendering tools of a wicked trade—their mantic touchstone gold, and the spices of séance. The Christian theology moved constantly toward minimizing all evil sacraments and scoffing at their power. St. Augustine worked to overthrow the dark god of Mani when he insisted the devil is a defeated creature, not God's timeless adversary —just as St. Francis, when he blessed fraternally death and darkness, only did it because all have the same Father of light. Christianity, far from loosing devils, bound them and put them in prison. And the devil's comeback between the Renaissance and the Enlightenment was encouraged in many cases by the heterodox. It is the "free thinker" of the sixteenth or seventeenth century who plays most recklessly with diabolic hypotheses. The tragedy of witch executions is that poor misguided women were killed instead of the real troublemakers. There was a better argument to be made for dragging some university wits to the scaffold than "Sairy" This-or-That who died in the off-glitter of their magic verse. Those who played at being Prospero helped those who looked like Sycorax toward horrible tortures. The devil-repressers were less tyrannous than devil-encouragers, and many a pyre was lit by the flames of "enlightenment." When the diabolist was not a free thinker, he was often a reformer. Luther was Lucifer's great rehabilitator, and it often seems that Antichrist was resurrected mainly to accommodate the

pope. It is marvelous how hatred of Rome insensibly exalted Satan—
which is one motive for Milton's obliquely majestic portrait of the devil.

If the Christian church did not so much create or encourage a belief
in devils as cut back that belief where it already existed, then Father
Schroth can cogently present his case: in a world that does not accept
the existence of devils as part of its religious language and inheritance,
true "exorcism" should mean a principled support for this rejection.
Earlier Christians had to drive out devils because they could not, yet,
drive out the idea of devils. But now the baptismal renunciation of
"Satan and all of his works" is a philosophical commitment to man as
freed from superstition's tyranny, exercising his freedom in God's
image.

The Father O'Malleys of the world answer that superstition is no
longer the problem—Christianity does not need to rescue us from that.
Here, at least, man was "saved" by science. The problem now is man's
incapacity to believe at all, to go beyond the immediate evidence of his
senses and the narrower calculus of reason. The devil is, thus, God's
last desperate proof that there is more in heaven and earth than IBM
has dreamed of.

Unfortunately, this approach makes God depend rather ignomin-
iously on his "bogeyman," scaring those he can no longer persuade.
This argument makes Jesus summon devils and open their mouths, instead of dismissing them under bans of silence. And, to crown all other
indignities, the treatment does not even seem to work. As Father
O'Malley himself admits, the devil gets his warmest reception these days
from those who are not what he calls "faith people." Indeed, what we
seem to be witnessing is a return to just that kind of idolatry Christians
did battle with in the Gospels' youth. Those looking for any old crack
in the walls of our secular factory are "syncretists" of the most slovenly
sort. Poltergeists, rappings, ESP, UFOs, drug visions, magic cards, dead
men's signals, prophets, devils, gurus, quacks—all find easy welcome
from those starved for the occult. Heaven and hell are interchangeable
"trips," and any Mesmer looks like the Messiah to a Colin Wilson.
What Jekyll once looked for in potions, students now look for in pills.
Anything unsecular becomes sacred; a deviation from reason is, by that
fact, holy. Upsidedown is rightsideup. Disease is health, and Satan
saves. Hell was Orwellian long before Orwell.

One's "fix" in this world depends on one's habit. Warm slurps of
communal gluing are enough to deliver Charles Reich from executive
suites. (But what do you do when executives slurp along?) Others need

dark rooms and astrological gobbledygook. (But what do you do when, as is happening, computers are rigged to serve up personalized "readings"?) Theodore Roszak needs the stronger medicine of Blake's "Satan" to break the stranglehold of logic. (But how do you grade pupils who cannot follow Blake's poetic logic?) Planchettes suffice for some; but others need Manson's affront to their imprisonment in mind. (And then what do you do when the cops arrive?) We have found our way back to old worships from which men thought they were delivered. So, by a roundabout history, it proves true once again that Beelzebub's house is not divided against itself—that those who summon demons are not of God. Those who summon spirits can hardly be considered exorcists; and those who try to make them do tricks for them, like trained dogs, hardly know the height or depth of evil's mystery. Our world will never be exorcized by the devil's PR men; and the more we attend to William Blatty, the less we shall be hearing from St. Francis.

—Esquire, December 1973

34

Why?

Answer to editors' poll of Christian writers:
"Why do you still believe?"

"Today we two shall be in Paradise." Only one man, I believe, could keep that promise, made to his fellow crook while they were being executed. How can I tell that? As I would gauge the size of a vessel that has passed, by the turbulence of wake it left behind. By the man's impact on the best and purest lives I know or have heard of. By the experience that the Gospel mysteries are the only language in which I can talk to the minds I most want communion with, from St. Augustine to Dietrich Bonhoeffer. By a consonance of mysteries, each incomplete in itself but the correlate of others—myself one of those mysteries; and you another; and he another (the darkest). By a sense that everything has a meaning beyond its meaning, reaching toward him. By a desperate process of guessing and hope. By prayer. By listening for a promise from my fellow crook.

—Harper's, April 1975

VIII
Athletes

The people in this section make art with their bodies. Their flesh is not only their instrument and their artifact, but a metaphor, finally, for spirit. They break down the ancient opposition of body and soul. When Beverly Sills effortlessly threaded a coloratura labyrinth of notes, or Ray Berry unraveled a tricky receiver's pattern (and there, at the end, was the ball waiting for him)—or even when Ali wove a crown of thorns around a man's head with fists moving faster than knitting needles—they had chastened their bodies back in time through the barrier of the Fall, and we glimpsed Eden. This was how things should be.

35

Muhammad Ali

Our hands I hope shall fence Our heads.
—Christopher Marlowe, *Edward II*

Boxing is a cruel art. Both the cruelty and the art are made possible by the modern sport's key feature—the cushioning of the hands. This is not done to protect the person being hit. Far from it. The human fist is a fragile little birdcage of bones. Impressive as Muhammad Ali's fist looked, photographed life-size for *Esquire,* it is not the "lethal instrument" outside the ring that legal fiction makes it. The more powerful a man's punch, the more body leverage he puts behind it, the worse would he shatter that bony wicker basket against any solid target—especially against the head.

Experiment with a skeleton. Fold that pterodactyl-webbery of hard bones, wired together like a Calder, then crunch it down on the holdable Yorick-ball of cranium, the attached helmet-guard flap of cheekbones, or the awning over the eye sockets. Admittedly the bare fist, perfectly placed, might get at the jawbone's juncture with the skull. But in trying to hit that precise joint, while ducking enemy blows and seeking the spot on a moving target, the fastest man would first disjoint and splinter his fingers two by two. The myth of facial vulnerability to the fist rests mainly on schoolboy memories of nosebleeds: nose cartilage is more friable, even, than knuckles.

A man who really learned to fight, in the first instance, as a boxer would be in very bad trouble if he took his acquired art into the nearest bar. My father was a melancholy example of this truth. He knew, in theory, that you cannot punch a man's head with your bare fist; but his fighting instincts had been so patterned by ring training that he always forgot it in the heat of battle. One time, when he had gone out to collect money from a resisting debtor, he came back holding the money in his broken hands, himself held up by the debtor: "I figured I better pay the little son-of-a-bitch, before he made me kill him." He regularly had to give up his *other* favorite sport, golf, until healing and a new grip made it possible for him to fold those sprung pterodactyl wings back around a club.

What my father did by force of training, some barroom brawlers do by movie-reflex. The great myth of the head punch controls all the hokey fight-out scenes on film. John Wayne, planted knock-kneed in his boots, fetches an Antaeus rocket from the dirt, and hits John Carradine "right on the button." The resulting sound is somewhere between the click of billiard balls and the crack of a rifle. I remember watching the movie *Mr. Lucky,* in which Cary Grant took a roll of quarters and wrapped his fist around it. A few seconds later we hear the rifle shot against the bad guy's jaw, and the quarters come rolling out under a door. In real life, it would have been the sound of all his little bones springing apart from each other in pain. The roll-of-quarters gimmick must have arisen from fantasies of the palm grip on "brass knuckles." Ancient boxing was done with hard leather prototypes of such "knuckles."

What on earth is supposed to make that pistol-shot sound when the hero hits another man's chin? The chin breaking? No, the villain gets up and takes twenty or so more shots to the same spot without any sign of fracture. The hand? That would make more sense—but the hero keeps swinging his fist, and never even loses skin off the knuckles. The clap of one surface on another? A slap *would* come closer to the sound, without getting anywhere near its plosive force. Go into the butcher's refrigerator, pick yourself up a plucked turkey, take its wing off, and punch it crisply in the shoulder socket. Even if you cradle it in your left arm while punching its shoulder bone with your right hand, you will not get much *sound*. Hang it up like a speed bag, where it has *give* when you hit it, and your efforts will be almost totally silent. The sound of the face punch in movies (or on radio, where it was even more dramatic) is as unearthly a pure artifact as Tarzan's yell.

Well, if bare knuckles are no good against the skull, what did bare-knuckle fighters hit? We get part of an answer from the reforms and rules that have grown up over the years—against things like kidney punches, rabbit punches, holding, butting. We get another clue from the stance of bare-knuckle fighters. The hands were held high, but with the fists turned back toward oneself. This let the fighter strike blows down backhanded when they were aimed at his kidneys, ribs, lower abdomen, testicles—after which his fist was in a position to ram the body. Holding and grappling were allowed even under London Prize Ring rules, the most restrictive of the pre-Queensberry era. Since fights were open-ended, and lasted often for hours, men were worn down by fatigue, and then, against a practically immobile opponent, the "chopper" could be used to bring about the knockout—the chopper was a hammerlike blow with the fist, where the real impact was taken by the fleshy part of the palm—very good for breaking noses and punching into eye sockets. (Oriental martial arts work from the obvious fact that the flat of the hand—or even the extended finger—is far more deadly than the clenched fist.) The bare fist can break a man's ribs precisely because the ribs are padded.

The Marquis of Queensberry did not introduce his rules to make boxing less brutal, but to make it more interesting. Most matches of his day degenerated into one long clinch, with men bear-hugging each other between flurries of blows to the back of the neck and the kidneys. A round was ended by a knockdown—which meant any man could buy himself a rest by slipping to one knee. The marquis wanted to make the fighting brisker, so he created the determinate round of three minutes; within that time, a man knocked down had to get up within ten seconds or forfeit the match. More important was the no-clinch rule; men were no longer *allowed* to hit (among other things) but *required* to hit (and nothing else). No grabbing, holding, gouging, elbowing. The only weapon left was to be the fist punch.

But to make that possible, the marquis had to introduce his most important innovation, the "muffling" glove. That meant not merely that a fight could be prolonged with the use of fists alone, but that the fist could be used repeatedly and hard against the head. The modern fighter's hands are not only gloved, but bandaged. The hand is bound in cloth wrappings, given a mummy's case knuckle by knuckle, the wickerwork extensions turned monodactyl. Evolution is reversed and the hand shrinks to a hoof. Fighters may not all be mumbling brutes, but they *do* all cease to be articulate at the finger tips; and the condition sometimes spreads. Only in this way can the fist become a weapon approaching the power of more obvious and effective parts of the body —the foot, which uses the much larger muscles of the leg for kicking; the knee and elbow, deadly in the right places; the head, used as a battering ram; the teeth.

The boxer's hand is wrapped so that the whole area between first and second joints on his fingers will present a single level plane, diffusing impact across its surface, not concentrating it on any one knuckle or letting finger slide against finger, fighting each other. Then the glove diffuses the impact even further, out beyond the original plane, and puts a cushion between one's fist and the other man's head. The effect is to put a shifting, invisible bag around the head, so it can be hit with impunity. Of course the gloves cannot be too large, or it would be hard to slip them past the other man's guard. The aim is maximum diffusion of impact on the hand's surface, and maximum concentration of impact on the head.

On the *head,* since the gloves make it not only possible to hit the head often, but desirable. That is where the art comes in, and the cruelty. A hit in the face is an affront. Slap a friend abruptly on the arm, and he will ask you what the hell you think you are doing. But slap him only half as hard across the face, and demons rise in him. There are many reasons for this. The face is the expression of our consciousness. It guards those vulnerable jellies, the eyes—which is why we flinch from a face blow. Even a slight swelling or discoloration affects facial expression immensely—contrast a huge bruise on the arm with a comparatively restricted "black eye." A hit to the head is far more dramatic and symbolically hostile than one to any other part of the body—which is the source of the anatomical absurdities risked to give us the movie punch-out.

And a hit to the head is more difficult than a blow to the trunk—just as it is harder to hit the speed bag than the heavy bag. The speed bag, a little head hung upside down, a bounceable leather scrotum, is a boxer's way of practicing his scales. (And remember that, even when hitting this airball, the fighter uses leather mittens, so as not to skin his knuckles.) The bag is swiveled to one spot, and even a slow fighter like Joe Frazier can make it stutter out patterns of imagined punishment to the head (the thing makes the kind of sound you would expect from a Claes Oldenburg "soft" pneumatic drill). But the head on a moving fighter is far more elusive—and it must be hit through a guard held high up.

The modern fighter no longer stands like the bare-knuckle fighter, ready to knock blows *down* with a backhand swipe of the forearm. That would expose the face to the gloves now able to attack it. Instead, the fighter guards his trunk with his elbows, so his hands will stay before his face. He catches straight jabs with his hand, or slips them by head movement; he blocks hooks with the back of his wrists. The effect is of fencing "in the ring." The modern fighter lurks behind a fluctuating cage of his own arms and gloves, and it is the job of the opponent to dart blows in through that fretwork of defensive battlements. Precisely because the head can be hit with gloves, it is carefully guarded. The smallest target has the most careful outworks. The skill needed to hit that head, and hit it solidly, is what the marquis added to the sport. He forced the boxer to throw away his pummeling blunderbuss and take up a rifle. The boxer's left jab arises within a constructed frame of artifice, like a dancer's *tour jeté* or a soprano's trill.

Muhammad Ali is the supreme marksman. Time after time he touches the target, so fast it is hard to count the blows. He washes his opponent's face in leather, raising a fleshy foam and red general swelling. Though he is very large, even for a heavyweight, his punch has been clocked, frame-by-frame, and found swifter than that of Sugar Ray Robinson, the fastest middleweight in anyone's memory. Ali does not, like many fighters, do speed-bag drills to show off in his camp—his punch is faster than the bag can show. There is an irreducible minimum of bounce-back time required between punches at the bag. But not in punches to the head. Ali disdains the dull but often necessary work of slowing men down with fatiguing blows to the ribs, midsection, and arms. Those fighting him know that almost every punch will go to the head, and guard accordingly—it is the equivalent of passing on every play in football, and never using the run. Certain fighters can take a tremendous amount of head punishment, so long as you leave their breathing apparatus alone. Ali does not adjust to this advantage. He just takes it as a greater challenge to his ability to hit even so impassive a head as Frazier's often enough and hard enough to chop him down.

Speed is nothing without accuracy, of course—and accuracy is the thing that sets Ali apart. Even Joe Frazier has an irregular little bobbing motion of his head that makes him hard to hit. But you would never know it from the Manila fight. Ali's accuracy is a technical accomplishment grounded in a moral one, in courage. The best rifle in the world is no good without the marksman's eye. Ali carries his head high and partly exposed, so he can see everything all the time. Even the coolest fighter flinches, closes his eyes, ducks his head while being punched (or a split second before). But Ali more than any other heavyweight keeps both eyes open and on his man all the time. He whips his head back just enough to escape a punch without losing sight of his man. His head seems to float above the fight, looking down on it. Occasionally a barrage will reach up to him, and he swerves away, but with that wide-eyed whinnying expression, as much of measurement as of wonder. He has the best eyes in boxing.

And his eyes are his principal defensive tool. The conventional supposition is that a stand-up "classical" boxer must depend on his legs to protect him—must stay on the move, circling away from the other man's strong hand, not letting him "set" for the big punch. That kind of mobility a Sugar Ray Robinson could retain through most of his career. And Cassius Clay had it as a light heavyweight winning his gold medal in the 1960 Olympics. He even had it as a very young heavyweight (twenty-two when he won the professional championship). But no man his size, having reached his mid-thirties, can dance all night with Joe Frazier. Ali has accepted the extra weight of his later years, conditioned it to take any amount of body punches, and set about protecting his eyes. He does this just by slight moves of his head—and mainly moves backward or to the side, so his vision is not blocked. He almost never ducks. He does not pay the price most fighters have to, of evading *this* punch by losing sight of the *next* one, momentarily. Ali's terrific speed and reflexes are not in his feet any more, but in his neck—guided by those undeflectable eyes. (Archie Moore, too, survived what seemed forever on his cool—by perpetual little defensive motions directed from a tranquil observation post.)

There is a quality of sheer concentration in Ali that would tire him if he attempted to sustain it through a long fight—a laser beam burning backward into its own generator. The effort of sighting, of throwing those fast punches, of hitting the target, is too much for him. He rests against the ropes, covering up his head and letting the other fighter punch himself out against his sides. In the thirteenth round of the Manila fight, Ali's quick flurry, that late in the evening, almost had Frazier out. But Ali could punch no more for a while. His foot slipped, and then he lounged around the ring, gathering resources. Rather than punch slower and slower through the night, Ali likes to deal out his lightnings in little packets, then save up for the next barrage. This loses him points on the scorecard, but he likes that too. He is in love with risk. He takes unnecessary chances.

He has everything—speed, size, strength, looks, charm, defiance. His very childishness takes the edge off the outrageous things he does—as was the case with Babe Ruth. Oddly enough, heavyweight champions often look less than imposing. They tend to be spindly-legged like

Dempsey or Jack Johnson; stump-legged like Marciano or Frazier; ape-armed like Liston; bony-framed like Patterson. But Ali looks as if Praxiteles had sculpted him from caramel. To get such balanced limbs, one normally has to go to smaller men, like middleweight Randy Turpin. Ali's body seems almost too symmetrical to be functional. He could earn an epinikion from Pindar any day. He is a superb athlete. I wonder why that is not enough.

It isn't, of course. Modern Pindars sing the weirdest songs about Ali. They cluster around him, trying to probe nonexistent mysteries. There is Mailer, under the impression that he is interviewing the Heart of Darkness. (What did Heart have to say today? "Being a fighter enables me to attain certain ends." Heavy.) Even Wilfrid Sheed asks Ali to recite the Muslim catechism. This buzzing of the literary gents around Ali calls to mind Nathanael West's description of Faye Greener:

> None of them really heard her. They were all too busy watching her smile, laugh, shiver, whisper, grow indignant, cross and uncross her legs, stick out her tongue, widen and narrow her eyes, toss her head so that her platinum hair splashed against the red plush of the chair back. The strange thing about her gestures and expressions was that they didn't really illustrate what she was saying. They were almost pure. It was as though her body recognized how foolish her words were and tried to excite her hearers into being uncritical. It worked that night; no one even thought of laughing at her. The only move they made was to narrow their circle about her.

Sheer physical beauty, of unusual degree, seems to become different in *kind,* to call for *complementary* significances. That can be bewildering to the beauty's possessor, who feels his or her power over others and recognizes that they want such power to include something *more*. More than mere animal glow. The magnetic "star" tries to supply this something extra—wit, say; getting primed with one-liners like Marilyn Monroe's answer to "What do you sleep in?" "Chanel Number Five." Or Muhammad Ali's "Float like a butterfly, sting like a bee." (I am surprised, with all these literary gents around, that no one has recalled James Whistler's graphic "signature"—a butterfly with a bee's sting added.) When the wit does not work, the star might try profundity—attend Actors Studio, marry Arthur Miller, read Dostoevski.

Ali cannot read books—he is barely able to make out newspaper headlines; but he had his gingerly affair with various causes in the sixties. His Russian novels became the emerging nations. The surest thing of all is to try religion—become a convert to one of Hollywood's theological extravagances. The star feels destined anyway; holds special powers. It is in the star's stars. Superstitions like astrology fill up any gaps in the stricter theology adopted. Ali, of course, had the best religion ever fashioned for a star—one that combined a high regard for money with black militance and Vietnam pacifism. For his draft-dodging he was punished by Spiro Agnew's Washington, just as Ingrid Bergman had been by Louis B. Mayer's Hollywood. It added a note of suffering to Ali's brashness, and made him a Cause somewhere down the line from boycotting grapes. By the end of the sixties, Ali was the intellectuals' catnip.

For some reason, people don't want fighters just to be fighters. They have to stand for an era, for the color of hope, for a metaphysics of the spirit. Virtue (Floyd Patterson) meets vice (Sonny Liston). Poor Jerry Quarry was, for a while, the voice of the new ethnics. Joe Louis, we were brought up to believe, dropped the first American bombs of World War II on Schmeling's head. It is as if the *art* were not enough to redeem boxing's violence, all that cruelty inflicted on the face—so we prefer to think the loser is being destroyed for some deeply ideological reason. Get the Nazi. Smash the traitor. Beat Whitey. This tendency, quite as degrading as it is silly, reached a new low in Manila, where people cheered for Frazier in the name of family life (Ali's current "fox" was more conspicuous at this fight than at others). Ali had finally to take a rap not analogous to Ingrid Bergman's, but identical with it. Our stars have to keep themselves pure. They are our own stars, and no one wants to read dirty jokes in his horoscope.

If all else fails, in our quest to have our heroes escape the confines of a merely physical luster, we can fall back on Hollywood's ultimate con —the claim that a starlet is really *bright* because she has the brains to recognize how *dumb* she is. This ploy turns publicity agents into scholastic disputants, and the star goes readily along:

> Had any other girl been so affected, he would have thought her intolerable. Faye's affectations, however, were so completely artificial that he found them charming. Being with her was like being backstage during an amateurish, ridiculous play. From in front, the stupid lines and grotesque situations would have made him squirm with annoyance, but because he saw the perspiring stagehands and the wires that held up the tawdry summerhouse with its tangle of paper flowers, he accepted everything and was anxious for it to succeed. . . . Faye did have some critical ability, almost enough to recognize the ridiculous. He had often seen her laugh at herself.

That is roughly the Wilfrid Sheed Thesis on Ali. The poor fellow may have a short attention span and low IQ; but he is on to his own hype, and graciously shares this awareness with us: "He grins a lot off camera, sheepishly, with an 'ain't I a devil? isn't this just too much?' quality. As if he wants *us* to be in on the joke on *them:* the basic principle of dramatic irony, whether he knows it or not." Marilyn, too, was always committing Ironies, according to her press agent. "He is, though he says so himself, a humble man: it is one of the weird secrets of his success. Once you notice it, his wildest boasts never bother you again." Sheed has gone backstage during Faye Greener's act, and seen the grips at work; so the lines no longer mean what they say for him.

The other Sheed Thesis is that fame on Ali's scale is a universal currency, convertible to any other kind of power or influence, good or bad. And he fears Ali's influence will be bad in time, out of mere quest for novelty, when he has worn out whatever good influence he cares to exercise. It is a more sophisticated version of the Ingrid Bergman problem —what if the children find out what Joan of Arc is *really* doing with her time? Sheed forgets how far Ali had dropped from view by the third year of his "exile." When he is not fighting, the magic drains from him. Then the doggerel is just doggerel. Ali will be a celebrity as long as he lives—like the Duke of Windsor. But he only *rules* from the ring. He has nothing, really, to say, except with his fists. Yet how eloquent they are.

—New York Review of Books, October 30, 1975

36

Shirley Verrett

The gypsy Azucena, by brooding on the past, twists the present. In the production of *Il Trovatore* that opened this year's Metropolitan Opera season, Shirley Verrett, singing Azucena's role, seemed to carry the past around like an invisible baby; curved over, curling in on her burden.

Restless, driven, beating her arms against the cold, she was a kind of singing vulture, circling to prey on herself.

When Miss Verrett came before the curtain for her bows, a lovely woman rose up from the witch—tall, young, laughing. Yet she had to be persuaded to make that stunning metamorphosis. Her agent had trouble getting her to smile when taking bows—for years she had "kept the character" even in curtain calls, as part of her fierce concentration on the role. She gained fame in Europe during the sixties as a sexily undulant Carmen, yet she has taken the part from her repertory because she thinks the opera is not performed as an integrated drama these days. She gave up the leading role in next week's Met premiere of Poulenc's *Dialogues of the Carmelites* in order to play the Second Prioress, who stays a nun throughout the opera.

Perhaps the extraordinary submersion in her roles explains why Miss Verrett still hovers on the very edge of that hysterical breakthrough that turns a career into a myth. Storybook things have happened to her, yet left her less than legend. In 1973, after singing the long lead role in the first half of Berlioz's musical epic *The Trojans*, she went on to sing the lead role in the second half (for an ailing Christa Ludwig). She opened La Scala's season last year as a spookily melodious Lady Macbeth, and opened the same company's first American tour in the same role. Yet six weeks after that triumph in Washington, she did not even fill the Kennedy Center Concert Hall for a brilliant recital. The underground trade in tapes of her performances is smaller and slower than one would expect for a singer of her achievements. Her commercially recorded "singles" are all out of print. She even plays down the important turn in her career that is making some critics dubious, when not openly critical. She is, in her mid-forties, changing from mezzo-soprano to soprano, and doing it without the help of a voice coach or teacher.

"I was always a soprano. I trained as one with my first teacher in Los Angeles, Anna Fitziu. I had scheduled a soprano aria for the Arthur Godfrey show, but changed to Delilah's *'Mon coeur s'ouvre à ta voix'* because it is a more popular melody." Miss Verrett won the Arthur Godfrey Talent Scouts show when she was twenty-four, and that led to a four-year scholarship at the Juilliard School of Music. This is very late for a singer to be starting. Born in New Orleans but brought up in Los Angeles as part of a black middle-class family of Seventh-Day Adventists, Miss Verrett had not been allowed to go to the theater, much less sing in one, as a child. She had to sneak over to an aunt's home to hear Billie Holiday records, and imitate them. She sang Methodist and

Lutheran hymns in her church, not "gospel" or jazz. At seventeen, she won a contest whose prize was the chance to study with the great diva Lotte Lehmann. She turned the prize down, embarrassed by her lack of training and background. Only after an early marriage failed, and she got bored as a real estate agent, did she start taking lessons on her own with retired Metropolitan singer Anna Fitziu, where the talent scout found her.

"At Juilliard, I entered every contest that came along, and won them all. I had to. I started late. I needed the money, which I plowed into extra coaching." While taking the full singing course at Juilliard, she used her prize money to hire special coaches outside the school—in voice, languages, drama. It is hard to imagine a more entirely *earned* career than Miss Verrett's. The contests brought her to the attention of musicians. Stokowski heard her, and wanted her to sing in Houston—but Southern opposition to blacks still blocked that kind of thing in the late fifties. She performed the three vocal episodes in de Falla's *El Amor Brujo* for Stokowski during his emotional return to the Philadelphia Orchestra, a performance recorded afterward. Stravinsky chose her to be Clytemnestra in his *Oedipus Rex*. Menotti cast her as Carmen for his Spoleto festival.

Yet the commitment to an opera career did not become final for six or seven years. The break with her church had been slow and painful. When she began her studies, the concert hall still seemed the safest route for black singers. And she did not like to be "typed" as a Carmen, a mezzo, a singer of accepted "black" roles like Ulrica or Azucena. She pushed out her repertory with recitals, still studying voice on her own. She turned down two offers to debut at the Met.

All that changed in 1967, when Covent Garden gave her a tumultuous reception in the difficult soprano role of Queen Elizabeth in Donizetti's *Maria Stuarda. Soprano* role. In 1972, she sang the soprano part of Princess Selika in Meyerbeer's *L'Africaine* for the San Francisco Opera. But everyone still thought of her as a mezzo-soprano—of the masculine recording of Orfeo, of the very feminine Carmen she sang at her Met debut in 1968.

The confusion began early. Madame Marian Szekely-Freschl, her teacher at Juilliard, heard her sing as a mezzo on the Godfrey show, and disapproved of the soprano exercises she was doing. Miss Verrett, regal on the stage, can be impish in private. A natural mimic, she tends to act out all the parts of any conversation she repeats. Her "Madame" is an act of very wide flourishes: " 'Something is funny going on,

Bootch'—she called everyone Butch. 'I gif you vun exercise, and you go and do something different. Vell, you vill gif that up, or you vill *brreaak* my heart. . . . You are very religious, so you go home this weekend, and you pray to your God. If you vant to be a soprano, I will be the first to help you find a new teacher. But it vill *brreaak* my heart.'

"So I went home and prayed. And paced the floor. And did exercises." Madame Freschl seemed to have a point. Shirley could not sing high notes *piano*. "Miss Fitziu had a natural *piano,* and did not know how to develop one in me." Shirley went back on Monday and capitulated. But the battle was not over. "Madame would still say to me, 'You take those notes too much like a soprano. I don't like that color.' But I said it was *my sound*. What she was asking me to do was uncomfortable, so it had to be wrong." Out from under the tutelage of Madame, she tried several vocal coaches without satisfaction, and became her own teacher. "I have read every book on the voice I could get hold of. Most were worthless. But perhaps there would be one thing in a book that seemed worth trying. If it didn't work, I put it aside. Maybe I would see its usefulness later." It all went into her vocal accounting book—a running record of her experiments, projects, achievements, goals. "I want to know everything I can about my instrument. I have always vocalized across my whole range, and then extended it *mezza voce*. I think every singer should do everything—trills, coloratura, everything: just to know, always, what the voice can do." She does not want to sing on the edge of her capability in the theater, but well within it. Yet she takes very good care of her voice: "The minute something hurts, you know you are doing it wrong, and must stop."

She brushes aside fears that she tries too many things without the continuing ministrations of a voice coach. "I am at the point where I want my principal teachers to be conductors. I love a man like Georges Prêtre—his hands just dance over the score, explaining all its possibilities. Some conductors simply tell you, 'That's fine,' and you learn later they thought it was *not* so fine. But they had no way of telling you. I want a man who *tells* me something. Teach me! Don't waste my time." She says she also learns from colleagues. When she was studying the part of Jane Seymour in Donizetti's *Anna Bolena,* she had trouble with Jane's high soft pleading "Hear my prayer," before Henry VIII. "I went to Giulietta Simionato, whom I heard sing a great Jane Seymour in a concert performance of *Anna*. She told me, 'On those high notes you cannot get a perfectly pure vowel sound and the right color too.' I was pushing the voice too far forward on the first vowel of 'Odi la mia preghiera.' By letting the voice go where it wants to go, ending farther back on the palate, I made 'Odi' sound more like 'Aw-di.' " A slight relaxation of diction freed the voice, to float. (Actually, on the record, her vowel sounds more like *Ohw*-di; but she was obeying the same instinct that made Caruso begin Handel's "Ombra mai fu" with an unashamed *Ahm*-bra.)

Miss Verrett thinks some voice teachers are just glorified babysitters, giving needed assurances to the stars, tending egos rather than technique. The teacher is also a legacy from older days, when the need for one was clearer. We do not, any of us, hear our own voice as others do. It comes to us resonated along and through our head, from inside as well as outside the hearing apparatus. That explains everyone's shock on first hearing his or her voice over a tape recorder. Shirley had just that experience while at Juilliard. "I thought, *wow,* if I don't even know what I sound like when I'm talking, what must my singing be like? I got a tape recorder right away, and I've used it ever since. The tape does not lie. If you listen honestly, you know that note wobbled, or that one did not emerge just right, or this one started flat." The recorder supplies the unbiased outside opinion that older singers had to get from coaches. Most singers now practice their roles onto tape, and take their recorders to rehearsals. Some even tape themselves as they are making recordings, so they do not have to wait for playbacks of the complete cast. They can go into a back room and play repeatedly a phrase, an entrance, a note-sequence, that is troubling them. What Shirley learns from the tape goes into her books, those accounting ledgers of vocal losses and gains she must keep on the credit side.

Her dramatic preparation is just as intense and thorough. She says she learns more at the ballet now than at operas—learning, always, how to move. "I begin with the walk. How would this person carry herself?" Even in a concert, she changes her whole body's carriage before each song or aria to suggest a character, a situation, a mood. Many physiological-psychological considerations go into this kind of total concentration. It is said that comedians who do impressions, like Rich Little or David Frye, are unable to catch a person's voice if they have not adopted a posture, a facial expression, a physical mannerism connected with that voice.

I ask her how she managed to *shrink* so as Azucena, and express some misgivings about making the role too haggish. Verdi said Azucena must never be crazy. She should have an air of power, like one of the Three Fates moving in disguise. "Yes, I overdid that a bit here, because the production was so static. I felt I had to move around; but I also had to look old enough to be Manrico's mother. There are several ways you can walk like an old woman. In London, I did it this way. . . ." She is up, leaning back, putting her feet hesitantly forward, afraid of falling off a cliff at every step. "The black costume made me slimmer; I was a mysterious gypsy, but the Met had all those veils and spangles, and I had to move faster." To do that and stay old, she went into her brittle crouch.

Her very carriage can be counted on to change the way a role sounds, as well as how she looks. And the bearing is chosen for each performance in terms of that particular setting. No singer is more interested in the entire production than she is. Her acting ability, as well as her beauty, made Sidney Poitier and Harry Belafonte ask her to try straight dramatic roles on film. She may yet do so.

When I first interviewed her, she was working on the part of the Second Prioress in *Dialogues of the Carmelites*. Her problem, she explained, is twofold: She has to move like a nun; but she also has to distinguish herself from a whole stage full of other women in nuns' habits. The Prioress is (by comparison) a peasant in an order of aristocrats' daughters; below them socially, yet placed over them in authority.

How to approach the role? One way that occurred to her was learning to say the rosary. Her happy second marriage is to a painter from an Italian Catholic family, so she asked her sister-in-law to teach her how to use rosary beads. (Shirley's mother was brought up a Catholic in New Orleans, and Shirley herself used to go to mass with her great-grandmother—a woman who was always quietly "saying her beads.") "At times, when I am not singing, my character would be likely to say the rosary to herself. I want to know what that would feel like." Perhaps the very act of saying the rosary would suggest a kind of bodily carriage as well as mental attitude: Where does one turn for strength *in* and yet *against* the religious community, in the *name* of the community?

Instinct is always at work behind Shirley's research. The rosary is simply a repeated "Hail Mary" punctuated into five "acts" on biblical themes. And the part of the Second Prioress has its finest lyric moment when her voice lifts effortlessly over her undeclared rival's as the community sings its "Hail Mary" in the chapel. Attempting that kind of lofty sweetness holds no terror for Miss Verrett now. She has developed on her own the *piano* high notes that her teachers could not find. Now, in Lady Macbeth, she ends the sleepwalking scene with that high D-flat "thread of voice" Verdi called for.

Her exercises gave her the facility in coloratura work that bel canto
mezzo-sopranos need, and this made people see she could sing florid soprano parts. There has been a tendency for mezzos singing Verdi to
darken their voices for contrast with the soprano. But that often makes
no sense in the earlier operas of Rossini or Bellini. In the latter's
Norma, an aspirant young Druid priestess named Adalgisa falls in love
with Pollione, the Roman occupier of England. She confesses this to the
head priestess, Norma, who has secretly borne children to Pollione.
Adalgisa, the mezzo, was first sung by Giulia Grisi, who had a lighter
voice than the original Norma of Giuditta Pasta. Adalgisa is allowed to
sing the verse of the famous duet, *"Mira, O Norma."* But when the
larger voice of Norma repeats the same notes for the second verse, the
effect should be one of growth and climax—Adalgisa thenceforth drops
a deferential third to harmonize with the woman in command. Shirley
says, "I played Adalgisa just as young and light as I could; otherwise
the story makes no sense." This is a far cry from the Adalgisas of Ebe
Stignani or Fedora Barbieri. Listening to private tapes of Shirley's first
Norma in the Met's 1976 touring company, I was struck again by all
the risks the part involves. Yet I see why she feels compelled to take
them on. As Norma she can change her whole vocal bearing in a
flicker: She is tender and harsh by turns; thinking of her country, risking her country; thinking of her children, and threatening her children;
sparing her rival, torturing her lover; caught on a spit that turns and
turns, and burns her with love and hate, of others and herself.

In the long final scene, when Pollione is brought on, and cries out
"Norma," she answers, *"Si, Norma."* It is a sour little cluster of notes—
halftone up, one and a half down—C sharp, D, B. Callas, in both her
recordings of the opera, makes the phrase threatening. Sutherland lin-
gers on the phrase and threatens to go to sleep. Shirley invests it with a
kind of resigned power, as if lifting the whole burden of her character
and fate in that name.

"Norma," as name and title, is invoked throughout the opera in a
haunting way, like the name of Euridice in *Orfeo.* Norma speaks in the
regal third person at her very first appearance, and after that the name
occurs often at the cadences of phrases. When, having promised to reveal the false priestess, she at last sings, *"Son io,"* the simple pronoun
has the effect of an uncrowning. The chorus asks, incredulous, *"Tu,
Norma?"* and Pollione says not to believe her. At that point, the queen
puts on her crown again, as a kind of torment: *"Norma* does not lie."
Miss Verrett gives that proud defeated phrase the same resigned maj-

esty she summoned to *"Si, Norma."* Just as she studies her bodily carriage for a new part, those who know her work can *see* how she would be standing as they listen to her records. It is the mark of a great singing actress.

—New York *Times,* January 30, 1977

AFTERWORD

Despite her intention to devote herself exclusively to soprano roles, Miss Verrett did sing mezzo parts after this, including more Carmens, while adding soprano triumphs like her Tosca. Today she says the whole compartmentalization of the voice in separate registers seems to her a sterile exercise.

37

Raymond Berry

Sherlock Holmes, given knowledge such a trade exists, could guess his trade. The little finger of the left hand is (a) not little and (b) as crazily jagged as one of his own zigzag patterns—the result of multiple dislocations in 1955, his first year in the pros. A football receiver is supposed to have "soft hands"; but the softness is all turned out to the ball, which does not answer in kind. The ball gets kissed; but kicks back. The backs of his hands are fleshily ridged, like tree bark reversing the Daphne metamorphosis. The spatulate fingers still fondle ball-shaped air as he gestures.

According to legend, Raymond Berry of the Baltimore Colts proved that art can be tortured from a body not naturally suited to high purposes. Here was a man who caught 631 passes between 1955 and 1967, gained 9,275 yards, ended up in the Football Hall of Fame—yet did this with one leg shorter than the other, only nine usable fingers, a bad back, and poor eyesight. It is a Cinderella story, often repeated; and there is only one thing wrong with it. None of it is true.

Berry, now the receivers' coach for the New England Patriots, just laughed his mild east Texas laugh when I ticked off the points in the legend: "It's a pile of bull. You don't get to play professional football without an extraordinary body—any more than you end up on the stage of the Metropolitan Opera with just an ordinary voice. What you do with the voice once you get it is a different matter. But first you have to have the natural gifts."

What were Berry's natural gifts? "Big hands and feet, and a sense of balance." He wears a size fourteen shoe. Why does that matter? "A receiver has to stop, cut, twist. He needs strong ankles and knees, and feet for planting." Berry appears slight, 180 pounds (his playing weight, maintained today) spread over a six-foot-two frame; but his training was so intense that none of it was fat. And his injury record shows that he had a remarkable toughness. The story goes that he built up this strength from the 92-pound weakling stage. But he denies that. "Look at Harold Jackson, a so-called small man who has played fifteen years in this league." Berry knows I just interviewed Jackson on his way out of the shower. He is only five feet ten, 175 pounds. "Harold has an extraordinary body for football—he's fast, but his joints are put together in a way that lets them take tremendous pressure."

I would find that Berry thinks "prevent" football—prevent injury, prevent interceptions and ball-drops and fumbles. Cut the odds against you. He only had one fumble in his pro career, and that was a disputed call. Others judge Joe Namath a great football player despite bad knees. Berry says good knees are a necessary gift, just like a strong arm or a sense of balance. John Unitas could stay "in the pocket," waiting for the last moment to release a ball toward Berry, because his body could take punishment. He was ready to get beaten up a second after the ball left his hand. Berry was ready to get beaten up too. Jackson told me, "This is my fifteenth year in the league, and Raymond played thirteen years. But my career has been extended by the rule that they can't manhandle you after five yards from scrimmage. Once I get my five yards, I'm free to run; but they were beating on Raymond all the way out."

The weakling Berry of sports legend could not have taken that punishment. Yet everyone believes the legend. Even Johnny Unitas told me one of Berry's legs is shorter than the other. This is the only myth Berry feels concern about. "I used to get letters from parents saying they had a child with one leg shorter, and they knew I was playing football in that condition. I had to write and tell them the story was false." How does such a story get started? "I was tested once by an osteopath for bruised nerves in my back; and one of the tests is to get the legs totally relaxed and keep bringing the ankles together to see if they meet perfectly—if not, you know there is tension in the back nerves. Some sportswriter got a garbled version, and all the others repeated him."

Well, what about his one unusable finger? "I use it all the time"—he flexes the weirdly askew little finger. "It got permanently dislocated, but I worked out all the stiffness and brought back the strength by kneading putty." I put my hand up against his—the longest finger is two-thirds of an inch longer than mine. "I could always catch the football. That was my natural gift. The reason I wanted to play professional is that college ball in my day was a running game. I was afraid I would never get a chance to do what I did best."

I was told by a well-known sportswriter that Berry had not caught many balls in college because of his bad eyes—Southern Methodist University played most of its games at night. Another myth. "Our games were in the afternoon, like most colleges'; I didn't catch many passes because not many were thrown." What about his eyesight? "I needed contact lenses. But the lenses would slip when I did rapid eye motions back toward the ball. If the sun caught a lens as it slipped, I was distracted from the ball. But I was lucky enough to find an optometrist willing to work with me for lens stability and comfort. I tried all kinds of lenses till we got what we wanted. I even had tinted lenses for sunny days, so I could watch the ball come right across the sun."

Far from being blind, he actually saw better than other receivers; but his constant changing of many contact lenses made sportswriters think he must have terrific eye problems. Didn't Berry have *any* of the physical disabilities he is now famous for? "I had a back injury from college; but by the time the osteopath figured out what it was, I could correct it with a canvas brace." Berry was known as a very slow receiver; Unitas told me he tended to overthrow to the left—but, he grinned, "Maybe that's because I had Lenny [Moore] on the right, and Raymond on the left."

Berry admits, "I was not very fast—I did the 40 in 4.8. But I improved my time by learning a fast takeoff, making my cuts without slowing. I turned just my head back, not my whole upper body, so I lost no speed watching the ball into my hands. If you do all that every play, you reduce the margin between yourself and the faster runners." Berry does not think of himself as overcoming any great handicaps. Then why did he make up all his famous drills for catching under duress?

Well, as you must suspect by now, some of those drills never existed. I had read in a Baltimore paper how he dealt with the problem of deflected balls—by getting friends to throw footballs against a basketball backboard, so he could catch them on the unexpected bounce. He just shakes his head: "Never did it." He was supposed to be so intent on catching the football that he had his wife throw to him when he went home. "No," he says, "it's bad practice to catch anything that's not thrown hard." Sally Berry threw some balls for the photographers, when they asked her to.

I was not surprised, by now, to learn that athletes have a low opinion of sportswriters. Berry, the least caustic of men, tries to be understanding. "I guess they don't get paid very much." Unitas thinks they have to sell papers by making up sensational stories. Both men believe that writers know little about the game. By the time I asked Berry these questions, I had followed him through two complete days of his coaching work. "You're the first writer that I know of who has done this," he said. "And I guarantee you they wouldn't say some of those things if they had." Berry, as usual, thinks prevention: "Actually, we should run clinics for writers, to help them understand the game."

Understanding the game has become an awesomely complex task. When I asked Berry if coaching presents him as many challenges as playing did, he answers: "Very few people know the problems a receiver coach faces every week—because, obviously, very few are directly engaged in it. You'll just have to take my word for it. I go to the Boston Pops here, and I admire the teamwork of all those musicians. Sally and I go to the ballet. But I've often wondered how well they would do if people were running through the orchestra grabbing their music off the stands, or wandering around the ballet stage trying to trip up the dancers. *We* could look perfect if we went out and did our kicking plays, passing plays, running plays, with no opponents on a Sunday afternoon. But we have eleven very skilled people out there trying to wreck our plans every time."

The way Berry describes the game, the natural odds are against any play's ever working. Mistakes on one's own side, a missed block, a tangled assignment, a bobbled ball, are constant dangers; and these are compounded by the efforts of the opposing team, which even accidents can make successful—the tackler ends up at the right place for the wrong reason. Modern defenses are increasingly sophisticated. Berry sees the job of a coach as cutting down the odds against any play's success—from, say, five to one, down to three to one. Out of the complex of designed and accidental obstacles, you have to succeed on just enough tries, react to enough breaks, learn to increase the odds against the *other* side.

And all this must be done on a schedule that is a nightmare. I had called Berry on a Tuesday. He told me, "Don't come by today or tomorrow. We don't have a spare moment either day. They're our worst." As one would expect, Mondays and Saturdays are the lightest days, bracketing Sunday's game. Monday is mainly retrospective. Coaches go over the previous day's game on film, early in the morning, then do a postmortem with the players, who suit up after lunch for a brief workout. "Players used to rest the day after a game, but that left them stiff. We have them work up a sweat and run off the soreness. It is also a good diagnostic tool—we see whether a sore spot is just hurting or indicates real injury."

Tuesday is the players' only day entirely off—and it is the coaches' longest day. They go over films of their opponent for the week, and read the computer printouts on what that opponent does—its statistics for defense lineups on third and long, on third and short, on plays inside the twenty, etc. These have already been broken down and put into a computer. Out of these figures and the films, the coaches construct their game plan—a single sheet of paper for the offense, top half running plays, bottom half passes. New plays are added, according to the strengths or weaknesses of the team to be faced. These are mimeographed. Defense formations are adjusted. The entire day is given over to intense debate and viewing, in various combinations of the coaching staff; and all the written material must be produced before they can go home—at eight or nine o'clock if they are lucky.

On Wednesday, the players arrive early and are given their packets of ten or a dozen sheets of paper. After lectures on the game plan, and films of the team they will be facing next, the players have lunch. Then certain combinations of players and coaches look at films again. Berry and his receivers gather in a little room ("About what an east Texas high school team would have," Berry laments) with the two quarterbacks and their coach, to look over the defensive backs they will be throwing or running against. The quarterback coach, John Polonchek, runs the projector with one hand and chain-smokes with the other.

Each play is followed to the point where receivers make their break and the defense "commits." Then the film is run back and forward several times. Receivers strain forward on the screen and pick up speed, then reverse their quick steps, tugged backward with increasing effort and diminishing result, till they settle once more in their takeoff position. Everyone is watching the way this back comes up or falls off, that one presses or "hugs" (plays outside the receiver). Players sprawled on the concrete floor talk sporadically of "double bumps." Berry, off and on, describes the defense's "philosophy," its basic patterns and tendencies, using a flashlight that casts an arrow-shaped beam at the key move on the screen.

Each receiver is watching his favored side of the field, the two or three men he must fake or outrun or outwrestle for the ball on Sunday. These films are not the TV kind, focusing on ball carriers or the ball's passage through the air. All twenty-two men must be kept in sight all the time. And each sprawled receiver in this room peoples the screen with his own ghostly self, trying to outthink the opponent, running it *this* way instead of *that,* succeeding where the offense of the prior week had failed. I asked, later, if Berry ever thought of himself as racing across that screen, trying his famous routines on a new generation of defenders. His answer is determinedly nonromantic: "No, it is just a different game now. Things that worked then would not now. Harold and Stanley [Morgan] have different bodies from mine, different gifts, and I have to concentrate on what *they* will be doing."

After a half hour or so of film, the players finish putting on their equipment and go to the field. The kickers are already there, warming up—as we come out into bright sunlight, a punted ball spins busily toward us, its head nodding lazily to this side and then that. I ask what Berry's responsibilities to Harold and Stanley, and to his tight ends, will be this afternoon. "I'll be giving no real attention to them. My job is to work with our defense so it looks like the Atlanta one." Offensive and defensive squads get only half the play period for their own practice. The rest of the time they pretend to be next Sunday's team, so the game plan can be practiced against actual formations.

I ask Berry how he gets the Patriot defenders to slip free of their own system and adopt another team's. "If their formation is enough like ours for any play situation, I just say, for instance, P-3 [Prevent-3] but *you* play up and *you* fall back [pointing]. For other situations, I use the cards." In his office, earlier, I had seen rows of big cards behind his desk, each pile with a specific offensive lineup printed on it. Before going to the field, Berry has drawn in the defensive positions of the next opponent, and between practice plays he goes into the defense's huddle with the appropriate card. All his thought about the opponent's defensive "philosophy," begun when he first started watching the films last week, leads to drawing up these cards and coaching the Patriots to pretend they are the foe.

Before practice can begin, the players do stretching and limbering exercises. Several lean their shoulder pads against the walls around the field and act as if they were Samson about to push over all the tiered rows of seats. The team then clots in little groups of specialists. Berry's receivers run back and forth across the goal line while the quarterbacks alternate throws to them. Berry has stationed himself at the approximate point where the balls reach the receivers, and he jogs in front of each, waving a hand in his face. In his own playing days, he used to get another receiver to harass him while he ran patterns, then return the favor. "Catching the ball is just half the job; you have to do it with every kind of distraction and opposition." Otherwise, you lead the tame life of a Boston Pops player whose music is not grabbed at by imps and goblins. During his practice with the Colts, Berry asked defensive linebacker Lenny Lyles to knock him around whenever he got in his vicinity, so he learned to live with (and ignore) the continual buffeting even Harold Jackson speaks of with some awe.

Berry, as Unitas's favored receiver, was a danger every time he got into the secondary; and he often had two players trying to hem him in, "to sit on me," as he puts it now. Yet, even when they knew it was coming to him, it tended to get there—three times in a row during the famous last drive of the 1958 overtime game that won the Colts their first championship. "John was the most accurate passer I have ever seen. He not only threw it to you, but away from the other guys." Berry says that what sometimes looked, to the layman, like a bad pass was deliberately thrown to the "wrong" side because Unitas saw a defender approaching the "right" side. Unitas earned his gambler's reputation when he threw on the Giants' six-yard line in the sudden-death overtime of 1958. He brushed aside objections that he was taking too big a risk by saying he threw behind Jim Mutscheller. That made it hard for Mutscheller to catch, but impossible for anyone else to.

Everything is done on the Patriots' practice field to a series of deadlines, sounded on an air horn by one of the training boys. After one-on-one drill, the receivers and defenders go into seven-on-seven patterns, one of the few times Berry will be able to watch his receivers run their patterns and make any suggestions. Berry was famous for the precision of his patterns—you could trace their shape almost as if he were skating and had cut his journey's record in the ice. He had the exact number of steps worked out for each variation of each pattern, and he timed himself over and over to make sure he had eliminated all waste motion and maintained full speed throughout. Unlike the speedsters, he would not vary his pace, to lull a defender with deceptive lagging and spurts. Maintaining full speed gave Berry time to add to the number of his fakes, the most famous of which left defenders tangled in their own feet.

Does he enforce a similar discipline, of running patterns "by the numbers," on his players? "No. By the time players get to this level, they have their own approach and style. You lose more by trying to change that than you could ever gain. My job is to know as much as possible about the people they will be facing, and to pass that knowledge on to them in a way that will do most good. You can't influence people unless you show that you respect them." The few times Berry has shown anger came when other coaches were driving or humiliating players.

At the sound of the air horn, Berry's players go from seven-on-seven drill to eleven-on-eleven. All the plays of this week's game plan must be run through, and they are all being filmed from the press box. The players look frustrated as the coaches call them back to scrimmage just as the quarterback is ready to release the ball, or a runner has broken through the line. The task is to get assignments in mind, test ability to read what is coming, adjust for it, complete the coverage. There is no time to let the play run its full course. Watching them, I get the impression I am back in the film room, seeing plays begin and never end, players run through the same motions backward and forward. Berry's own drills, run through after his practice sessions with the Colts, came from his shocked realization, on joining the pros, that a receiver rarely catches the ball when plays are being rehearsed.

When the players trot off to their rather cramped locker room, and strip off the heavy padding, I search for Harold Jackson, but don't catch him till he comes out of the shower. What's it like to be coached by Berry? "I wouldn't have felt my career was complete if I never had a chance to work with him. After all, when it comes to wide receiving, he wrote the book. I had heard from other receivers he coached around the league how helpful he was to them." But Berry came to the Patriots after Jackson had played in Foxboro over a stretch of ten years. What more did he have to learn? "You get into bad habits with your patterns.

It's good to have somebody knowledgeable look at them from the outside, and sharpen you up. He also taught me to study a field beforehand." Berry used to pace off an entire field before playing on it—testing for soft spots, loose dirt, mud, slippery grass—so he could arrange his special combinations of cleats, to go along with his layout of contact lenses. But now most fields are artificial. What is there to study? "They each have different slants." Artificial turf is installed with slopes for drainage. A receiver, who lives much of the game with his head screwed around the wrong way, should know whether he will be running "uphill" or "downhill" in various parts of the field. He should also check which way the sun will hit the field at different parts of the game, or how close the wall is at both end zones. When I told Berry that Beverly Sills always checked out an auditorium she was going to sing a recital in, getting to know the stage geography and sight lines as well as the acoustics, he nodded enthusiastically. He likes the comparison of singers to athletes—both have to use their bodies as instruments, and no one can know that instrument as well as the performer.

I ask Jackson if Berry is a hard taskmaster. "No, he is always asking me how my legs are feeling. If there is any doubt, he has me shorten up the practice. He says an athlete's life is all a matter of pacing."

It is a lesson Berry learned the hard way. "When I went to the Colts, it was an obscure team, and John [Unitas] and I were hungry young guys willing to work hard. So I had him throw to me after practice. I thought I could run as long as he was willing to throw. I didn't know yet that John was a horse for work and his arm never got tired. I was doing too much, leaving part of my game on the practice field. When Bobby Shaw came to the Colts in 1958—the only receiver coach I ever had—he persuaded me that I should take it easier, to have more bounce in the game. My father [a high school coach] had told me you should always quit exercise with energy *left,* not totally drained; but I needed Shaw to make the lesson take hold." The myth is that Berry was indefatigable in his endless drills. Actually, he prepared them to a tight schedule so he could do a great deal in a little time, without wearing himself out. He organized his drill so he could catch sixty balls in fifteen minutes, three or four of each kind he worked on (high and behind, low-front-scoop, sideline toe dancer, etc.).

By the time I left the Patriots' locker room, the coaches were back at their desks, the lights dimmed as they watched more films on the wall nearest each one's chair. But they were no longer watching this Sunday's team. Next week's players were flitting about the wall like crazed mosquitoes in the cones of light. I ask Berry if it isn't a distraction to look at two teams on the same day. "Not really. We're not *studying* these films yet, we're just making up the pads to feed the computer."

What they are giving the computer is a "call" on each play of the game—for Berry, what the defensive formations were in each situation. This time has to be "stolen" from preparation for *this* Sunday, since there is no place else for it on the schedule. The computer printouts have to be ready Monday for the coaches to start work on their next game plan. The authority on each "call" is the specialist on the coaching staff for that area—so Berry has final say on what the defense formation is. This ensures uniformity of terminology and stability of result.

By now it is six o'clock. Berry is hungry, and the coach he is "calling" with, Tom Yewcic, begins to yawn. "How many more plays?" Berry moans. "Just a few." They jump up at the end of their chore, rubbing their eyes and making motions to close their desks for the day. But the head coach shouts in from the other room that a problem has arisen. He is watching this afternoon's practice films, already developed in-house, and on a double blitz no one is picking up the weak side safety. Several plays have to be reformulated, to free the tight end for this chore. The debate over which plays to change, and how, goes on for thirty minutes or so. We leave the head coach still pondering the troublesome sequences. Tomorrow morning he must report on them to the players, to correct the matter on the field. Friday's practice is reserved for such problem areas, and for special parts of the game that need work or are likely to be stressed on Sunday. Saturday is comparatively relaxed. The players come in for some running and stretching and more films, but they do not suit up.

Over dinner I ask Berry what he will do during Sunday's game. "I sit up in the box and analyze their defensive backs' moves. When I see anything important, I phone it down to the field." Again, he does not even see his own receivers—he will watch them Monday morning, on film. "It takes all my concentration to keep seven men in view, and record their motions. I am more and more convinced that people who say they know about a game just after it is played cannot realize what they are saying. If, with my experience, I find it an absorbing task to watch seven men, how can anybody keep track of twenty-two of them all at once?"

In his playing days, Berry rated his own effort, not on the basis of his feelings on the field, but after watching the game film on Monday. He set two team records during the sudden-death championship game of

1958—twelve catches for 178 yards—but rated himself only 45 percent. "If I hadn't missed a downfield block, [Alan] Ameche would have scored in the fourth quarter and we would not have had to go into overtime." I ask what was the highest rating he ever gave himself. "Earlier in that 1958 season, against the San Francisco Forty-Niners, I earned an 85 percent." In that game the Colts came back from a halftime score of 27–7 to win 35–27, with Unitas hitting Berry like a drum.

What was the basis for Berry's ratings? "Playing full-out every time I left scrimmage." He ran as fast and far on a fake pattern as a real one, threw blocks as hard when he knew the ball was going elsewhere as when it came his way. He tried to achieve a uniformity of intense performance every minute he was in action. "What can you say?" Arty Donovan told me. "Raymond was just a perfect football player."

Then Donovan, the jovial all-time all-pro guard from the Colt championship teams, told me The Pants Story in one of its many variants. "He washed his own pants, so they'd fit perfectly. Underpants, too, I guess." Arty seldom leaves a story unimproved. The sportscasters told how Berry, in his emphasis on smooth performance, finally got a pair of football pants that fit perfectly, then lovingly laundered them himself to prevent shrinking or stretching. Any truth to that? "No," Berry tells me. How did the tale arise? "When I went to the Colts, they used the same dirty pants every day in practice and rarely washed them. I wasn't brought up to wear dirty clothes, so I washed my own—to get them clean, not to make them fit. Later, I bought my own pair of game pants because the Colts' cheap canvas felt stiff. I always believed in wearing the best equipment." Like the very best contact lenses. In a mimeographed set of tips for receivers that Berry put together in the early seventies, he advises players to get the best in helmets and shoulder pads even if they have to buy their own. The cost is nothing compared with an injury.

By now Berry's wife has joined us at the restaurant, and she laments the long hours of the playing season. "For six months," he nods, "we work seven days a week, twelve hours a day. I cannot see my daughter compete in gymnastics because our seasons overlap." But doesn't that leave a whole half-year of off-season for relaxation? Husband and wife laugh together, she a little less heartily. "There is no off-season. We just go to a five-day-a-week nine-to-five job." What is there to do? He looks at me with something like pity. "During the season, we are living from day to day. I'm focused on next Sunday's defensive backs, rarely on my own receivers. Our *defensive* backs' coach is watching the opponents' *offense*. We never get a chance to compare notes or look at each other's areas during the regular season."

In the actual season, each week is spent developing and practicing a *game* plan aimed at one team. The off-season (so called) is spent developing a *season* plan—basic changes in the system, analysis of long-term strengths and weaknesses, changes to take advantage of new players. A new playbook is devised and printed—the repertory of moves from which the game plan will be drawn, week by week. The game plans will only be as good as the season plan from which they are carved in little sections. And, again, time presses. "We get our vacation in June, just before training camp opens." It is the coaches' relaxed "month of Saturdays" before the long Sunday of the whole season.

Training camp is hectic from the outset. "We only have a month before the exhibition season opens. It's not enough time to look at rookies, get veterans' skills sharpened again, teach the playbook." When the exhibition games start, the coaches have to juggle their time between looking at new talent in action and giving veterans practice in the unduplicatable conditions of a real game. Berry, the stickler for detail, throws up his hands in despair at the acceleration of pressures to perform without adequate preparation. In his day, there were twelve teams in the NFL, and they played twelve regulation games. Now there are twenty-eight teams, and they play sixteen regular games, along with televised exhibition games. Owners and fans want results, even from the exhibitions; both, after all, are paying for the spectacle, even that early. In the old days, each of the teams had thirty-three players. Now the twenty-eight have forty-five. Simple math (Berry loves numbers) gives the game 1,260 players to keep track of instead of 396. And one must keep track of them over a period one-fourth longer than it used to be, even if exhibition games are not counted. The computer, Berry says, saves him about two hours a week—important hours, on the most harried day, Tuesday. But what is that against the accumulation of problems to be coped with? "You sit down with the figures on Tuesday morning, and it is like tinkering with an atomic bomb. Most often it is going to explode in your face."

The play is more complex now, but also more diffused. "I think there is more football talent around now, but it is spread a lot thinner." There was perhaps no denser concentration of sheer football talent than the 1957–60 Colts team. Baltimore has put only six players in the Hall of Fame, and they all played on that team. Gino Marchetti was the man whose preeminence is least likely to be challenged—the Platonic ideal of defensive ends, violence artfully mobilized: *mens agitat molem*. Right next to him swarmed Arty Donovan, less mind than mass, but enough of the latter to get the job done.

Move over on that line, and you ran into Big Daddy Lipscomb, who competed with Sonny Liston to challenge Joe Louis's kempt performance with "Bad Nigger" flair. Berry says of Lipscomb: "He could stretch his arms and touch the walls of most rooms he went into. His was the biggest talent I ever met, and he was the dearest man"—till the dear man overdosed himself with heroin.

On the offensive line, Jim Parker was all-pro for eight years, and went into the Hall of Fame in 1973 with Berry. For runners, Alan Ameche and Lenny Moore were Mr. Inside and Mr. Outside. Berry says of Ameche, "We never had a running game after he left." The six Hall of Famers—Berry, Unitas, Marchetti, Donovan, Parker, and Moore —had ample help from people like Ameche and Lipscomb. I asked Berry if he thought that team, as many do, the best in football's history. "We didn't have secondary defenders like those on the modern Steelers, or linebackers like theirs. Of course, with Marchetti and Donovan coming at him from one side, no quarterback had much time to look at the secondary. But I have to think Unitas, given the 2.8 seconds he asked for and got from his offensive line in those days, would find a way to beat any team. He just didn't believe in losing. I never saw a more competitive man."

What made for their legendary teamwork? It was written, at the time, that they thought on the same wavelength, there was a kind of telepathy telling John where Raymond would be, Raymond where John would throw. Even after twenty years, that legend seemed confirmed in 1978, when the retired players of the Giants and the Colts played a game of touch football in Central Park to recall the first sudden-death match for a championship. From a jovial start, the blocking and running turned serious. Arty Donovan shakes his head in wonder at the folly of taking on Unitas and Berry again. "Those crazy guys—[Kyle] Rote and [Frank] Gifford especially—were out there to wipe out our 1958 win. They thought they should have had that game. They were out for *blood*. Modzelewski knocked me over and laid on me. But Raymond and John were still in terrific shape. Raymond could still get out there, and John could still throw the bomb." By the end of the afternoon, Unitas was hitting Berry at will.

Did the two men think on the same wavelength? "No," comes Berry's answer, very fast. "We didn't think alike at all." His wife is nodding her head emphatically. "Every season we had to start all over on our timing, especially for the long ball. He knew he had to release the ball when I was eighteen yards from scrimmage for me to receive it thirty-eight yards out. I knew I had to make my break in those first eighteen yards and get free within 2.8 seconds." Berry numbered the steps of all his patterns, and knew, for each one, on what step the 2.8 seconds ran out.

I wondered at his emphatic rejection of the "thinking alike" line. Unitas and Berry make up the most celebrated passing team in the game's history—the perfect mating of skills, more finely tuned to each other's motions than pairings like Isbell to Hutson, Graham to Lavelli, Waterfield to Fears, or Bradshaw to Swann. It would pierce the ultimate myth if Berry now admitted impediment to that "marriage of true minds." I ask Berry why he and Unitas did not think alike. For a man with a near-kamikaze candor, he shows a reluctance to answer; he speaks in parables. "When I found that I could not get free of a defensive back with direct moves, I had to use indirect ones." Was he telling me that he had to use indirect moves on Unitas? (He was certainly using them on me at this point.) "You only get a few chances to win some people's confidence, and your facts better be right on those first tries." It was hard to follow the talk-pattern Berry was running on me. I asked what kind of information he had to be certain of, to win Unitas's confidence. "Well, if I told him a defensive back was going to act a certain way, I'd better have my facts straight."

I wondered if Unitas would answer more directly when asked about disagreements with Berry. I should have known he would. He talks the way he played, with no time for nonsense like tact. "Raymond had sixteen ways to run every pattern. He wanted to tell me which foot he was going to cut on. I said I didn't have time for things like that. I had to watch other receivers, and defenders. I didn't care whether he caught the ball on his right foot or his left foot."

Unitas brought to each game a fierce concentration and resourcefulness, and showed his contempt for anyone less combative than he. If he lost, that just increased his appetite for winning; but he did not go back over the loss, again and again, thinking how he might have won by improving his own performance. He clearly thought the other players had lost it for him. When Berry wanted to talk over each play of the game, Unitas got bored and turned him off. The most direct thing Berry would say about their differences was: "John did not have my sense of detail."

In Tolstoy's scheme of things, Unitas was a Napoleon, pushing boldly ahead, seizing the moment, startling others with his willingness to take risks. Berry, by contrast, was Tolstoy's General Kutuzov, cautious, outwaiting an opponent, making few mistakes, letting the other man's mistakes undo him. Berry spent most time watching the films of losses and defeats. When he made training films for receivers in 1970, he began each segment with a shot of himself dropping the ball (though for some types of reception he could not find an example of that). Then he would show the drill he used for that particular kind of catch—with a final game film of a successful catch. For Unitas, such plodding routine would take the excitement from the game, dull the all-important competitive edge. Yet Berry, however thorough, does not lack imagination. Some of his drills were attempts to keep himself interested in the dreary work on details. "My father had told me no player is really in shape until he has played a game or two. So, for my second year in the pros, I watched a game film with stopwatch and pad, and put down the motions of the wide receiver throughout the game—time between plays, time on the bench, time spent running the plays. Then I tried to repeat those motions to that timetable. Though I thought I was in good shape, I couldn't get much beyond the first quarter, running each play full-out. But I kept trying, every couple of days, and I was able to 'play' the full game before I got to training camp."

Such dogged, repetitive thoroughness would in most men become rigidity. Eddie Rickenbacker became a World War I ace, not out of romantic flair for flying, but from a literally endless tinkering with his plane, making it an ever more efficient instrument of his will; but Rickenbacker was a martinet in his dealings with others, trying to force them all into his own mold. Berry is, if anything, too aware of his players' peculiarities, of the futility of changing others from outside. It is part of his resignation. He knows one can only beat the odds part of the time, and not for long. He plays a "prevent" game with life, knowing that death will win in the end. It was at the peak of his playing time, the year of his best season (1960), that he felt life's futility most and underwent a born-again conversion. Art Donovan says, "He used to have a few beers with us till [linebacker] Don Shinnick convinced him it was the devil's brew." God offers the ultimate in prevent games.

Berry began his coaching career with another born-again Christian, Tom Landry of the Dallas Cowboys, and was happiest in that job. But he quit it to make his training films, which turned out to be a bad investment. Hush Puppies used the films to sell an ill-fated football shoe with cleats on a revolving plate. Berry broke even in the long run, but could not pay his bills in 1971 without taking the first job at hand—with a college team. Then he went to Detroit, to work with a coach who promptly died; and to Cleveland, where the whole coaching staff got fired by an irascible owner. Now, in his fifth year with the Patriots, his wife hopes they are settled for a while. "No one can stay in coaching for the pay," Berry grimaces. Unitas, the winner, has become a very successful businessman. Thwarted of a chance to coach the Colts, he turned his concentration and resourcefulness to investment. I asked Berry if he regrets not opening a business, as Marchetti and Unitas and Donovan have done. "No, I think that everything that has happened to me did so because the Lord is trying to teach me something—though I must admit, I don't know what it could be," he says with his contagious laugh. It seems safe to assume that Berry will stay at the assignment until he learns it, down to the last detail. In *War and Peace*, after all, Napoleon, the winner, gains victory after victory and loses. Kutuzov suffers defeat after defeat and wins.

AFTERWORD

This article was written for *Inside Sports,* but was finished too late for the football season; and by the next year the editor who commissioned it was gone. I placed it elsewhere; but before it could appear, Berry was fired, along with the whole coaching staff, after the Patriots' dismal 1981 season. The receiving of the team was not its problem; but Berry is stoic: "It is a good idea to start with fresh people all around. The trouble is they should get rid of *more* people, not fewer." Offered several promising business executive posts, Berry took the one that would not entail a move from the Boston area. He is now a real estate developer, thankful for release from the pressures of coaching. "It is not a job a sane man undertakes," he says now.

38

Beverly Sills

It is a Holbein painting, with everyone in Tudor styles—women bared or tightly bound above, but with legs sunk in vague mushrooms of skirt; men in sleek stockings below, but puffy above under sixteenth-century shoulder pads. The parts, so strategically cushioned, will have trouble combining.

And then the painting comes to life; well, half-life. To a liquid dither in the strings, the men engage in elaborate sidling. A goes to B with a secret, who carries it to C. When C tells it back to A, A is not surprised to be hearing the same thing over again. It is all a hoax, you see, to break the secret out toward us; and just in case we did not get it, they soon turn and "whisper" out loud, in tune, together, that this is the secret: The Queen is in trouble.

The Queen, tonight, is Beverly Sills. Backstage, not long ago, she was noticing that the periodic fade and flare of her red hair rinse made the appended "fall" strike discords now—artifice at one remove fighting with slightly more distant corrections of nature. But a third device, she is confident, will rescue her, the stage lights that lave her in a blending final compensation. Reality moves up, on nights like this, by a thousand small ratchets, adjusting live men and women into the Holbein picture's intricate machinery.

It is a process that both enlarges and diminishes. Caught here in a cool, too brightly lit, modern dressing room in Lincoln Center, she looks a captive from some other time, ahead of ours or behind, at once too large and too small to be talked with in a normal way. The gown she will be wearing hangs dead behind her—it, too, needs stage lighting to come alive. She is wrapped in a workaday robe, her Supphosed feet in slippers; but her flat-looking eyes are hedged already with the spiky outworks of her makeup, and false lashes. The eyes are mild animals caged—unless the barbed wire is more a weapon than imprisonment.

"One-half hour" comes over the intercom, as from a submarine commander; and she scurries through hygienic corridors to her very own control panel—at least she thinks of it as hers. Her fellows, in on the conspiracy, cheer this act of insubordination: it is Beverly breaking the rules, turning the house lights up, tier by tier, and lowering hundreds of pounds of gold curtain with the flick of a button. Only proper union members should do this, but the proper ones in this house have made her an honorary exception. When all the ratchets are crowding live flesh into place, minor rebellions become major symbols. She returns to a scatter of applause, the only kind she will hear backstage.

In her costume, as it is buttoned, she does a mock-hysteric scale, then squeaks Mickey Mouse imitations. Nerves matter more than vocalizing now, and she is bluffing others as a way to bluff herself. "Miss Sills," on the intercom: and she goes to her entry point on an offstage scaffold. Jane Seymour has already gone onstage at a lower level, and told the audience that she will not tell the chorus what she is nervous about. Beverly, Anne Boleyn for tonight, enters fast, with little pretension, teasing the spotlight after her; comes down the stairs, killing applause with her quick opening lines. Once down, after twitters of accord with Jane, Anne takes a chair at center stage with her back to us. Donizetti wrote in the first great prima donna era, after the castrato's days of fame; and he liked to tantalize the audience with delay in using the soprano. *Smeton dov' è?* "Where's Smeton?" The contralto page must cheer the Queen in her melancholy: he/she launches into a big romanza that means, among other things, that Beverly is denied us for a while.

Much of the diversion is a matter of waiting, of waiting others out—though there is not a great deal of time to wait. Withholding is bluffing. That is how Beverly became a Queen. In the game played for the high stakes of big roles newly produced for the big houses, she lost *Daughter of the Regiment* to Joan Sutherland; but that gave her the jump on Montserrat Caballé for the three Tudor Queens of Donizetti. It is Lincoln Center poker, not the most pleasant game in the world; but one that spectators are more interested in, tonight, than in sixteenth-century English politics. The Queen who is endangered is Beverly, and her rivals are not on this stage or in this story. But their emissaries are here, out in the audience, behind that carefully turned back. These have weighed her opening conversational notes, and found them wanting;

watched each move and considered it unregal; played every turn on the theme of Rudolf Bing, who as director of the Metropolitan lost the Queens and took his revenge with a famous bitter crack: that Miss Sills, by virtue of her birth in Brooklyn, must hold "priority in the portrayal of British royalty." Those, and not the court's rumors about Jane Seymour, are the whispers she must face tonight, and every night she sings.

Smeton, the page, has been making airy passes at a classical-looking lyre, to incongruous huge swells and subsidings of harp in the orchestra pit. He is delivering a consolation song in a Renaissance style borrowed from the ancients; it has echoes of Sappho and Catullus, since Donizetti's teacher, Giovanni Mayr, was noted for his learning, and wrote an opera on Sappho. Smeton's romanza is stilted, as befits a song within a song, the equivalent of the *Hamlet* play within a play. It is a melody in four couplets, of which the second imitates the first; but the third one pushes off from them and begins to wander, and the fourth recovers only with a lurch and final swagger. Donizetti is setting us up. Second stanza; mere repeat, but with florid ornament, written out. Another signal. Pause, for the harp bridge—where to go after that thickly encrusted last bit of song?

"Taci!" Beverly erupts from the chair, tortured by praise of her beauty (since she knows how she betrayed it, marrying Henry VIII). She has fluttered the dovecote of choristers, and strings now plunk out the expectable aria background. Donizetti has us set up. Ducks in a row.

And what happens? She starts singing, in transmuted but obvious form, a mere variant of Smeton's song. She even has her own classical reference—hers to Vergil: *Son calde ancor le ceneri.* . . . But just where Smeton's song went wrong, things go magically right—into a new third phrase and up to a triumphant fourth one that comes and comes and comes, on which she dwells and dwells. The change from one song to another is as much social as emotional—a mark of caste. The Queen gets to sing things "right," where a page can only stutter melodically. Anne's aria is actually an aside; it is what she was thinking while the ardent Smeton (in love with her, of course) sang the romance. In a different place, with a different composer, Anne's musings would have intruded on, been interspersed between, and finally blended with Smeton's phrases in a duet. Donizetti cannot do that for several reasons: a duet does not belong here—it should come when the two characters are already established, and even then it must be with an equal (vocally, dramatically, socially). She must wait for Percy to sing her amorous duet.

Donizetti was one of those nineteenth-century romantics who got into trouble with real kings out of his Walter Scott love for evanescent dream kingdoms. It is hard for us to imagine this museum picture come to life, with its deceiving Holbein costumes, as a live political statement; but Mazzini considered it that. In his view, Donizetti's *Anna Bolena* struck a blow for freedom; but his was still a political game in which all the players, no matter what cards they held, knew the difference between a queen and a knave. Windsor Castle, even when Walter-Scottized, is no Brooklyn.

Once we get this dose of Beverly, we know we can expect more right away—the cavatina leads necessarily to the cabaletta, fright song to fight song, reflection to resolution. That is another reason Donizetti followed Smeton's romanza with a variation which becomes Anne's cavatina—the customary two-part pattern is thereby made a three-parter: expectation, to be played on effectively, must also be played against.

Men's cabalettas are often sword-drawing songs—all about dashing off to battle or vengeance or death as soon as I dash off this stanza or two of pyrotechnic vocalism. Since Anne has been musing to herself how she forfeited her real love (Percy) through ambition for the crown Henry could give her, she turns decisively to Jane Seymour and tells her not to let ambition fool *her*. Anne's tune is rather soupy at the outset, but quickly reaches one of those lovely phrases—"Don't let it fool you" —that Donizetti has his heroines hug to themselves, and caress, and then fling over and over to the audience as so much vocal largesse. Beverly is down by the footlights now, doing what she does best, making tough show off phrases very touching. Her voice, even in full prance and canter and leap, has a hurt sound, almost a wince in it, that her critics call a whine. It is a light voice, good at teasing and tantrums, coquettish as Manon, lachrymose as Elvira, and skittish as Rosina. Its critics—and, opera being largely an athletic contest over known obstacle courses, many come just to criticize—call the voice thin, and contrast it with the big ceaseless flow of Joan Sutherland's sound, which hits fast high notes like a focused water hose knocking bottles off a fence one-two-three. Beverly, though she hits the same notes hard and clean, without the "white" (unresonating) sound of so many coloratura sopranos, seems to put them together in a wispier way, as with a silvery tinsel hung from note to note.

She is called too damned ingratiating to take on the role of a tough schemer like Anne Boleyn, here revealed in midconversion back from Queen to woman. Though Beverly is a big woman in her mid-forties, she seems much younger on the stage. Her face under the vivid red hair is all sharp curves blanked by the lights onto a single plane, a cubist child's face overdrawn by makeup to Picasso jumbles. She takes such obvious delight in singing that Donizetti's mournful droopings of the voice, to the faint sighing of horns, can seem incongruous. She even bounces to the meter, in an unqueenly way, when the music is getting to her. "I'm a real foot stomper," she admits. Once, when she was rehearsing a recital with Charles Wadsworth of the Chamber Music Society, Mr. Wadsworth warned her of this "tendency to bounce." Beverly winked at the clarinetist, Gervase de Peyer, and they did an impromptu dance, bobbing up and down around Wadsworth's piano, during their next song. "Charley, you're getting too stuffy," she concluded.

But none of these criticisms seem to matter now as she turns and turns, entranced with her own phrase, entrancing us: "Don't let it fool you." Her admonition to Jane has taken Anne back, behind ambition's fooling action, to the earlier love follies of her own girlhood. She always, even under her tons of Bette Davis makeup as Queen Elizabeth in *Roberto Devereux,* seems to reveal the little girl buried in each Queen.

"She never had a girlhood," her brother Stanley told me. "She was always going to lessons. On Saturdays. After school. Voice lessons. French lessons. Italian lessons. Dancing lessons. Elocution. She had no friends her own age." Her first piano teacher used to take her to the movies. From the age of five she was determined to be a singer, and her mother was determined to get her the best training available—voice from Galli-Curci's teacher, Estelle Liebling, piano from Paul Gallico's father, stagecraft from the Metropolitan's Désiré Defrère. In all this, Beverly is not very different from the girl who preceded her by several years on the Major Bowes Amateur Hour, and who was born across the East River from Brooklyn as Maria Anna Cecilia Sofia Kalogeropoulos, but graduated from P.S. 189 as Mary Anna Callas.

Callas, after imitating phonograph records and doing the same kind of children's shows that Beverly did, went to Europe for better training. Beverly did not have to. Estelle Liebling was the last great exponent of Europe's high Marchesi school of bel canto singing. Callas and Sutherland both began as dramatic sopranos and trained up into the stratosphere of coloratura. Beverly studied the arias of Donizetti and Bellini from the outset, long before there was any bel canto "renaissance." She

was always a Donizetti heroine, though it took people an unconsciona-
ble time to find that out.

Donizetti made his first great impact on the international music world
with *Anna Bolena* in 1830, and it has remained an era-shaping opera
ever since. Callas's 1957 revival of it in Milan—in a truncated form, but
brilliantly staged by the moviemaker Luchino Visconti—had as much to
do with the bel canto rebirth as any single event. One hearing of that
performance sent Herbert Weinstock into the archives to produce his
book on Donizetti. Beverly took a calculated risk in saving it for last in
her series of Queens. A chronological sequence—by either the order of
things portrayed, or Donizetti's order of composition—would give us
Bolena, Stuarda, Devereux with the strongest opera first and the
weakest following it to suggest that all the best music was used up. "So
we led off," she says—"we" being Beverly and Julius Rudel, the director
of the New York City Opera—"with *Devereux*. The sheer drama of
Queen Bess could carry us." It carried her, among other places, onto
the cover of *Time,* made up as the old Queen. "*Stuarda* had to come
next, so we could finish strong. The problem with *Stuarda* is that Mary,
who is supposed to be sexy, never has a chance to sing with any lover—I
had to suggest her sexiness at long distance." The first two productions
had been mounted in 1970 and 1972. *Bolena* was added in 1973, as the
crowning work. She was to be seen as all three Queens in this last season, but the musicians' strike made it impossible to fit in *Devereux,* and
she will sing all three Queens in a single New York season for the first
time next year. Perhaps the cancellation was a service to her. The strain
of three such big and drama-laden roles was telling, all the critics
agreed, by the end of the 1972–73 season. Her voice bounced back ad-
mirably for last year's *Bolena* and Bellini's *Puritani*. She has made the
Queens her own property, and in the process made herself a permanent
part of opera history.

Bolena has been central to many careers, beginning with that of Donizetti himself. By 1830 he had already, with awesome facility, produced
thirty-four operas to run-of-the-mill libretti. At age thirty-three he was
settling into a rut as a talented second-rater. Vincenzo Bellini, four
years his junior and with only six operas to his credit, had surpassed
him in reputation, and was about to follow his appearance in Milan
with a new opera on the admired Victor Hugo play *Hernani*. At this
trying point things began to break in Donizetti's direction. First, he was
given a good book by the leading librettist of the day, Felice Romani.

Second, he was assigned for his heroine the great Jewish soprano Judith Negri, known as Giuditta Pasta. Seven years earlier, Stendhal had predicted that the next figure to command the opera scene after Rossini would be the one who tailored a great part to the odd but unforgettable voice of Pasta. Donizetti fulfilled the prophecy. He composed *Bolena* at Pasta's villa on Lake Como, where she could try on each phrase for size.

Pasta did not have the big voice that has come into style since her time—theaters were smaller and more intimate then, small enough for Donizetti to score the mad scene of *Lucia* to the weird breathings of a "glass harmonica" (water glasses played by friction on their rims, which simply would not be heard in a modern house—the flute fills in that part). Besides, Pasta's voice had a clear "break" in registers between the high head tones and her dark contralto "from the chest." Instead of trying to disguise this "seam," she played the two vocal qualities against each other, and Donizetti exploited this in *Bolena,* making savage leaps down to a voice of anger or outrage. Some of Pasta's famous "moments" were almost straight dramatic declamations, like her shout at Henry, *"Judges,* for *Anna?"* or her taunt, "Justice is silenced by Henry." The drama of these lines was perfect for Callas, whose voice, too, had breaks between registers (and gave out early, like Pasta's). Beverly cannot weight such phrases with the Callas scorn. Her best lines come at times of near despair, when she is not hurting but being hurt, as when it dawns on her that Jane Seymour is the one who has replaced her in Henry's bed. She wanders the stage, trying to pound that into her mind—*La mia rivale!* over and over—unable to grasp the enormity of the betrayal; her head will finally break in the effort. She credibly prepares us for her later mad scene (the first great example, in Donizetti, of what became almost his trademark). Beverly's strength is her weakness, her ability to take wounds on the stage.

That very fact tells against her with some intenser types of fan, who liked to be kicked and mistreated by their operative mistresses. In all the craziness that surrounds the hypercharged world of opera, none is crazier than the adoration of a prima donna, which sometimes reaches masochistic extremes. Opera fans were the first groupies, and there is a disturbing resemblance between a farewell concert by Maria Callas (her voice a creaky ghost of its former self, but the crowd still calling for more of it) and the velvet-and-leather legions who have turned out for the later appearances of Judy Garland, Mae West, or Marlene Dietrich. Beverly has never been a favorite with this crowd; she is the heterosexuals' soprano, revoltingly normal and earthy. It is hard to be "La Divina" when anyone who is around you for two minutes starts calling you Beverly, either by instinct or by invitation. With Callas, "Maria" itself became a title. "Beverly," by contrast, is always a familiarity taken, not far from the name she was born with, "Belle." A beaut.

Given the pressure of opera's insaner fans, the stories enacted, the rigorous training involved, the high competitiveness between performers, it is not surprising that a Donizetti performer becomes, all too often, an hysterical woman playing hysterical women before hysterical people. Even if *she* keeps sane, there is no final protection from her own crazies, not to mention her rivals' various self-appointed spokesmen. Joan Sutherland is a very gracious gentlewoman, but Beverly, and Beverly's people, bump more often into raging Sutherlandites than into the diva herself. And vice versa.

The fan of a voice is bound, I guess, to go overboard. Look at Stendhal; he was watching a precious commodity, a once-in-a-lifetime miracle, go to waste: Pasta. He pleaded with Rossini to give this instrument its proper music. When Rossini did not respond, Stendhal looked around for another musician worthy of the organ; found none; raved a bit—and prophesied.

Imagine a harpsichord of bewitching sweetness, like nothing created before; perhaps with slight flaws that serve only to endear it to those who see its unique quality. Then add an essential point: this harpsichord is disintegrating at a very rapid pace—it virtually fumes away before your eyes. A Stradivarius plays on, mellowing, through centuries. But this instrument, of unique quality, has perhaps two decades in it—it must catch the right people's attention, get a house to play in, get music fitted to it, achieve a repertoire in a quickening race with time. A thousand things distract from this one instrument's glory, drawing people back toward other instruments already fading or faded, to musicians working in a different vein, to houses commercially sewed up and audiences deaf to this sweet new tone, the last of its kind. Is it any wonder Stendhal was desperate? The only wonder is that his prophecy was, just in time, fulfilled.

It goes to the head, this kind of adulation. How is the possessor of such an instrument to deny it? A young violinist, lucky enough to buy or receive his or her own Stradivarius, will hug it, treasure it, interpose his/her very body to protect it from danger. Yet one Stradivarius more or less makes a small dent in the totality of music's history. What if a girl's body, uninterposable between danger and her dearest musical possession, *is* the instrument, the evaporating harpsichord never heard before? Should she not take care of it? Yet *she* is it, and the idolatry a young pianist can direct toward an exceptional Steinway looks like mere egotism when it is directed to a person's own respiratory system.

Great sopranos must undertake a kind of protective custody over their musical instrument of flesh and bone—swathing, spraying, steaming, pampering; and they compete energetically for attention, respect, and opportunity. Each year that passes without a major role added, a record made, a triumph repeated, is a loss of one tenth or one fifth or one half of the instrument's life span—one never knows exactly how much time is left to a voice daily put to the strain. of utmost musical athleticism.

Beverly makes a great, almost too great, effort to avoid such protective custody, as exercised by herself or by others. She is accessible, unpretentious, and makes no display of caring for her voice. But one can never become totally unconscious of the fact that it is a priceless human Stradivarius being bumped in the elevator, splashed on the street, exposed to a thousand germs as wheezy fans press up to kiss her. When, as a child, she went to baseball games with her brothers, they would tell her, "Stop shouting, Bubbly. You'll hurt your voice." Her mob-scene receptions in the narrow corridor backstage sometimes unnerve her husband.

Then consider the problem of teeth. Every part of a singer's body is important—part of the instrument; not only the chest and lungs and throat and cords, but anything having to do with general health. Flying affects the sinuses, or ears, of some opera stars; air conditioning dries others up—or, in New York air pollution irritates those who cannot stand air conditioners; Joan Sutherland has back trouble in long roles, or in her long rides to the long roles in new cities; even a woman's monthly period can affect the voice. Anything having to do with the head is especially important. The head is the last box through which the opera singer's prodigious quantities of breath reverberate. Caruso used to credit much of his tone to the big mouth cavity in which he could secrete an egg with his teeth firmly closed. The palate and teeth affect the shaping of tone. A whole different set of considerations arises, therefore, when singers have tooth trouble. A dental plate could change the placement of notes, affect diction, or be blasted loose on the stage. Opera singers live in a bombardment of sound, their own and other singers' and the orchestra's, that can be, in a literal sense, teeth-rattling.

So it was no ordinary accident for Beverly's car, some years back, to skid on a wet Manhattan street into the back end of a braking truck. "I literally bit the steering wheel," she remembers. "It broke all four front teeth right off." A plate was out of the question; as little alteration as possible should be risked. A dentist in this situation is like one of those experts called in to patch Michelangelo's *Pietà*. The roots were kept, and false teeth built around implanted poles. By the spring of 1974, the outer caps were to be replaced and matched to her other teeth. While the four individual teeth were being cast, her dentist gave Beverly two temporary caps of two teeth each. A singer's schedule does not adjust itself around dental appointments. She was singing her last Queen, Anna, and the temporaries seemed to be holding—until one Sunday night in March, when one fell out onto her tongue in midnote.

"Luckily, I had time to turn around and stick it, hard, back into place." But it tends to distract, not knowing where one's teeth will show up next. In fact, they showed up again on the steps leading up to Henry VIII's chair. She had just spat out her "Judges for Anna?" (*Giudici ad Anna*) when she spat out one of the double caps. "It was bad enough to be without the two teeth. Not only was I funny-looking, I had to lisp until I got them back." Then she saw the fulfillment of all her nightmares taking shape. Henry, with his gouty limp, was about to climb the stair; and his course was all too clearly charted in her mind. She waited until the very last moment, when his foot was almost on the cap, about to crunch. With a leap, she grabbed it, shoving Henry off with her left hand. Even so, his boot drew blood from her right hand, firm in its grip on the precious cap. Henry (Richard Gill that night) staggered down the steps, wondering what the hell. But she had done it so deftly that no one else, onstage or in the audience, noticed anything amiss—not even Julius Rudel at the podium, with whom she claims an almost telepathic communication when she is singing. Only one close Beverly-watcher backstage realized what had happened, and ran for the proper first aid. As she went off, he pressed a tube of Duco cement in her hand and whispered, "This will fix it for the next act." Fix it, it did. The next day it took her dentist twenty minutes to chisel the cap loose again.

Opera performance is a horror tale of such minor crises, adding to the general ordeal. One of the nights when I saw Beverly perform *Bolena*, a new tenor was struggling with the virtuoso demands Donizetti made on his own first Percy, Giambattista Rubini. "I've been thinking," she told me between acts, "what I could do to help him out. But there's little we can take it easy on in this opera." Rudel was back and forth to the man's dressing room, giving aid and comfort; it demoralizes the whole troupe for a singer to be having a bad night. Beverly was tired, too; her speaking voice was almost gone. "I lose that first; I talk unnaturally low, since my father did not like the high chatter of women." She is not afraid, so far, for her singing voice. That stays strong if anything does. She is talking up into a falsetto to make herself audible. But fatigue gets to the breath support, which takes great expenditures of energy. "If you can't give them full support, the damn trills stick." And there are some fiendish trills, coming up in the very last moments of *Bolena*. "That's the killer. And it ends with a high E-flat, the very last note people hear in the whole opera. [The City Opera omits some final bars by the chorus and ensemble.] A whole performance can be soured by a slip at that point. A knowledge of that haunts you all night. No matter how good you have been up to then, you are never home free until you make that last note." Turning her attention back to the tenor, she reflects that there is no way a singer can walk through a performance as some actors do—slowing the pace and reducing the intensity, even omitting things at will. Singers must keep their part and time in the entirety of orchestral and ensemble music. "You see why we play so many crazy ladies. We have to be crazy to start with."

In such a world, Beverly does not seem crazed enough for some; she disappoints by placidity and lack of "temperament"—and may have forfeited some opportunities by her accommodating ways. But she persisted; and when the chips were down, she could be firm. Rudel learned that. By 1966, she had become a mainstay of the City Center, rather taken for granted, living in the shadow of the Met, where she had turned down an unflattering contract. It was the City Opera or nothing for her. And she had reason to hope. Julius had been a pianist coach when she first tried out at City. He took pity on her about midway in her eight unsuccessful auditions and started accompanying her (which saved her a fee for bringing her own pianist). The two had risen together—he invited her to solo at one of his first star conducting turns in Chicago; both of them made Douglas Moore's *Ballad of Baby Doe* the one great success in a City Center season of American works.

But now, in 1966, the big one came along. Lincoln Center was opening its stamped-tin palaces. City Center, rechristened City Opera, had edged its way into the project against heavy opposition from the Met. Bing had commissioned a new Samuel Barber opera on Cleopatra for the Met's opening, to star Leontyne Price. Rudel did not have the resources for such a commission, but he could stage an expensive revival, and he chose Handel's *Giulio Cesare.* This happened to contain a big coloratura role for an eighteenth-century Cleopatra, to contrast with the twentieth-century one being crafted just a carom shot off Lincoln Center's fountain wall. Rudel claims that this was accidental, and the heavy secrecy around the early stages of a new commission tends to confirm him. But Bing, at the Met, was miffed: "What are you trying to prove—that Handel is a better composer than Barber?" he asked Rudel.

The crusher, for Beverly, was this: Rudel, who meant to highlight the American debut of mezzo Maureen Forrester, gave the soprano's Cleopatra part to Phyllis Curtin. And Beverly dug in her heels. "Phyllis was not even part of the company. [Neither, of course, was Maureen Forrester.] I went to Julius and said he just could not open the new house with a soprano not even belonging to the company. He owed this to me." It was eyeball to eyeball, and Julius blinked. Miss Curtin had to be, as gently as possible, nudged out. "Are you and Miss Curtin still friends?" I asked Rudel. His shrug was resigned: "I hope so."

Giulio Cesare was Beverly's major breakthrough—after a long series of muted triumphs, small glories, the accumulation of a devoted band of followers. The Barber opera flopped. The Handel, kookily staged (almost camped), caught fire. Hubert Saal of *Newsweek,* sent to cover the Forrester debut, came away talking nothing but Beverly, and became her most ardent promoter in the press. He had heard her a hundred times—even in a twelve-tone rarity in Boston—and had never heard her. What the Cleopatra role gave her was a chance to show her coloratura technique in conjunction with her vocal acting and stage presence. The three, noticed peripherally by many people, had never fused in their minds as they did now.

At last, after three decades of professional singing, she was an "overnight" sensation. Westminster wanted an album of her coloratura arias. She worked on the ornamentation with her musicologist friend Roland Gagnon, who suggested various embellishments in the style of Donizetti's period. He would propose two for her to choose between, and she kept incorporating both. At last he broke it to her: "You are going to be criticized by the reviewers for overornamenting." "I know it. I don't care. This may be the only record I ever make, and I want them to know I could do it." She had "arrived" in her late thirties, and had no further time to spare. We mortals all have an inbuilt clock, and we can feel it running down to the last alarm, which rings in silence. But opera singers have a clock within that clock, with a far shorter span; and they are reminded of its message whenever they test their voices in the morning or try to lift them, above the dragging of fatigue, at night. Beverly, for all her relaxed self-mockery, knows exactly what time it is on that clock. Mr. Gagnon, with whom she studies scores by the hour when she is at home in New York, says, "She judges each score in terms of the time left to her. How many more coloratura roles can she play before the top goes? What can she still sing when the breath begins to go? She is already looking at the French repertoire." Estelle Liebling was insistent on perfecting languages, and Beverly's French diction won exceptional notice when she performed in Paris. Her Manon is called by many her best role, and she wants to find other French parts.

She can be so calculating about her career because she knows it will end, and does not intend to be one of those sad women nursing relics of a beautiful lyre now broken and quavery. Too many faded opera stars try to build a more imposing nightingale cage and attract the world to it —only to watch the singers writhe there. A great part of the problem in having a great voice is learning, ahead of time, to survive it. Her mother says, "I always told Beverly, when the curtain comes down, there must be a woman left who can go to her home." It is the woman, more than the singer, that "Mamma," Shirley Silverman, is proud of—and with reason. Beverly gave up her career almost totally for two seasons in the early sixties. She had discovered that her infant daughter was deaf and her newborn son severely retarded. She comes from a European-style family with very close ties. "Both my sons and my daughter never left the house," Mamma says, "without kissing their father on the forehead. Even when they were grown up." Beverly never smoked or drank simply because her father told her she must not. Beverly and her brothers are very close; they vacation together and almost share the three separate broods of children; and Mamma is always with them—her grandchildren treat her with the same affectionate respect she describes as directed toward her husband. When Mr. Silverman died, Beverly was working her way down to a low-paying opera engagement in South America, singing for her fare on shipboard. News of her father's death reached the captain, but he held it from her until they reached New York. Mamma says, "She just sat around the house for months after that, not practicing, not going out to find engagements. At last I told her I had promised her father that his sons would finish their schooling and she would finish her training, and that I must keep that promise." Beverly went back to Estelle Liebling.

Against the family background we must estimate the blow of learning she could have no more children. Her son was so retarded that he had, eventually, to be boarded at a special school. Mamma also stepped in then: "Beverly could not even teach him to go to the bathroom. At last I said, 'I gave my children an education. What are you teaching him? If all he can learn is to keep himself clean, at least send him where that can be taught.'" This broke at last her daughter's resistance to surrendering her "Bucky." (I fear I may be giving an impression that "Mamma" is what people call a typical Jewish mother, but Mrs. Silverman is a saint, and saints are not typical anythings. Everyone who knows them both agrees that Beverly's most endearing qualities—her warmth, thoughtfulness, and genius for affection—are borrowed directly from her mother.)

While Beverly was trying to teach Bucky, and helping her daughter Muffy lip-read, Julius Rudel kept pestering her pleasantly to take up her singing again. "Bubbeleh, Bubbeleh," he would write her, with tempting talk of new roles, till at last he used the brusque approach, reminding her she had a contract to fulfill. When she did return, people took her more seriously. She was no longer just "Bubbly" with the self-deprecating humor and wide-eyed amiability. Tragedy had touched her, through her children; and many claimed to find signs of that in her singing. Mamma scoffs at the idea: "She always had deep feelings. She was just not given the parts in which to show them." Any woman of deep feeling who "never had a girlhood," out of devotion to her craft; who had been disappointed in her career over and over again (starting with those eight failed auditions for the City Center); who lost her father and saw her mother undergo economic hardships for years, working her children through school—any such woman does not need a retarded child to be taught that life is dark and not often easy. Besides, the proffered explanation is too like a Hollywood story about musicians—break your heart, and your voice improves—to be taken seriously. Beverly's delayed fame came from people's final discovery of that secure technique which was the very thing she started with. She had matured in her artistry, of course; but that was a continuing process, taking place long before the children made people look past the protective manner of a "cutup" and discover she had always been a Queen.

Beverly is a brilliant woman. A mathematics teacher claims she has an extraordinary gift for abstract problems. He once saw some old 78-RPM records she used to carry around in her high school days. On the stiff cardboard jackets with the big round windows, she had worked at her geometry assignment. "One problem she solved by a brilliant shortcut that she could not have been taught at that stage. She just saw it on her own." This side of her mind is fed, now, mainly by puzzles, word games, Double-Crostics, which she works at high speed on planes and backstage. Her study of musical scores, alone and with Gagnon, absorbs her scholarly time. Gagnon says, "She sight-reads like lightning, and can memorize a score in no time, even on a plane." One reason she takes criticism with such equanimity is a tremendous assurance on the technical side. "There are very few people who have spent as much time as I have studying the music of Donizetti's period. I cannot bother with the judgment of people who are not my peers." Her ego stands very nicely on its own; which means she can ignore it and give real, rather than feigned, attention to others and their problems.

Her work for the retarded is becoming legendary, and she is joined in it by her husband, Peter Greenough, a patrician Bostonian who was, before his retirement, financial columnist for the Boston *Globe*. She shares many interests with him, and he provides a saving distaste for much of the opera world's nonsense. When a Beverly addict chided him for not applauding his own wife vigorously enough, he snapped, "I am her husband, not a cheerleader." When the head of the Naples claque came to bicker with Beverly for his "pay," he was sent off empty-handed. When he returned after a rehearsal and said, "I apologize; you need no claqueurs," Greenough said, "Now you get a tip."

There have been Greenoughs at Harvard for about as long as there has been a Harvard, but Bucky will never go there. Although Beverly has given up receiving most honorary degrees, she accepted one from Harvard last June, on one condition—that it be awarded her in her married name. Despite all the Greenoughs who have gone to Harvard in the past, there has never before been a *Dr.* Greenough, until now—when there's Beverly.

Even at the pinnacle, with her career assured and all the summits conquered, it is easy to underestimate Beverly. She protects herself from her status with comic mugging and an impish way even onstage. Called in at the last minute to do *Traviata* in a production she had never seen before, she played the first scene tipsy, to cover any difficulties others might have with unrehearsed moves—and those who saw it believe this might have been her best performance as Violetta. She likes the stage, and is inventive in her use of it. When I asked her why she never plays Galli-Curci's great part, Verdi's Gilda in *Rigoletto,* she sighed, "Gilda's such a sap." Sappier than Elvira? But Gilda has only one solo, and is overshadowed by her father in the great ensemble scenes. Beverly likes big juicy roles with herself at the center. She also likes them to have a hint of naughtiness about them, a touch of the flirt or the urchin. She plays her Manon, all fresh from the country at the outset, with a *calculated* naiveté: "She's been tumbled in the hay already," she confided one afternoon. "I shocked Roland this morning. We're working on *Lucrezia Borgia* for next season, and I played the scene with her son in an incestuous way. It may work." Mr. Gagnon was not enthusiastic about the interpretation. When I asked him if he thought she would play it that way onstage, he said, "I don't think she can resist it."

Even her Anne Boleyn tries to scheme as long as she can. When Percy reveals to Henry his earlier betrothal to Anna, the Queen, though she loves Percy, is not yet willing to die with him. She asks *Io?* "Who, *me?* Never heard of the guy." Anna has good cause for fear. Percy's news seals her fate—it gives Henry just the excuse he needs for annulling his second marriage and moving on to Jane. It even allows him to indulge in a bit of self-righteous moralizing, in his briskly thrumming vengeance song, *Coppia iniqua*—"Evil pair!" From Henry VIII, that's a bit much. Enough to drive a woman mad.

Which, of course, it does. For her last scene, Anna comes in aimless tunefulness down the same stairs she descended with her quick opening lines two hours ago. Smeton, the poor lad smitten with her, has been tortured into informing on her. The council has condemned her. The chorus is still saying, but even more piteously, that the Queen is in very *bad* trouble. She has the smile of all these operas' predemented Ophelias at last gone round the bend. It is her wedding day, after all—not to Henry, of course; to Percy. But Percy is angry at her for something, she cannot remember what. Woman's hysteria was holy to the generation brought up to think of Pre-Raphaelite ladies as always walking on some cliff of utmost vitality and nervous awareness and insight—and always on the verge of falling off.

At one point in this reverie, Anne leads her invisible child-self back into the scenes where she grew up: *Al dolce guidami castel.* . . . It is at moments like this, as she leads in dumb show a mute alter ego, her own childhood, up the stairs, that Beverly sentimentalists feel her own real life is enriching (at too great cost) her art. Cooler judges may note that Joan Sutherland, for all her vocal perfection, or even Maria, with that compelling hauteur, cannot achieve the pathos of this hurt voice coaxing at itself: "Give me back girlhood, if only a day. . . ." Because she never was a girl, she will always be one. Art creates what nature lacked. This *is* real life.

She turns to poor bloodied Smeton, and asks for a song again. But no: the lyre, overstretched, has destroyed itself with song, just as her mind disintegrates in music. "Sweet bells jangled." At last she kneels, for a prayer spun weirdly out of the old tune we know as "Home Sweet Home"—praying for death as her homecoming, to a hymnlike blending in of others' prayers. It is a mad scene far more tranquil, almost happy, than disturbed. Lucia wanders the stage still dabbling in her husband's blood; but Anne is released from marriage by the king's evil act (one in the snoot for *tyrants,* Mazzini thought) and her doom is an escape.

But then the last stroke of the unexpected, of form played against form, Donizetti proving his stature once for all. This scene, which has floated along as melodiously as a flower-petal Ophelia in the Millais picture, ends in a fierce vendetta song, cabaletta to the prayer cavatina. Anne even begins her outburst with a deliberate echo of the king's curse song: *Coppia iniqua!*—the wicked pair now being Jane and Henry. It has the beat of a sword-drawing song, a prelude to action. Yet it is ending the opera, along with Anne's life. She is a victim, not the avenger. Anne has gone out beyond madness to some otherworldly point of equanimity, and she prays forgiveness for the evil pair. The song is a formal paradox: *"Vengeance!*—is what I do *not* pray." Rising up through the vendetta strain comes trill after trill that speaks of pardon. To complete the paradox, Donizetti has the trills climb, step by step, on the word *scende*—she prays as she descends to death. These are the "damn trills that stick" after the fatigue of a long tour de force like Queen Anne's. And the E-flat is coming any second. All her affecting dreams of childhood had lived under the gallows threat of that symbolic test. It is too much to ask of a soprano after this long night of miraculous Pasta roulades and raucous outcries, the Queen's constant commerce with the stage and other actors, her comings and goings that trail big traceries of sound behind her. No time, even now, to think of the E —trills of forgiveness have to keep boiling up phoenix redemption from this deceptive last hate song.

Maria Callas, recording this scene in her prime, makes the cabaletta ring with a hellish mockery. For one thing, the trills are beyond her, and without their lightness defying the song's drive, Donizetti's paradox fails. At this point, technique *is* the meaning. Callas grinds out her trills like a gnashing of teeth. "Forgiveness" such as this were vendetta indeed! Pasta could growl and stab like Maria; but she could also do quick runs upward with a shaped trill on each step—Donizetti gave her a first-act chance to prove that in the quintet that brought Percy and

Anne together in Henry's sight. Maria cannot do that. No Queen is perfect. Beverly, "So they'll know I could do it," ornaments the song's repeat by *doubling* the trills as they rise.

And now the E. She hits it clean, and holds it—and damned if, just slightly, she doesn't bounce the meter out as she holds. "No class," a Callas advocate told me later in the lobby. "Too shrill," said a Joan fan. But she had just bounded, like her own vengeance song of forgiveness for her rival, beyond such categories. One might even guess what Stendhal felt when his prophecy came true.

—*Esquire,* September 1974

AFTERWORD

Beverly retired at fifty, as she said she would, and became the manager of her own company, the City Opera, where her many skills are tested raising money, picking singers, choosing repertoire, soothing temperaments. Her husband is a steadying presence, encouraging others as well as he does her. After I turned down an award offered by the New York Sons of the American Revolution (the group had just given a standing ovation to an hysterically right-wing speech), he wrote a letter of support which said I had "created a wave of nostalgia in me; you took exactly the same route with that organization as I did back in 1950." At a dinner where he was to be received as a member, the same kind of speech was given and applauded; so he rose and withdrew his application.

Boxing: A Palinode

Clifford Geertz has a lot to answer for. Ever since he published his essay on Balinese cockfighting fifteen years ago, sportswriters, who used to be (mainly) fans, have increasingly been social historians (or fans disguised as social historians).[1] They use sports to tell us about everything in society except sports—about the relations of the classes or the sexes, about the community's legal machinery, about its political values. Geertz found that cockfighting in Bali was both a protest against the legal order (first imperialist, then puritanically nationalist) and a replication of it, partly inverting and partly enforcing ordinary divisions in the society. This makes cockfighting "a metasocial commentary upon the whole matter of assorting human beings into fixed hierarchical ranks," with the fixed betting orders put under strain by the bettors' identification with the cocks, who revert to harsher standards, whereby survival is the only privilege and death the only deprivation.

Well, you can say that sort of thing about tennis, too—and sociologists are now ritually saying it. The Olympic Games are more intensely scrutinized for international political significance, at the moment, than is the United Nations. But Geertzism is nowhere more at home than in the boxing ring. Boxing, through most of its modern history, has been illegal, like cockfighting—in fact, early prizefights were often paired with cockfights or rat fights. The early organization of prizefighting into an unofficial hierarchy of champions, the enforcement of progressively more intricate rules, the uncouth exclusivity of "the fancy" (regular fight followers, a club made up of all the people who passed the Groucho test against "better" clubs)—we can find in all of these an inversion-replication of "ordinary" civilized life.

Boxing in the nineteenth century had all the thrill of fashionable outlawry, even before one bare fist dug into a scarf-wrapped abdomen.

Only those in the know (an admittedly larger number as the century wore on) could even *find* the fight, and then it might be moved in mid-course if the cops caught on, or were observed to have caught on, or were prodded on the scene into reluctant action. Fights were held in no man's lands—on islands between states or just over borders. The fancy was a floating world, reassembled from match to match, obeying its own half-articulated code, much like Clopin's ragged kingdom of thieves and beggars in Victor Hugo. There was a special *frisson* for the aristocrat or intellectual who won acceptance in this underworld, like Gringoire brought to the Court of Miracles. Byron at ringside felt like a mini-Conrad, his own corsair. Geertz felt he was truly accepted by the Balinese only after police tried to arrest him for being at an illegal cockfight. Though he was not arrested, he shared the camaraderie of those "caught, or almost caught, in a vice raid." Tex Rickard would later make that same *frisson* available to women when he created a "Jenny Wren" section in the arenas where he held his fights, so ladies bold enough could break the taboo on their presence.

The struggle of early fight promoters with the law has a comic air, like the tales of moonshiners outracing revenuers. Tremendous ingenuity went into the logistics of quick getaways, with the winner of a bout pulling on clothes as he raced to a specially rented train already steamed up on the track. Even after boxing became legal in some states, boxing films were outlawed in them all (largely to prevent the showing of black fighter Jack Johnson's prowess). When customs officers seized prints of the Johnson-Willard fight, which had been held in Cuba, a crazy ingenuity gained possession of the image without physically moving any film over a border:

> A movie camera was set up on the American side eight inches from the New York–Canada border, with the lens directed to the north. At approximately the same distance on the Canadian side rested a box with an electric light; through this was run an original positive film taken from the negative film made in Havana. Unexposed film was moved along a reel and through the camera on the American side; the two reels were connected in a way that allowed both to turn simultaneously, and thus a negative reproduction was made in America of the positive film in Canada. Pantomimic Corporation then rephotographed from the secondary negative and prepared to distribute positives of the contraband film.

That story is one of the many interesting ones told in Jeffrey Sammons's *Beyond the Ring*. But for a full treatment of boxing's early days—and for the most overtly Geertzian account—we must begin

with Elliott Gorn's *The Manly Art.* For Gorn, the bare-knuckle era of
the nineteenth century was the only pure age of boxing—an expression of "the free and easy cultural style" of the working class, which
was later taken over and manipulated by middle-class evangelical reformers. The fancy was a set of Robin Hoods resisting the onslaught
of industrialized capitalism. The goldenest age within this happy time
was the early part of the century, the romantic era when high-culture
misgivings about the disappearance of ancient ways took the concrete
form of toffs attending prizefights. For Gorn, Lord Byron at ringside
was engaged in essentially the same enterprise as when he went to lib-
erate Greece. At a time when the industrial revolution was consuming
workers in large schemes of production, men "took control over their
lives" by resorting to the ring. For a while, they had some heretical
upper-class support for this rebellion against the "moral arrogance of
the middle class," with its Bibles and its timetables preaching abstinence and productivity. Taverns were the lairs in which people could
hide out from Mrs. Grundy: "Saloons were at the heart of the working-class life. Cliques of men created informal but stable brother-
hoods in particular bars," which served as home base for the fancy.

Gorn deliberately emphasizes gender in brotherhood. Men went to
the bars to hide from their wives as well as their bosses, since Bible
and timetable found their allies in the feminine part of the working
class. Male rites and bonding were at the heart of bare-knuckle boxing, according to Gorn, who quotes lavishly from "homoerotic" descriptions of boxers' bodies in the sporting press of the time.

The heroes in older books on boxing, those written by mere fans,
were people who tried to legitimize the sport, to make it respectable
and regulated. These are the villains in Gorn's book, the cat's paws of
the reforming busybodies, those who wanted to get control over the
workingman's recreation as well as his work time. John L. Sullivan,
who made the sport widely admired if not entirely respectable, is
treated partly as a dupe of the commercial world and partly as a be-
trayer of his own class. Sullivan was the one man of his time who could
introduce the Marquis of Queensberry rules into general use without
seeming to "sissify" the sport. Those rules called for gloves, which
seemed more gentlemanly (amateurs used gloves in their sparring
matches at athletic clubs), though professional use of gloves is actually
more deadly than bare fists.[2] For Gorn, the most subversive of the new
rules allowed fights to be held on a stage rather than the traditional
"turf." That, along with indoor lighting, enlarged the audience that

could see a fight. (Many early bouts had been fought in the dark to evade the law.) Indoor fights let promoters set conditions on attendance, ranging from admission charges to the suppression of drinking and violence outside the ring. The collaborative crowd became the regulated audience.

Michael T. Isenberg, in *John L. Sullivan and His America,* shows that Sullivan preferred Marquis of Queensberry rules because they fit his style of fighting better than the old bare-knuckle (London Prize Ring) rules. Sullivan had a reckless energy of total assault; but as an alcoholic from his youth, he fought usually out of condition, and sometimes drunk. Though he was able to endure incredible amounts of punishment, his ability to deliver it flickered on and off after his crushing first charge. Under London Prize Ring rules, a round ended with every knockdown, throw, or fall to one knee, and the downed man had thirty seconds to make it back to scratch. So there is a falsely heroic air about old accounts of fights that lasted eighty or ninety rounds and went on for two or three hours: much of that time was consumed in the thirty-second rests a fighter could buy for himself at any moment simply by taking a fall. That was particularly frustrating for Sullivan, who could not rain thunderous punches on a person who went down at the first tap when Sullivan caught up with him. Just fighting "the Great John L." down to exhaustion, and avoiding being knocked out by him, was a victory for some of his opponents, who lived to mock his fabled strength.

Queensberry rules forced any downed man to get up within ten seconds *and continue the round,* coming back again and again to the fire blast of Sullivan's fists. Furthermore, since Sullivan fought in berserk seizures of destructiveness, he was not careful enough about where he landed his blows, and his hands were damaged under London rules. He remained a rather bloated bare-knuckle champion as long as he did by challenging foes to a preliminary Queensberry fight before the official (but illegal) championship fight; then he finished them off so thoroughly with the gloves that all interest was lost in a subsequent bare-knuckle event.

Queensberry rules confined and intensified what had been open-ended and slipshod. No more wrestling holds or throws. No more staggering around in the mud for hours. No more injuring an opponent with one's shoes rather than one's fists. Since the "turf" afforded poor footing, often at the beginning of a bout and always by the end, fighters improvised cleated shoes, with which they gouged the opponent's feet, ankles, and calves. Sullivan seemed to be wading through blood in his international fight, held in France, with the British challenger Charlie Mitchell; but it was mainly his own blood, shed from his lower extremities.

Though Isenberg is writing a complete biography of Sullivan (a difficult task given all the intervening legends), he too is primarily interested in social history, placing Sullivan in the world of vaudeville, celebrity, and excess that marked the Gilded Age's climax. Where Gorn thinks of Sullivan as a tool of the genteel moralists, who tamed the raw vigor of the unauthorized ring, Isenberg finds in him a product of an age that openly glorified greed and ego, the reign of robber barons and the raw nobility of the yellow press. Theodore Roosevelt idolized Sullivan, much as Byron had doted on John Jackson. In Veblen's chosen decade, Sullivan was as conspicuous a consumer as one can imagine of booze, women, money, fame, and physical punishment.

Isenberg, like Gorn, studies boxing's connection with the working class. But where Gorn glorifies a tavern society that forged bonds of brotherhood, permitting a spontaneity banned elsewhere in society, Isenberg finds a world of men beating their wives, children, and friends. Sullivan, sober, had a kind of Babe Ruth childishness and generosity. Drunk, he pummeled anyone near him, including his wife. Like the "Raging Bull" in Scorsese's movie, Sullivan punched the world with his face, amazing the world and destroying his face. Emblematic of the many evenings he ended up comatose, sometimes in jail, sometimes in the hospital, most often in a strange hotel room, was the night he wandered out onto a train's caboose to urinate from it, pitched face forward into the gravel, and was resuscitated only when people missed him on the train, had it backed up, and found him inert on the tracks. The showbiz side of Sullivan did not prevail until he could no longer fight professionally, at which point he took to the vaudeville circuit as a temperance lecturer, heroically if belatedly practicing what he preached.

Boxing of the post-Sullivan era is covered by Jeffrey Sammons. Though he concentrates on the years 1930–1980, he picks up the story by going back to Sullivan, "the Father of American Prize-fighting" in his account, and—more significantly for his book—continuing through the first decades of this century, which was the era of Jack Johnson. If boxing was an underworld reflection of (and on) the larger society, black boxing was a reflection of that reflection. The idealized brotherhood of Gorn's white working class was nowhere more authentic than in its opposition to interracial prizefighting. Sullivan's famous boast that he could knock out any man alive was based on the tacit thesis that Peter Jackson, the great black fighter from Australia, was not a man.

Sammons is a black scholar young enough to have adored Muhammad Ali in his childhood. It is not surprising that he treats more centrally the racism that all authors of the books under review admit has been a permanent feature in the fancy. The dubious social standing of boxing was given a supreme test when a black man finally became the heavyweight champion in 1908. As we have seen, that development led to the banning of fight films, even in states where boxing had become legal, to prevent people from seeing the revolting spectacle of a black man beating a white. But, on the other hand, it led to a worldwide hunt for a "great white hope" to reclaim the crown, giving society a more widespread and intense concern with boxing. Jack London called for James Jeffries, the undefeated former champion, to come out of retirement to "remove the golden smile from Jack Johnson's face." Johnson mowed down Jeffries and several hulking white hopes (including Victor McLaglen, later part of John Ford's repertory group) before a circus strongman, Jess Willard, was found and hoisted up, like a tower of flesh, to fall on Johnson.

Tex Rickard, who had exploited the great white hope in setting up the Johnson-Jeffries match, waged a brilliant campaign of publicity, bribery, and blackmail to turn the vindication of the white race into a larger legitimation of boxing. He helped create the myth of Jack Dempsey (who defended his title only six times in seven years and fought only one superior opponent, Gene Tunney, losing to him both times). Part of Rickard's attempt to "elevate" boxing was the use of his "Jenny Wrens," women whose presence was a kind of blessing on the sport. Katharine Fullerton Gerould even reported the first Dempsey-Tunney fight for *Harper's,* treating it as a Greek tragedy. In state drives to legalize boxing, the presence of women at bouts would be used over and over to prove that boxing was not brutalizing.

Meanwhile, the racist aspect of the sport continued on an international scale. When "Battling Siki" of Senegal became the light-heavyweight champion of the world, American newspapers rebuked the French for the political risk of putting a colonial in the same ring with Georges Carpentier of the colonizing country. (Ironically, Siki had been a war hero fighting for the French during the Great War.)

Dempsey refused to fight Harry Wills, as Sullivan had avoided Peter Jackson. Another group discriminated against in America was admitted to the ring before blacks were. "By 1928 there were more prominent Jewish boxers than there were boxers from any other single ethnic or racial group." Even Max Baer, the "sometimes Jew," fought with a Star of David on his trunks. But Sammons shows that the press treated Jewish boxers with condescension. I could add to his examples a sports column written in the *Des Moines Dispatch* (February 5, 1937) in which "Dutch" Reagan rebukes "Bob Pastor, a young Jewish lad," for running away from Joe Louis in the ring rather than standing and taking his beating like a man. This made Pastor a disgrace to his mother, who had "spent her life battling slums and social injustice." According to Reagan, Bob Pastor was a quitter—the charge that had echoed since 1935 around Max Baer, who also refused to be beaten any further by Louis. Jewish fighters were either quitters like Baer or clowns like "Slapsie Maxie" Rosenbloom.

Gates dwindled, along with the perception of the sport's seriousness, during the period of Jewish ascendancy (the late twenties and early thirties). But then promoter Mike Jacobs put together Mrs. Hearst's Milk Fund, as a sanitizing charitable cause for bouts (accompanied by good Hearst paper coverage), with a milk-brown fighter given humility drills, to revive the sport. What Jacobs realized was that patriotism could be used to temper racism. He matched Joe Louis against Primo Carnera on the eve of Italy's invasion of Ethiopia. A clever cartoonist drew Louis as the little African country and Carnera as the elongated "boot" of Italy. Black youths in Harlem shouted after the fight, "Let's get Mussolini next." Then Louis repeated his victory over fascism in the form of Carnera by beating the Nazis in Max Schmeling. (The fact that a Jew, Max Baer, had already defeated Schmeling in 1933 was not taken seriously by the press.)

Despite continuing resistance in the South, Louis integrated boxing a decade before Jackie Robinson could do the same thing for baseball. And in both cases, the rules were the same—a demonstrative gratitude for the opportunity. But before blacks could dominate the sport, Italians got their turn. Sammons is careful to point out that the criminal influence on boxing had been there long before the era of Marciano and Graziano and LaMotta, and that it was a multiethnic mob that preyed on gambling profits. But Graziano and LaMotta make clear in their own books that part of their working-class culture was the pervasive Mob.

• • •

Sammons's book reaches its climax in the rise and decline of Muhammad Ali. Here all his themes combine—race, politics, art, money, theater, and physical destruction. When my family was watching Ali fight Ken Norton on television in 1973, my twelve-year-old son went upstairs before the end, saying he could not bear to see Ali lose. (He was fighting, we did not then know, with a broken jaw.) The last part of Sammons's book is the scholarly equivalent of that reaction. In Ali, lethal agility was perfected to destroy similarly perfected bodies. It took all his skills to let him prevail long enough for those very skills to be dramatically obliterated. The result is that tragic caricature of himself that walks about as a standing rebuttal of all boxing's claims. Sammons concludes his book with a massing of evidence from economic studies, physical tests, demographic tables, and the history he has told himself, to prove that boxers are particularly exploited members of the generally exploited classes in our society. The few who succeed delude the many who try into short efforts that unfit them for other walks of life. Even those who succeed usually end up damaged, in debt, or impoverished. In a survey of bouts done during the 1970s, there was an average of 21 ring deaths per year (3.8 deaths per 1,000 participants, compared to college football's 0.3 deaths per 1,000). And the deaths are just part of the toll for people who suffer brain damage and all the other debilities that testify to the deadly improvement of the human physique when it is turned into an implement for its own demolition.

Sammons should be required reading for anyone who takes up Joyce Carol Oates's *On Boxing*. Oates continues the tradition of her fellow Jenny Wren, Katharine Fullerton Gerould, finding classical Greek antecedents for "America's tragic theater," "not an altar of sacrifice solely but one of consecration and redemption." Like Elliott Gorn, she sees class bias in any opposition to boxing, which is "particularly vulnerable to attack by white middle-class reformers (the AMA in particular) who show very little interest in lobbying against equally dangerous Establishment sports like football, auto racing, thoroughbred horse racing." She agrees with him that boxing is "a celebration of the lost religion of masculinity." She rightly notes that boxers are as proud of being able to take a punch as of delivering one. Jake LaMotta, in his autobiography, has his best friend hit him over and over as hard as he can (in the movie, this task is given his brother)—the bond with this friend is the true love story followed in the arc of the book, as wives come and go.

The brotherhood of pain makes Oates feel a teensy bit more brave herself as she contemplates it: "To move through pain to triumph—or the semblance of triumph—is the writer's, as it is the boxer's, hope." It is amazing how much boxing reminds people of what *they* do, only more so. Norman Mailer tried to capture the tremendousness of it all by comparing a boxing match to his having to debate Bill Buckley for twenty-four hours without a break.[3] Geertz boasts that, at the pit where cocks fought to the death, he and his wife, facing the police, "had not simply pulled out our papers and asserted our Distinguished Visitor status" (anthropology as a blood sport).

Oates admits the dangers of the sport, only to belittle them: "Surely it is championship chess, and not boxing, that is our most dangerous game—at least so far as psychological risk is concerned. Megalomania and psychosis frequently await the grand master when his extraordinary mental powers can no longer be discharged onto the chessboard."

And if killer chess is mowing down bright people, the streets will get lower class ring fodder: "So frequently do young boxers claim they are in greater danger on the street than in the ring that one has to assume they are not exaggerating for the sake of credulous white reporters." This statement reveals a blindness to the differences between moral situations strikingly like that displayed in the argument that more people died of traffic accidents during the Vietnam War than from battle wounds.

At least Oates is honest about what appeals to her in boxing. It is the blood. Like "Vicki" in Scorsese's movie, she is ready to kiss the cuts above the fighter's eyes. She glories in

> the electrifying effect upon a typical fight crowd when fighting suddenly emerges out of boxing—when, for instance, a boxer's face begins to bleed and the fight seems to enter a new and more dangerous phase. The flash of red is the visible sign of the fight's authenticity in the eyes of many spectators, and boxers are justified in being proud, as many of them are, of their facial scars.

For many fans, Muhammad Ali was not a "real" fighter so long as he could boast of being pretty, of not having been bloodied. For Mailer, Ali could not truly win until he had been brutally beaten by Joe Frazier. Only that would "show America what we all had hoped was secretly true. He was a man."[4] The blood lust was sharpened in the days of bare-knuckle fighting when bets were laid on "first blood" as well as on the final outcome.

Oates's description of the electrifying effect of blood is uncannily like the words Saint Augustine used of his friend Alypius, a fan of the gladiators who thought he had overcome his addiction till friends dragged him back to a match. "At sight of the blood, he took a sip [*ebibit*] of animality. Not turning away, but fixing his eyes on it, he drank deeper [*hauriebat*] of the frenzies without realizing it and, taking complicit joy in the contest, was inebriated [*inebriebatur*] by his delight in blood" (*Confessions* 6.8). For Augustine, the real harm was done in the crowd rather than the arena. Alypius "was wounded deeper in the soul than was the gladiator in his body." I had read that passage dozens of times, but it did not come home to me till, five years ago, I stood talking with Muhammad Ali, embarrassed by his inarticulateness, and deeply ashamed. It was not only his own superb body that had done this horrible thing to his own superb mind. I had done it too, as part of that crowd urging him on, applauding the blood. I have not watched a boxing match since then.

—*New York Review of Books,* February 18, 1988

NOTES

1. "Deep Play: Notes on the Balinese Cockfight," *Daedalus* (1972); reprinted in Clifford Geertz, *The Interpretation of Cultures* (Basic Books, 1973).
2. See my article "The Sporting Life," *New York Review of Books,* October 30, 1975.
3. Norman Mailer, *King of the Hill* (New American Library, 1971), p. 19.
4. Mailer, *King of the Hill,* pp. 92–93.

Index

Garry Wills, one of our most distinguished historians and critics, is the author of numerous books, including the Pulitzer Prize–winning *Lincoln at Gettysburg, Saint Augustine,* the bestsellers *Why I Am a Catholic* and *Papal Sin,* and, most recently, *"Negro President."* A regular contributor to the *New York Review of Books,* he has won many awards, among them two National Book Critics Circle Awards and the 1998 National Medal for the Humanities.

NIXON AGONISTES

"Astonishing . . . a stunning attempt to possess that past, that we may all of us escape it." — *New York Times Book Review*

Wills offers this classic dissection of Richard Nixon that is also an incisive and provocative analysis of the American political machine. With a new introduction.   ISBN 0-618-13432-8, $15.00

■

INVENTING AMERICA

"A scintillating tour de force of historical detective work." — *Time*

An adept and controversial analysis compares Jefferson's original draft of the Declaration of Independence with the final, accepted version, thereby challenging many assumptions. With a new introduction.

ISBN 0-618-25776-4, $15.00

■

THE KENNEDY IMPRISONMENT

"The ultimate Kennedy book." — *New Republic*

This is the definitive historical and psychological analysis of the Kennedy clan and its crippling conception of power, from one of America's foremost historians. With a new introduction.   ISBN 0-618-13443-3, $14.00

■

WHY I AM A CATHOLIC

"Exhilarating . . . Garry Wills reminds us of what Catholicism should be." — *Los Angeles Times*

A *New York Times* bestseller, this book offers a historical look at how the papacy has reinvented itself in times of crisis and an argument for its future relevance. With a new introduction.   ISBN 0-618-38048-5, $14.00

■

"NEGRO PRESIDENT"

"A bracing study of the mischief wrought by the Constitution's so-called three-fifths clause." — *Wall Street Journal*

In this provocative new book, Wills explores a controversial and neglected aspect of Thomas Jefferson's presidency: it was achieved by virtue of slave "representation."   ISBN 0-618-48537-6, forthcoming in paperback